```
581.634 Mor          199513
Morton.
Major medicinal plants.
```

The Lorette Wilmot Library
Nazareth College of Rochester

Major Medicinal Plants

Upper photo.

Back row: Desenex® (undecylenic acid — from castor oil), Atropine Sulfate, Beladonna tincture, Cas-Evac® (cascara sagrada), Quinine Sulfate, Quinidine Sulfate, Agoral® (agar, tragacanth and acacia), Metamucil® (psyllium hydrophilic muciloid), Senna Laxative.

Second row: Colchicine, Quinine Sulfate, Velban® (vinblastine sulfate), Ephedrine Sulfate, Tincture of Benzoin Compound, Senokot® (senna), Thymol.

First row: Ouabain, Pilocarpine HCL, Reserpine, Colchicine, Podophyllum Resin, Juniper Tar (cade oil), Papase® (proteolytic enzymes from *Carica papaya*), Caroid (proteolytic enzyme), Donnatal® (hyoscyamine sulfate, atropine sulfate, hyoscine hydrobromide).

Lower photo.

Back row: Digoxin®, ergot, menthol, camphor.

Front row: Tylenol® (codeine), ipecac, Empirin® (caffeine), cocaine, menthol, glycyrrhizin, Vicks® (ephedrine sulfate, camphor, menthol, methyl salicylate, eucalyptol).

MANY DRUG PRODUCTS ON THE SHELVES OF MODERN PHARMACIES CONTAIN ACTIVE INGREDIENTS WHICH MAY BE DERIVED FROM PLANTS.

Photographed at the Rivera Pharmacy, Coral Gables, Florida, through
the courtesy of Nathan Tartak, M.S. Pharm., Proprietor.

Major Medicinal Plants

Botany, Culture and Uses

By

Julia F. Morton, D. Sc., F. L. S.

Director
Morton Collectanea
University of Miami
Coral Gables, Florida

CHARLES C THOMAS • PUBLISHER
Springfield · Illinois · U.S.A.

Published and Distributed Throughout the World by
CHARLES C THOMAS • PUBLISHER
BANNERSTONE HOUSE
301-327 East Lawrence Avenue, Springfield, Illinois, U.S.A.

This book is protected by copyright. No part of it may be reproduced in any manner without written permission from the publisher.

© 1977, by CHARLES C THOMAS • PUBLISHER
ISBN 0-398-03673-X
Library of Congress Catalog Card Number: 77-3287

With THOMAS BOOKS *careful attention is given to all details of manufacturing and design. It is the Publisher's desire to present books that are satisfactory as to their physical qualities and artistic possibilities and appropriate for their particular use.* THOMAS BOOKS *will be true to those laws of quality that assure a good name and good will.*

Library of Congress Cataloging in Publication Data

Morton, Julia Frances
  Major medicinal plants.

  Bibliography: p.
  Includes index.
  1. Botany, Medical. 2. Materia medica, Vegetable.
3. Botany, Medical—United States. 4. Materia medica, Vegetable—United States. I. Title.
QK99.A1M67   581.6'34   77-3287
ISBN 0-398-03673-X

Printed in the United States of America

Foreword

It has always been my habit over the past twenty years to cast a jaundiced eye on virtually any book bearing the title, "Medicinal Plants," either directly or indirectly. The reason for this is that, to authoritatively treat the subject, one must have a broad understanding of at least botany, chemistry and pharmacology in order to place the information in proper perspective. With the exception of college textbooks on the subject, during the past twenty years, there has been only one such book printed in America that could be considered as authoritatively written, and that was *Green Medicine*, by Margaret Kreig, published initially in 1964. We now have what I would consider to be the second authoritative book written on the subject of medicinal plants during the past two decades, i.e. *Major Medicinal Plants: Botany, Culture and Uses*, written by Doctor Julia F. Morton.

One might ask why such a book is needed, and what advantages does it offer over textbooks used by pharmacy students that surely must cover the material in this book. In order to answer these questions, it is necessary to digress somewhat from the central theme of this introduction.

Pharmacognosy is the study of drugs and other economic products obtained from natural sources. Fifty years or more ago, the subject was centered around the theme of utilizing morphological and histological features of plant and animal drugs as aids in their identification. This was during the presynthetic drug era, when the medical profession was completely dependent on drugs from plants and animals, or those of mineral origin. One needs only to look through some old pharmacopoeias to determine how true the statement is. Thus, it was important for the pharmacist to know, at that time, how to determine the purity and identity of crude drugs, or powdered crude drugs, since often they were subject to adulteration, either accidental or intentional. There was no other means available at that time to determine this, thus pharmacognosy as a descriptive science realized its most prestigious era.

However, as synthetic organic chemistry came into being and the first synthetic drugs were introduced into medicine, the use of crude drugs and extracts of crude drugs became less important. This lessening of importance of crude drugs was paralleled by a decreased need for the pharmacy student to study traditional pharmacognosy. However, as in any changing science, the practitioners of pharmacognosy began to realize a necessity for change in their discipline in regard to approach and outlook.

It should be pointed out that initially what we know now as pharmacology, pharmacognosy and medicinal chemistry were formerly taught collectively under the title of Materia Medica. As each of the three sciences developed, they branched out and became independent entities.

Thus, pharmacognosy has developed slowly into a dynamic science based on botany, chemistry, biochemistry and pharmacology. It now includes information on the sources of major natural products, methods for isolating the active principles, the chemistry of the active principles, the biosynthesis of these principles and their uses, side effects and interactions with other substances that may be administered concurrently with subsequent detrimental effects. Even though pharmacognosy has changed its face completely, it is a subject in the pharmacy curriculum that is perhaps berated more often than any other, has consistently had to defend its existence in the curriculum over the past twenty years, and indeed has been eliminated from the curriculum of many colleges of pharmacy in the United States. Initiation of this de-emphasis has been primarily through the efforts of a handful of pharmaceutical administrators and non-academic bureaucrats who have proclaimed that the practice of pharmacy must change from a product-oriented outlook to patient-orientation. This has resulted in the inclusion of a large component of "clinical" instruction in the curriculum that in many places has replaced pharmacognosy as well as other basic science courses. The ironic nature of the situation is that the "clinical" instruction forced on the students, while probably justified in one sense, is oriented toward the hospitalized patient (or institutionalized patient) and not the ambulatory patient who will be seen most frequently in the community pharmacy setting by 80 to 90 percent of the practitioners. It will be the larger latter group who will use the material in Doctor Morton's book to greatest advantage in maintaining their professional competence in natural drug products.

Concurrent with the de-emphasis on teaching students about natural drug products has been an alarming increase in the awareness and concern of many people in this country about the side effects of synthetic drugs and a desire to learn more about the efficacy and safety of "natural drugs." With the current state of affairs of pharmaceutical education, our students in most instances are not prepared to advise their patients on such matters. How then will our students, who become pharmaceutical practitioners, prepare themselves to effectively do this?

The most obvious place to seek such information is in textbooks. At the present time, there is only one current American textbook on pharmacognosy, and from a quick glance through this book, it becomes obvious that it is not written to answer many of the questions that will be asked the practitioner regarding natural products. To ask the practitioner to take the time to search the scientific literature for the required information would be to ask an impossibility. Thus, it would appear that Doctor Morton's book will fill this void and will be an invaluable ancillary textbook for any pharmacognosy course; it should be a required reference book in every pharmacy in the United States.

Perhaps at this time it would be of interest to point out to the reader just why the material in this book, or information on natural drug products in general, is important to the practitioner or pharmacy as well as medical practitioners in general.

Most people think that practically all drugs currently used are prepared synthetically. Such is definitely not the case. For example, it is now known that of the 1.532 billion new and refilled prescriptions dispensed from community pharmacies in the United States in 1973, 25.2 percent contained one or more active constituents still extracted from higher plants (seed plants). Microbial products accounted for 13.3 percent of the prescriptions and animal-derived drugs for about 2.7 percent. What is truly amazing is that during the period 1959-1973, drugs from higher plants maintained an amazingly consistent figure of about 25 percent of the total prescriptions dispensed in the United States. It has been calculated that, in 1973, drugs from higher plants alone cost the American public some three billion dollars. Further, as we add up all of the prescriptions derived from natural sources, they account for almost 50 percent of all prescriptions dispensed. It is difficult to comprehend, then, why the subject taught in pharmacy schools that have been entrusted with educating future pharmacists in this important aspect of their profession is being relegated to obscurity.

But even perhaps of more importance is the recent surging interest on the part of the American public to learn more about the use of natural crude drugs, and it is reasonable that they will turn to either the pharmacist for this knowledge or seek out knowledgeable books on the subject. For this reason, it becomes more important that this authoritative treatise on medicinal plants has been painstakingly written by Doctor Morton. In effect, she has taken up where pharmacognosy textbooks have left off. Virtually every plant drug in use today or from which useful drugs are being derived is discussed, as well as those plant drugs which were once used extensively but are of lesser importance today.

Finally, not only should this book be available in every pharmacy in the United States but many physicians will want to include it in their library of reference materials. It should be a required secondary textbook in all pharmacognosy courses taught in colleges of pharmacy. I suspect, however, that it will be most popular with the laity who desire authoritative and accurate information on plant drugs, which will now be available to them for the first time.

Norman R. Farnsworth, Ph.D.

Professor of Pharmacognosy
Department of Pharmacognosy and Pharmacology
College of Pharmacy
University of Illinois at the Medical Center Chicago, Illinois

A great deal of useful information on the subject of currently employed medicinal plants will be found within the covers of this book. Particularly useful will be the sections dealing with the botany of each species considered, as well as the production of crude drugs from the whole plant, seeds, roots or other parts.

Over the years, many plants have "supplied" medical materials for use by mankind. A list of such plants would be an almost endless one. From time to time, certain plant drugs have lost favor or, in some cases, their uses have changed. Some drugs of plant origin have been abandoned because they did not prove as valuable as was hoped. However, there are outstanding examples of newly-discovered plant medicines which have proved to be safer than and/or superior to those displaced. Reserpine, the tranquilizer from *Rauvolfia,* is a prominent example of a plant drug that came into wide use some years ago. Another genus that now plays an important role in modern medicine is *Catharanthus*. The search for new medicinal plants goes on and on as it has for centuries during man's history. Exploration and investigation have broadened to cover the globe.

During recent decades, emphasis in research and teaching of material on natural products has undergone gradual change until, at the present time, natural product chemistry seems to dominate the subject. Of course, the biological and biological-activity phases are also considered, but with less emphasis than in the past.

The widespread attention being given drug plants in modern times is evidenced by the extent and diversity of scientific reports—the bibliography of this volume consists of over six hundred references. Students and others will find this feature a valuable guide to the literature. Throughout the book, the user will glean much timely information pertaining to the technology of the production of medicinal and related materials of natural origin as well as by-products of economic botanical interest.

In the category of pharmaceutical aids or adjuncts, the marine algae are presented, followed by those higher plant forms—embracing many *taxa* (categories) of seed plants—which, while not strictly medicinal, make significant contributions to health care. The numerous illustrations add greatly to the book's instructional value.

Though this volume is not a textbook of pharmacognosy and is not intended to be, it embodies essential, up-to-date, practical knowledge of the botany and culture of plants studied in pharmacognosy, and the environmental conditions and systems of handling and processing which determine the availability, quality and utility of the ultimate healing agent. To any person who is likely to be approached concerning information on any of these matters or is professionally or non-professionally interested in this field, this book will be invaluable.

Maynard W. Quimby, Ph.D., F. L. S.

Professor Emeritus
Department of Pharmacognosy
School of Pharmacy
University of Mississippi

Introduction

This book is not meant to be a textbook in pharmacognosy but is intended to supplement such texts. Its aim is to give to teachers and students, growers or potential growers, dealers and consumers more information on the botany, culture, harvesting and handling of medicinal plants and their products than can be found in any other single source.

There are many factors in drug crop production that determine the feasibility of cultivation, economic importance to the producing areas, variation in quality and potency, availability on the market, competitive status and fluctuations in selling price. This statement would apply, of course, to any agricultural endeavor, but the medicinal plant industry is probably subject to more variable conditions than any other. In spite of the impact of this industry on the well-being of mankind, there is little popular communication of its problems and methodology—in contrast to the wealth of readily available information on food plant cultivation, for example.

Much of the professional and technical knowledge of medicinal plant exploitation is inaccessible, and that which has been reported represents a lag behind developments within the industry. It is clearly impossible to keep fully abreast of the findings of phytochemists and other investigators whose latest discoveries may be presented at scientific meetings or submitted to technical journals even as manuscripts such as this go to press, for chemical studies and clinical trials of traditional or new natural products go forward every day in an ever-continuing effort to improve health care and reduce hazards of drug administration.

Therefore, no great claims are made here for scope and currency. This book embodies an effort to bring together a concise account of the physical aspects of each of the major medicinal plants currently in use in the United States, along with a brief outline of their chemical constituents and their past and present therapeutic uses. Where applicable, toxicity is explained and also other economic uses and by-products. In some few cases, it will be seen that the medicinal "crop" itself is a by-product of a plant which is grown for an entirely different purpose, e.g., steroid extraction from the fiber plant, *Agave sisalana*.

Plants which have, in relatively recent years, fallen into disuse, appear in Table I, in the Appendix. Many others, employed in the distant past, have been abandoned or unrecognized officially for many decades and need not be mentioned. It should be realized, though, that numerous "obsolete" medicinal plants are still used in formulating patent medicines, some persist today in domestic or "folk" use in rural areas, and quite an array is being distributed by "health food" outlets as beverage or curative materials.

Plants which are solely or mainly pharmaceutical adjuncts—functioning as lubricants, vehicles or flavors—are relegated to Table II. Various spices are included in pharmacognosy textbooks as having carminative activity, but this is not sufficient reason to classify them here as medicinal plants; to include all the aromatic natural materials capable of such action would be unreasonable.

As this work is focused on the plants which provide therapeutic agents rather than on their applications, the species are presented in natural botanical sequence by family. This arrangement is desirable for all readers because it keeps together those plants which have or should be expected to have similar properties and cultural and other requirements.

Inasmuch as most of the commercial growing of medicinal plants takes place abroad, much of the field and processing detail has been derived from foreign sources. European chemists are very active in medicinal plant research and in patenting improved methods of extraction and purification of active principles. It will be noted that, in many instances, I have referred to and listed in the Bibliography only the English abstract of foreign articles as published in *Chemical Abstracts* (CA) or *Biological Abstracts* (BA), for I have no facility for translating the languages of central and eastern Europe or the Far East. Nevertheless, all of the literature—complete texts, reprints or abstracts—drawn upon in the preparation of this volume is presently in the subject files or on the reference shelves of the Morton Collectanea of the University of Miami.

J. F. MORTON
October 17, 1976

Acknowledgments

Grateful acknowledgment is made of the research conveniences afforded by the Calder Memorial Library of the University of Miami's School of Medicine; also of the encouragement and helpfulness of Doctor Maynard W. Quimby, Professor Emeritus of the Massachusetts College of Pharmacy and School of Pharmacy of the University of Mississippi, and of Doctor Norman R. Farnsworth, Head, Department of Pharmacognosy and Pharmacology, College of Pharmacy, University of Illinois at the Medical Center, Chicago, both of whom have given of their valuable time to read and comment on the manuscript.

Deep appreciation must be expressed for the courtesies and cooperation extended by the Directors and staff members of Longwood Gardens, Kennett Square, Pennsylvania, the Morris Arboretum of the University of Pennsylvania, the National Botanic Garden, Washington, D.C., and the Fairchild Tropical Garden, Coral Gables, Florida; also to Doctor Franklin W. Martin, Mayaguez Institute of Tropical Agriculture, Puerto Rico, and Doctor James Duke, Chief, Plant Taxonomy Laboratory, USDA Agricultural Research Center, Beltsville, Maryland. The generosity of others who have loaned photographs or specimens is recognized in the legends of the respective illustrations.

I am heavily indebted to Doctor Robert J. Knight, Jr., Research Horticulturist, USDA Subtropical Horticulture Research Unit, Miami, Florida, who has nobly and expertly shared the burden of proofreading.

Contents

Foreword v

Introduction xi

Acknowledgments xiii

Major Medicinal Plants
    Clavicipitaceae (fungi) 3
    Pinaceae 11
    Cupressaceae 25
    Ephedraceae 31
    Bromeliaceae 37
    Liliaceae 45
    Agavaceae 65
    Dioscoreaceae 73
    Podophyllaceae 85
    Menispermaceae 91
    Lauraceae 101
    Papaveraceae 109
    Hamamelidaceae 127
    Leguminosae 135
    Erythroxylaceae 175
    Rutaceae 185
    Euphorbiaceae 191
    Rhamnaceae 199
    Malvaceae 205
    Sterculiaceae 209
    Styracaceae 215
    Caricaceae 221
    Apocynaceae 231
    Labiatae 269
    Solanaceae 281
    Scrophulariaceae 311
    Plantaginaceae 323
    Rubiaceae 333

Appendix

    Table I—Medicinal Plants no longer official in the United States of America but still mentioned in the U. S. Dispensatory and/or American textbooks on pharmacognosy.......................... 359

    Table II—Plants which serve as pharmaceutical aids or adjuncts....... 372

Bibliography 383

Index 413

Illustrations

Figure

1. Ergot, *Claviceps purpurea* 4
2. Ergot, *Claviceps purpurea* 8
3. Slash pine, *Pinus elliottii* 12
4. Slash pine, *Pinus elliottii* (lightered stumps) 14
5. Slash pine, *Pinus elliottii* (logs) 14
6. Slash pine, *Pinus elliotti* (tree base and cross section) 15
7. Slash pine, *Pinus elliottii* (high-gum seedling) 15
8. Longleaf pine, *Pinus palustris* 16
9. Prickly juniper, *Juniperus oxycedrus* (rooted cuttings) 26
10. Prickly juniper, *Juniperus oxycedrus* 29
11. Prickly juniper, *Juniperus oxycedrus* 29
12. Gerard ephedra, *Ephedra gerardiana* 32
13. Pineapple, *Ananas comosus* 38
14. Pineapple, *Ananas comosus* (harvesting) 42
15. Pineapple, *Ananas comosus* (harvested fruits) 42
16. Mediterranean aloe, *Aloe barbadensis* 46
17. "Bitter aloes" (inspissated latex) 49
18. Autumn crocus, *Colchicum autumnale* 52
19. American hellebore, *Veratrum viride* 58
20. White hellebore, *Veratrum album* 62
21. Sisal, *Agave sisalana* 66
22. Sisal, *Agave sisalana* (bulbils) 70
23. Sisal, *Agave sisalana* (inflorescence) 70
24. *Dioscorea composita* in bloom (female plant) 74
25. Mexican yams in cultivation 78
26. *Dioscorea composita* in bloom (male plant) 79
27. *Dioscorea composita* in bloom (female plant) 79
28. Mexican yam tubers: *Dioscorea floribunda* (left); *D. composita* (right) 81
29. *Dioscorea floribunda* (harvested tubers) 82
30. *Dioscorea elephantipes* 83
31. May apple, *Podophyllum peltatum* 86
32. Levant berry, *Anamirta cocculus* 92
33. Pareira, *Chondrodendron tomentosum* 96
34. Camphor tree, *Cinnamomum camphora* 102
35. Camphor tree, *Cinnamomum camphora* (bark) 106
36. Camphor tree, *Cinnamomum camphora* (young) 106
37. Opium poppy, *Papaver somniferum* 110
38. Poppy pods, *Papaver bracteatum* or *P. pseudo-orientale* 122
39. Sweet gum, *Liquidambar styraciflua* 128

40. Oriental sweet gum, *Liquidambar orientalis* 132
41. Gum acacia, *Acacia senegal* 136
42. Gum acacia, *Acacia senegal* (tree) 139
43. Gum acacia, *Acacia senegal* (oozing gum) 140
44. Tragacanth, *Astragalus gummifer* 142
45. Indian senna, *Cassia angustifolia* 146
46. Alexandrian senna, *Cassia senna* 150
47. Licorice, *Glycyrrhiza glabra* 154
48. Balsam of Tolu, *Myroxylon balsamum* 160
49. Balsam of Peru, *Myroxylon balsamum* var. *pereirae* 164
50. Calabar bean, *Physostigma venenosum* 170
51. Coca, *Erythroxylum coca* (foliage and flowers) 176
52. Coca, *Erythroxylum coca* 179
53. Paraguay jaborandi, *Pilocarpus pennatifolius* 186
54. Jaborandi, *Pilocarpus jaborandi* 188
55. Castor bean, *Ricinus communis* 192
56. Cascara sagrada, *Rhamnus purshiana* 200
57. Buckthorn, *Rhamnus cathartica* 203
58. Marsh mallow, *Althaea officinalis* 206
59. Karaya, *Sterculia urens* 210
60. Sumatra benzoin, *Styrax benzoin* 216
61. Papaya, *Carica papaya* (oozing latex) 222
62. Papaya, *Carica papaya* (ripe fruits) 227
63. Arrow poison tree, *Acokanthera schimperi* 232
64. Periwinkle, *Catharanthus roseus* 236
65. Serpent-wood, *Rauvolfia serpentina* (in flower) 242
66. Serpent-wood, *Rauvolfia serpentina* (in fruit) 247
67. Serpent-wood, *Rauvolfia serpentina* (roots) 247
68. American serpent-wood, *Rauvolfia tetraphylla* 250
69. African serpent-wood, *Rauvolfia vomitoria* 254
70. Smooth strophanthus, *Strophanthus gratus* (flowers and pods) 258
71. Smooth strophanthus, *Strophanthus gratus* 260
72. Green strophanthus, *Strophanthus kombé* 264
73. Japanese mint, *Mentha arvensis* subsp. *haplocalyx* Briquet var. *piperascens* Holmes 270
74. Thyme, *Thymus vulgaris* (in bloom) 276
75. Thyme, *Thymus vulgaris* 279
76. Belladonna, *Atropa belladonna* 282
77. Indian belladonna, *Atropa acuminata* 288
78. Corkwood, *Duboisia myoporoides* (flowers) 292
79. Leichhardt corkwood, *Duboisia Leichhardtii* (flowers) 298
80. Leichhardt corkwood, *Duboisia Leichhardtii* (tree) 300
81. Henbane, *Hyoscyamus niger* 302
82. Egyptian henbane, *Hyoscyamus muticus* 306
83. Foxglove, *Digitalis purpurea* 312

84.	Grecian foxglove, *Digitalis lanata*	318
85.	Psyllium, *Plantago ovata*	324
86.	Black psyllium, *Plantago psyllium*	330
87.	Ipecac, *Cephaelis ipecacuanha*	334
88.	*Cinchona calisaya*	340
89.	*Cinchona officinalis*	344
90.	*Cinchona succirubra*	345
91.	*Cinchona pubescens*	346
92.	Coffee tree, *Coffea arabica* (in bloom)	352

COLOR ILLUSTRATIONS

Frontispiece. Pharmaceutical products

Color Illustrations

1. Pineapple, *Ananas comosus*
2. American hellebore, *Veratrum viride*
3. May apple, *Podophyllum peltatum*
4. Great scarlet poppy, *Papaver bracteatum*
5. Great scarlet poppy, *Papaver bracteatum*
6. Talha, or suakim, gum arabic tree, *Acacia seyai*
7. Senegal senna, *Cassia obovata*
8. Coca, *Erythroxylum coca*
9. Castor bean, *Ricinus communis*
10. Papaya, *Carica papaya*
11. Periwinkle, *Catharanthus roseus*
12. Belladonna, *Atropa belladonna*
13. Henbane, *Hyoscyamus niger*
14. Foxglove, *Digitalis purpurea*
15. Coffee, *Coffea arabica*
16. Arrow-poison tree, *Acokanthera schimperi*

Major Medicinal Plants

Clavicipitaceae

Figure 1. Ergot (*Claviceps purpurea* Tul.) infestation of a head of rye, with two ergotized seeds at right. This photograph by H. Garman appeared as Fig. 9 in his *Kentucky Weeds and Poisonous Plants*, Bull. 183, Kentucky Agricultural Experiment Station, Lexington, Kentucky; 1914.

Ergot

Botanical name:

Claviceps purpurea Tul.

Other names:

Ergot of rye, smut of rye, spurred rye, spur kernels, blight kernels, horn seed, mother-of-rye, muttercorn.

Family:

Clavicipitaceae, a family of sac fungi (formerly called Hypocreaceae).

Description:

A parasitic fungus, developing from spores into mycelia (fine, threadlike filaments), excreting honeydew filled with microscopic bodies (conidia) which are carried by insects from one host plant to another, where they reproduce. The mycelia from whence they came continue to grow in the ovaries of the flowers of the host and become sclerotia (fruiting bodies)—each sclerotium a cylindrical, slightly three-sided body, tapering to rounded ends, usually slightly curved, hornlike, hard, grooved on one or both sides, dark-purple to nearly black externally and white or pinkish within; ⅓ to 1½ in. (8.5 to 38.1 mm) long and ⅛ to ¼ in. (3 to 6 mm) wide. The odor is unpleasant, somewhat "fishy." These sclerotia, which seem to be oversize seeds to the uninitiated, actually take the place of the seeds which the fungus has prevented from developing. The sclerotia finally fall to the ground and, the following spring, there emerge from the soil tiny mushroomlike forms (ascospores) which infect the next crop of grain.[197, 214, 624]

Origin and Distribution:

Ergot exists as a parasite on the heads of rye (Secale cereale L.) wherever this grain is cultivated and is also found on wheat, barley and certain other grasses. Heavy infestations occasionally occur in periods of humid but cool weather.

Constituents:

Over fifty alkaloids have been reported in ergot sclerotia, mainly ergonovine (also called ergometrine, ergobasine, ergotocine and ergostetrine) ($C_{19}H_{23}N_3O_2$), ergotamine ($C_{33}H_{35}N_5O_5$), ergocryptine ($C_{32}H_{41}N_5O_5$), ergocornine ($C_{31}H_{39}N_5O_5$), ergocristine ($C_{35}H_{39}N_5O_5$), ergosine ($C_{30}H_{37}N_5O_5$),[116, 117] and ergovalide.[61] One of the most publicized properties of ergot alkaloids is lysergic acid diethylamide, a potent hallucinogen, which is linked with groups of amino acids varying according to the alkaloid.

The alkaloid content differs with the host plant, and pharmaceutical ergot in the United States is specifically that which infests rye. Certain rye cultivars produce ergot of higher alkaloid content than others.[214] There is also much variation in alkaloid content of different strains of ergot.[435, 478]

The physiologically active alkaloids all stimulate the smooth muscle of the uterus, inducing rhythmic contractions. Ergonovine is the most effective, whether given orally or intravenously. Ergotamine must be given intravenously and is slow in action. Ergotamine constricts cranial blood vessels, modifying their pulsations.[428] The biosynthesis and metabolism of ergot alkaloids has been the subject of much study by pharmacognosists at Purdue University[193, 267, 323, 478] and the University of Texas at Austin.

Amino acids that have been isolated include tyramine, thiolhistidine betaine, histamine, ergotic acid, agmatine, putrescine, cadaverine, *iso*amylamine, trimethylamine, choline, acetylcholine betaine, clavine, tyrosine, histidine and tryptophane. Also present are ergosterol, fungisterol, clavicepsin, sclererythin, ergochrysin, ergoflavin, inorganic salts and complex proteins.

A chemical test for the presence of ergot in food products activates the color reaction of sclererythin and produces a deep-violet hue in the solution.[20]

Propagation, Cultivation and Harvesting:

Most of the ergot in the drug trade is that which has naturally formed in fields of rye grown for cereal use. In 1936, the successful production of ergot on cereal culture medium was reported, but this method has not proved satisfactory for commercial purposes because of the hazards of contamination by other fungi and bacteria. In 1942, experimental field culture of ergot was undertaken in India because of wartime suspension of imports. By 1950, 100 acres were in production in the Nilgiris. The rye seed was sown in April, and the flower heads formed in early July. These were sprayed with conidial spores of the fungus as many as six and eight times to assure adequate infestation. Sclerotia were fully formed in fifteen days. Surveys were made to locate other suitable regions for ergot production, considering not only the climatic conditions but also the hazard to other cereal crops and to humans and livestock. In 1960 and 1961, test plantings were made in Jammu. Best results were obtained by sowing rye in mid-September and infecting the heads throughout January and the first half of February by spraying spores suspended in a 7.5% cane sugar solution, which was two to three times more effective than water as it did not run off nor did it evaporate quickly. Spraying began when 10 percent of the field was in bloom and was performed fifteen times on alternate days between 10:00 and 12:00 A.M.[214]

In Switzerland, Germany, Hungary and Czechoslovakia today, much ergot is obtained by deliberate inoculation of rye crops grown for the purpose. In these countries, inoculation is performed with batteries of needles which are dipped into a spore solution and then brought into contact with the inflorescences manually or mounted in the front of tractors for rapid mass application.[122] Ergot cultivation is also carried on in New South Wales.[20]

The mature sclerotia may be harvested by hand, by especially designed machines, or may be sorted from the harvested grain by various hand or mechanical processes.[122] In the United States, ergot was almost entirely imported from Spain, Portugal, Germany and Russia prior to World War II. During the war, American pharmaceutical firms purchased ergot gathered from domestic rye and wheat fields.[631] Now, much is recovered when electrostatically screening rye (mainly from Minnesota and neighboring states) for contaminants. The imported supply comes primarily from Canada.[618] In the northwestern states, the sclerotia are fully formed in mid-August and persist on the plant until late fall.[114]

A yield of 95 lbs. (43 kg) of ergot per acre (approximately 190 lbs or 95 kg per hectare) was realized in India.[20] The harvested sclerotia must be dried and kept in airtight dry storage. There is constant improvement of techniques for determination and separation of the usable alkaloids.[434, 442, 594, 612] From 595 lbs. (270 kg) ergot, Ram *et al.* obtained 0.17 percent crude ergometrine, 75% of which was later purified as ergometrine tartrate.[464]

Medicinal uses:

Ergot has been employed by midwives in China for many centuries. It was not adopted by the western world until the 17th century.[440] Late in the 18th century, it was introduced into obstetrical practice, and it became official in most Pharmacopoeias early in the 19th century. It is now no longer an approved drug in the United States, having been replaced by certain of its alkaloids.

Ergonovine is administered primarily to women in the final stage of labor and immediately following childbirth, especially if hemorrhage occurs. It may be given, too, after cesarean operations.

Either ergonovine or ergotamine may be prescribed for migraine sufferers, ergotamine being the more effective but more apt to cause digestive

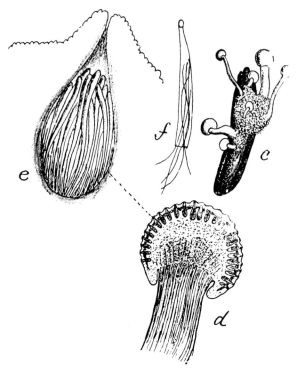

Figure 2. This drawing, a reproduction of Fig. 11 (c, d, e, f) in Garman's *Kentucky Weeds and Poisonous Plants*, shows: "(c) An ergotized seed with small mushroom-like fruiting bodies of Ergot; (d) a section through fruiting body, showing developed spores about margins; (e) perithecium with enclosed spore sacs; (f) an ascus or spore sac, with thread-like spores partly discharged. After Tulasne."

upsets. Dihydroergotamine (the methanesulfonate salt) is less likely to produce side effects. Small doses bring relief if taken early; later, larger doses are required and the response is delayed.

Ergotamine has been given with caffeine in tablet form or in rectal suppositories in some migraine cases. It is not advisable to employ any of these ergot derivatives frequently in therapy or as preventatives.[423]

Dihydroergotoxine (of which hydergine is the methansulphonate salt) is a mixture of dihydroergocornine, dihydroergocristine and dihydroergocryptine, in equal parts.[213] It serves as a vasodilator, and it has been in use for many years in treating peripheral vascular diseases and hypertension.[423] It has recently received acclaim for its favorable effects in some cases of senility.[418, 196]

Ergot alkaloids or preparations may not be given in early stages of labor, or to patients with sepsis or impaired kidney or liver functions, or peripheral vascular disease.[423]

Toxicity:

Accidental intake of considerable ergot or overdoses of ergot alkaloids may result in acute poisoning. Symptoms include pallor, coldness of skin, staggering, partial paralysis accompanied by tingling and numbness in the limbs, weak pulse, thirst, burning in the mouth and stomach, fever, nausea, vomiting, diarrhea, dizziness, drowsiness, loss of consciousness, prostration and often impaired vision, delirium and convulsions.[20, 423]

Chronic poisoning from ingestion of small amounts over an extended period results in digestive reactions and the development of gangrene, sometimes causing loss of hands and feet. Dreadful convulsive and gangrenous ergotism frequently occurred in Europe in olden times from eating bread made from grain contaminated with ergot. Occasionally entire villages have been affected, and the mysterious affliction was called "Holy Fire" or "St. Anthony's Fire." There was a severe outbreak in Pont St. Esprit, France, in August, 1951, and in Calcutta in October of the same year. Cases of ergotism occurred in the central and midwestern United States from 1820 to 1885 and in Russia in 1926 and 1927.[631] Ergot deliberately taken to achieve abortion has had fatal consequences.[423]

In grazing animals, accidental intake of ergot in quantity generally brings on depression, acute pain in the extremities, lameness, then dry gangrene of feet, tail and ears, sometimes also of lips and tongue and shedding of hair and teeth, followed by death from exhaustion. Paralysis, beginning in the tongue and muscles of the throat and quickly generalized, has been observed in horses. Lowered temperature and a weakening of the pulse has preceded death, which occurred within a few hours.[114] In the spasmodic type of reaction, there is muscular trembling, incoordination, spasms, delirium and sometimes abortion.[156] Experimental feeding of sheep has produced ulceration of the tongue and inflammation of the stomach and bowels,[197, 216] but sheep are apparently not subject to accidental poisoning by ergot. Investigators maintain that the milk of animals that have consumed ergot is not toxic to humans.[130]

Poultry are highly susceptible to ergot, and gangrene occurs in the comb, wattles, beak and toes.[197]

Pinaceae

Figure 3. Slash pine (*Pinus elliottii* Engelm.) cones and needles. Courtesy Charles S. Bush, Nursery Grades Supervisor, Division of Plant Industry, Florida Department of Agriculture, Gainesville, Florida.

Slash Pine

Botanical name:

    Pinus elliottii Engelm. (*P. elliottii* var. *elliottii* Engelm.)

Other name:

    Swamp pine.

Family:

    Pinaceae, the Pine Family.

Description:

    A fast-growing tree, developing from an erect, slender seedling to a height of 50 to 100 ft. (15 to 30 m), with short, thick branches, the trunk attaining a diameter of 24 to 40 in. (.6 to 1 m). Outer bark is dark gray, furrowed, breaking into oblong plates; inner bark is red-brown. Needles, in 2s or 3s, are 7 to 12 in. (17.7 to 30 cm) long. The tree begins to bear when ten years old but not abundantly until over twenty years of age. Strobiles (in spring) are rose-purple and ¼ in. (6 mm) thick, the male 1½ to 2 in. (3.8 to 5 cm) long, densely clustered; the female, ½ in. (1.25 cm) long. Cones, red-brown, maturing and dropping in the fall of the second year, are 3 to 5½ in. (7.6 to 14 cm) long, narrow-ovate, becoming broad-ovate when open, each scale tipped with a short, straight or recurved spine. Seeds are ovoid, ¼ in. (6 mm) or more in length, smooth, gray, with a glossy, brown, membranous wing ⅝ to 1⅓ in. (1.5 to 3.5 cm) long, which becomes detached.[32, 341]

Origin and Distribution:

    Native to the southeastern United States from southern South Carolina to central Florida, southern Alabama, southern Mississippi and southeastern Louisiana. From central Florida southward, it is replaced by var. *densa*, which has remarkably heavy, hard wood. Slash pine is cultivated and naturalized as far west as eastern Texas and, since 1950, has been extensively planted in Brazil.[395]

Constituents:

    See under LONGLEAF PINE.

Propagation, Cultivation:

Seeds germinate quite readily beneath the parent tree without burning of ground litter. This pine is not very fire resistant, and burning for any reason is avoided until the trees are at least 12 to 15 ft. (3.6 to 4.5 m) tall. Disking the ground will increase the chance of seeds making good soil contact. Sprouting occurs in about two weeks. The tree is also successfully propagated by grafting and air-layering. In the wild, slash pine favors sandy soil underlain with poorly drained hardpan, generally in wet flatwoods, on the margins of ponds, creeks and bays where it is safe from fire. With fire protection, it has spread to sandhills and scrub-oak ridges and even invades stands of longleaf pine. Seedlings are prey to Pales weevil (*Hylobius pales*), the pine webworm (*Tetralopha robustella*), and young and old trees are subject to fusiform rust (*Cronartium fusiforme*) and pitch canker (*Fusarium lateritium* f. *pini*). Mature trees are attacked by the fungus diseases red heart (*Fomes pini*), butt rot (*Polyporus schweinitzii*) and recently by root rot (*Formes annosus*). Bark beetles (*Ips* spp.) cause much damage, especially during droughts. The black turpentine beetle (*Dendroctonus terebrans*) breeds in stumps and attacks some sound trees but mainly those that have been tapped for gum. Cones are attacked by the rust *Cronartium strobilinum* and the cone moth (*Dioryctria* sp.).[32]

Additional characteristics and uses of the slash pine will be discussed in the following section on the longleaf pine.

Figure 4. Lightered stumps of slash pine (*Pinus elliottii*) being delivered by railway cars for destructive distillation and production of pitch, turpentine, rosin, tall oil and various derivatives. Courtesy USDA Forest Service, Southeastern Forest Experiment Station, Olustee, Florida.

Figure 5. Slash pine (*Pinus elliottii*) logs are transported to paper mills for pulping and extraction of tall oil from the waste materials. Courtesy USDA Forest Service, Southeastern Forest Experiment Station, Olustee, Florida.

Figure 6. Base of a slash pine (*Pinus elliottii*) which was treated with an herbicide (Paraquat®), causing the tree to "lighter" (become soaked with gum) as shown by the darker portion of the cross-section. This artificially stimulated storing of gum results in an increased yield of tall oil of which turpentine and rosin are important products. Courtesy USDA Forest Service, Southeastern Forest Experiment Station, Olustee, Florida.

Figure 7. Slash pine (*Pinus elliottii*). Seedlings of a high-gum strain in a plantation, Osceola National Forest. Seeds from 100 acres of seed orchards are being distributed for commercial production. This strain, yielding twice as much gum as an ordinary slash pine, was developed at the USDA Forest Service, Southeastern Forest Experiment Station, Olustee, Florida, the source of the photo.

Figure 8. The longleaf pine (*Pinus palustris* Mill.) is distinguished by its spreading tufts of needles—to 18 in. (45 cm) long, grouped in 3s—and its cylindrical cone—to 10 in. (25 cm). The gum has run down in streaks from a gash in the trunk.

Longleaf Pine

Botanical name:

    Pinus palustris Mill.

Other names:

    Southern pine, southern yellow pine, turpentine pine, yellow pine, Virginia pine, broom pine, fat pine, Georgia pine, hard pine, heart pine, longstraw pine, pitch pine, rosemary pine.

Family:

    Pinaceae, the Pine Family.

Description:

    Seedlings in the "grass stage," normally the first three to seven years, are stemless, merely a tuft of needles rising from a taproot; after a period of root development, the tree begins to grow, at the rate of 1 to 3 ft. (.3 to .9 m) per year and eventually reaches to 80 or even 100 ft. (24.3 to 30.4 m), with a straight trunk to 2½ ft. (.75 m) thick and an open crown of short branches. Outer bark is dark, furrowed, separating into irregular plates; inner bark is reddish. The drooping needles, 3 or rarely 4 or 5 joined at the base, borne in dense tufts at the ends of the branches, are 8 to 18 in. (20.3 to 45.7 cm.) long. The tree generally does not begin to bear until the trunk reaches a diameter of 6 in. (15 cm). Male strobiles, in spring, are clustered, red-purple, 2 to 3 in. (5 to 7.6 cm) long and about ½ in. (12.5 mm.) long, ¼ in. (6 mm) thick. Female strobiles are purple, cylindrical, ½ in. (12.5 mm.) long, ¼ in. (6 mm.) thick and develop into cones which mature and are shed in the fall of the second year; the cones are stalkless, 6 to 10 in. (15 to 25 cm) long, slightly curved when immature, each scale bearing a recurved spine at the thickened tip; when mature, the opened scales release the mottled, ridged, winged seeds, ½ in. (12.5 mm.) long, the wing glossy, dark-brown, 1½ to 2 in. (3.8 to 5 cm) long.[32, 601]

Origin and Distribution:

    Native and abundant in the southeastern United States from southern Virginia to central Florida and west to eastern Texas. This is the largest and formerly the most important of the southern yellow pines.

It is said that it was once the most abundant tree in the United States.[250] Its numbers were greatly reduced by logging but, in recent years, it has been reestablished on a vast scale through reforestation programs.

Constituents:

Natural oleoresin exudate from pine resin ducts contains about 66% resin acids, 25% turpentine, 7% nonvolatiles and 2% water.[311] The main components of turpentine vary in quantity according to the source of the material: alpha-pinene represents 60 to 65% of gum turpentine, 75 to 80% of wood turpentine, 60 to 65% of refined sulfate turpentine and 50 to 65% of crude sulfate turpentine; beta-pinene represents 25 to 35% of gum turpentine, 0 to 2% of wood turpentine, 25 to 30% of refined sulfate turpentine and 20 to 30% of crude sulfate turpentine. Dipentene and other monocyclic terpenes constitute 5 to 8% of gum and refined sulfate turpentine, 15 to 20% of wood and crude sulfate turpentine; camphene is lacking in gum turpentine; the amount varies from 4 to 8% in wood turpentine and from 0 to 2% in refined and crude sulfate turpentine.[304] Turpentine from slash pine contains *l-a*-pinene, while that from longleaf pine contains some *d*-pinene.[311] In the production of synthetic camphor, the pinene is converted into bornyl chloride by the action of hydrogen chloride; thereupon, treatment with sodium acetate yields isobornyl acetate. Isoborneol results from hydrolysis, and camphor is formed by oxidation.[594]

Rosin consists primarily of diterpene resin acids of the abietic type (abietic, neoabietic, palustric and dehydroabietic) and pimaric type (pimaric, isopimaric and sandaracopimaric).[634]

Crude tall oil contains 40 to 60% resin acids, 40 to 55% fatty acids and 5 to 10% neutral properties. The fatty acids consist primarily of n-C_{18} acids, of which 75% are the monoenoic (oleic), and the dienoic (linoleic) acids, with only small amounts of the trienoic and saturated acids.[634]

Pine tar, a dark-brown, viscid fluid, contains turpentine, resin, guaiacol, creosol, methylcreosol, phenol, phlorol, toluene, xylene and other hydrocarbons.[415] It becomes granular and opaque with age.[122] Rectified tar oil consists mainly of pinene, phenolic compounds and acids, including acetic acid.[122]

Green needles from young longleaf pine yield .42% (by weight) of a balsam-scented oil. Those of slash pine yield .27%. The oil consists primarily of camphene, beta-pinene, cadinene and borneol.[289, 601] The phenolic constituents, catechin and leucocyanidins, are found in the sapwood of pine.[311]

Propagation and Cultivation:

Longleaf pine trees may be grown from seeds planted in nurseries, but stands are more often increased by encouraging natural regeneration. This pine is fire-resistant after its juvenile stage, and the forest floor is prepared by burning off grass and litter just before seedfall. Vegetative propagation trials have been much less successful with this species than with slash pine. The tree thrives in highly acid, dry, sandy soil or rocky gravel, sometimes in red clay, and flourishes best in regions having heavy summer rains. It is fast-growing, as stated, after the first seven years, though, in unfavorable situations, it has been known to remain in the grass stage for as long as twenty years. Seedlings face a number of hazards. Wild hogs uproot them to reach the starchy root-bark; the buds are eaten by cattle, sheep, cotton rats and rabbits. White grubs feed on the roots, and defoliation may be caused by the sawfly (*Neodiprion lecontii*), the pine webworm (*Tetralopha robustella*) or the Texas leafcutting ant (*Atta texana*). Longleaf pine is less susceptible to pine beetles than other southern pines. The most serious disease of seedlings is brown spot needle blight.[332] Many cones are destroyed by squirrels.[32]

Status and Oleoresin Extraction from slash and longleaf pine:

Today there are nearly 19 million acres of longleaf and slash pine, and these two species are the sources of most of the resin products and turpentine marketed in the United States. Faster-growing strains, resistant to pests and diseases and high in yield of oleoresin, have been and are being developed, and the trend is toward cultivated and fertilized plantations for more efficient management.[332] Two seed orchards of improved strains were established in Florida and Georgia in 1969. It was estimated that these and other orchards would not make an impact on the pine extractives industry for forty or fifty years.[304]

Until about 1935, the longleaf pine was the leading source of oleoresin, formerly called "naval stores," a term based on the use of pine pitch for caulking wooden ships. At that time, the slash pine assumed the greater importance and has continued to dominate the industry.

Pine extractives are obtained by three different processes. The original practice of cutting deep "boxes" in the base of the tree to catch the pine gum was recognized as destructive and was gradually abandoned after the introduction, in 1903, of an improved method of tapping, less injurious to the tree, less laborious and giving higher yields.[166, 231] In this shallow tapping process, a notch is made with an axe a few inches above the base of the trunk, and a metal apron is nailed on to lead the flow into a cup (or disposable paper bag) beneath. Horizontal incisions are made weekly, removing a strip of bark and a thin layer of sapwood, starting just above the apron and continuing higher and higher for several months to stimulate the flow of oleoresin—a somewhat translucent fluid gum—yellowish-white, aromatic, pungent, bitterish. The cup or bag is emptied monthly. Tapping may be carried on for five to eight years. Incisions can be made biweekly or monthly if the flow is extended by applying a spray or paste of 50% sulfuric acid to the cut "face," which is now generally done. A single tree 11 in. (27.9 cm) thick can yield 50 lbs. (22.6 kg) of gum over a period of four years.[139] Tapping requires two man-days of work and ten miles of walking for each barrel of gum. Up to 1930, 80 percent of the pine gum produced in the United States was collected by tapping; by 1950, only 40 percent; and 20 percent in 1960.[251] Because of the manual labor requirement and the scarcity and cost of labor (which represents 60% of total production costs), only 5 percent of pine extractives are now obtained by this method.[139]

Since 1910, pine oleoresins have been derived from heartwood chips (by-product of lumbering) and also from stumps and roots.[250] The stumps, which were formerly left to rot or were pulled up and burned, were the subject of much research prior to and during World War II which led to their full utilization. They are left in the ground long enough to thoroughly dry (a period of years), then are mechanically taken up, transported to the mill, washed, ground up and shredded, and then subjected to hot, pressurized solvent extraction in sealed tanks (which has replaced the original procedure of steam distillation). When the liquid extract cools, the resinous pitch which settles out is sold to foundaries as a core binder. The rest is distilled and the petroleum solvent drawn off to be used again. Next to be separated is turpentine and then the pine oil. The residue is a dark-red fluid resin that hardens as it cools. By purification processes, this material yields a decolorized wood rosin and a dark pitch.[634] The resin-free wood chips serve as fuel for the boilers in the processing mill.[517] On the average, 4,000 lbs. (1,814.3 kg) of wood will yield 7 to 12 gallons of spirits of turpentine.[601] In 1969, 20 percent of southern pine turpentine and 50 percent of all the rosin came from wood processed in this fashion or by the older method of steam distillation.[251]

The most recent development, which began in 1928, is the recovery of turpentine, rosin and other products of southern pine "tall oil" from the waste material of Kraft paper manufacturing by the sulfate process. When the pine wood is pulped, sulfate turpentine is condensed from the cooking vapors. The fats and resins from the wood are converted into soaps, which are skimmed off the spent liquor. The addition of sulfuric acid to the skimmed soaps causes crude tall oil to separate and it is drawn off. The name is a corruption of the Swedish word *tallolja*, meaning "pine oil," and is used to avoid confusion with the essential oil of pine.[634] Tall oil is utilized as such or is fractionated into rosin and fatty acids. Each ton of sulfate pulp yields 80 to 90 lbs. (36.2 to 40.8 kg) of tall oil. In 1969, 40 percent of all domestic rosin and 70 percent of domestic turpentine resulted as by-products of sulfate paper production.[304]

Medicinal Uses:

In the past, an oil, steam-distilled from the needles of longleaf pine, was used to a limited degree as an antiseptic, and the fiber extracted from the residual "pine straw," being still aromatic, was employed, at least experimentally, as a disinfectant surgical dressing and maggot-repelling bandage for wounds.[157]

Turpentine was formerly administered internally in the treatment of gonorrhea, leucorrhea, various urinary diseases, catarrh, rheumatism and chronic inflammation of the bowels.[305] A few drops of turpentine on sugar has been a popular home remedy for colds. Externally, turpentine and rosin (colophony) were used as rubefacients and stimulants in rheumatic and chest complaints and ointments and plasters containing rosin were applied to boils and ulcers.[624]

Pine tar was formerly an ingredient in expectorant sirups taken for the relief of bronchitis. Pine tar is mildly stimulating and antiseptic and is today prescribed only for external use in cases of chronic, and especially parasitic, skin diseases. It is usually employed as a 50% ointment or lotion.[423]

Terpin hydrate is the main synthetic product of turpentine used in pharmaceutical preparations.[634] The *cis*-form is employed as an expectorant. Apart from administration to humans, it is commonly used in treating chronic bronchitis in horses, cattle and dogs.[552]

Synthetic anethole is given as an intestinal stimulant and expectorant.[552] Among other pine-derived pharmaceuticals are *dl*-menthol and synthetic camphor, produced on a small scale, which are employed for the same purposes as the natural products.

Tall oil rosin contains sterols,[290] mainly sitosterol, a potential precursor of cortisone.[634] In 1968, the first factory for the production of steroids from pine pulp extractives was built in Russia.[311] Russia is also pioneering in the commercial production of vitamins A and E from pine needles.[634]

Toxicity:

Pine tar may cause skin irritation unless diluted.[423] Turpentine is a skin irritant and tends to produce allergic contact dermatitis, often manifested as dryness and cracking of the skin. When fresh, it is less irritating and sensitizing than after storage. Household products containing turpentine are common sources of skin difficulties.[190]

Other uses:

Harper wrote concerning longleaf pine in 1928, "This species probably has more uses than any other tree in North America, if not in the whole world ... the exploitation of its products has furnished the principal source of income for millions of people at one time or another. Its lumber and naval stores have been exported to all parts of the civilized world, over a million dollars' worth in a year going out from the port of Mobile alone."[250]

In the late 19th century, when the price of jute for baling cotton became exorbitant, a million yards of bagging was manufactured from pine-needle fiber as an emergency substitute.[157] The fiber has also been used in upholstery. Longleaf pine needles are much utilized in basket-making. Machine-chopped needles serve as litter for poultry sheds, bedding for horses and cattle, and also as fertilizer in cotton fields.[601]

Formerly much used for ships' masts and for paving blocks, pine wood has been highly prized for construction since colonial times. It is particularly valued for framing, laminated beams, plywood and for creosoted or pressure-treated poles; vast quantities are utilized for paper pulp.

Turpentine was previously most important as a solvent for resins, waxes, varnish and paint. Its major role now is as a chemical raw material, primarily in the conversion of α-pinene with aqueous mineral acids to synthetic pine oil (consisting mainly of α-terpineol), which is being increasingly used as a bactericide and scent in household cleansers and in treating textiles. Secondarily, turpentine is a source of polyterpene resins used in adhesives, chewing gum and sizing of paper and fabrics.[634]

Among the various fractions of turpentine now marketed are α-pinene, β-pinene, dipentene, terpinolene, camphene and other monocyclic terpenes,[32, 575] which are utilized in a diversity of manufactured products. Synthetic materials derived from turpentine include essential oils used extensively in perfumes and others which impart licorice, *Citrus*, lemongrass, nutmeg, peppermint, spearmint, cinnamon and other flavors to food products. Menthol from turpentine is added to tobacco in cigarettes and to cosmetic and toilet products.[634] Toxaphene (insecticide) is a complex mixture of at least twelve compounds formed by the chlorination of camphene from southern pine.

Whole tall oil is used in making crude soap, cleansers, disinfectants and asphalt emulsions and serves as a flotation oil in mining. Much is fractionated into rosin and fatty acids.[304]

Rosin is mainly employed in sizing papers and in the making of pressure-sensitive adhesives; it is used also in the compounding of natural rubber and the emulsifying of synthetic rubber. It was formerly much used in laundry soap. Minor rosin products now include chewing gum. Too, rosin is applied to violinists' bows, rubbed on athlete's hands and used for coating the inside of beer casks.

The fatty acids of tall oil are made into protective coatings, epoxies, inks, soaps and detergents.

Related species:

Pharmaceuticals are derived also from the Loblolly pine (*P. taeda* L.), now the dominant pine species from New Jersey to central Georgia and west to eastern Texas.

Cupressaceae

Figure 9. Prickly juniper (*Juniperus oxycedrus*). Rooted, growing cuttings collected by Sylvester G. March of the National Arboretum, May 1973, in the Juniper Forest under protection of Nikitsky Botanical Garden, Yalta.

Prickly Juniper

Botanical name:

　　Juniperus oxycedrus L.

Other names:

　　Prickly cedar, sharp cedar, brown-berried cedar, large-fruited juniper, brown-fruited juniper, cade-oil plant.

Family:

　　Cupressaceae, the Cypress Family.

Description:

　　A shrub, low-domed and compact or prostrate and spreading, 3.3 to 6.6 ft. (1 to 2 m) high, or a small tree to 33 ft. (10 m). The thick trunk of a specimen which had attained extraordinary size—62 ft. (19 m) in height—was displayed at the Exposición del Mueble, Barcelona, in 1923.[196] Branchlets are fairly slender, angular; leaves, borne in whorls of 3, are evergreen, spreading, stiff, linear, the upper half tapering to a needle-like point; ½ to ¾ in. (12.5 to 19 mm) long and .04 to .06 in. (1 to 1½ mm) wide, with 2 lengthwise bands of silvery-white stomata above, all-green and convex beneath. Male catkins ovate, axillary, nearly sessile, solitary, ⅛ to ¼ in. (3 to 6 mm) long. Fruit (cone), ripening the second year, axillary, solitary, short-stalked, globose, composed of 3 to 6 bracts or scales; ¼ to ½ in. (6 to 12.5 mm) wide, glossy, red-brown, sometimes partly coated with bloom; enclosing sometimes, 1 usually 2 to 3 triangular seeds. Variety *brachyphylla* has shorter, less sharp-tipped leaves; variety *maderensis* has very narrow leaves rounded or short-pointed at tip; variety *umbilicata* differs in the shape of the cone; variety *viridis* bears green cones.[115, 454, 474]

Origin and Distribution:

　　Native to the Mediterranean region of Europe from Portugal and Spain to the Balkans; also the Aegean Islands, Cyprus, Morocco, Algeria and Tunis, Syria, Lebanon, Turkey, Caucasus, northern Iran and Iraq, occurring in dry pine and oak forests at elevations of 1,475 to 5,640 ft. (450 to 1,750 m). It was introduced into England late in the 18th century and is occasionally cultivated there, as well as in North America and Argentina as an ornamental.

Constituents:

The chipped heartwood and large roots yield by steam or vacuum distillation a product which, after standing two to two and one-half weeks, separates into three layers, the lower being tar-like, the center watery, the upper a transparent, orange-brown or dark-brown oil (cade oil or juniper tar oil) with a tarry, smoky odor and bitter flavor. The main constituents are the sesquiterpene, d-cadinene, ($C_{15}H_{24}$); also l-cadinol and dimethylnaphthalene;[122, 210] and guaiacol, ethyl guiacol and cresol.[584]

Propagation, Culture and Harvesting:

This plant, like all junipers, can be grown from seeds or from cuttings of young growth taken in the fall. The seeds remain viable in cool, dry storage for a number of years. Germination may take several months or as long as one year.[155] There are no commercial plantations. All of the raw material for pharmaceutical purposes is furnished by the abundant wild stands. As soon as possible after harvesting, the wood is chipped and distilled in factories in Yugoslavia, southern France and Spain to obtain the essential oil. Annual production is between 125 and 200 metric tons of crude oil.[41, 583]

Medicinal uses:

Centuries ago, cade oil was employed to treat corneal opacities, was placed in dental cavities to allay pain, was rubbed on the head to kill lice and their eggs, was applied to snakebite and was applied to the penis before intercourse as a contraceptive. It was administered internally and externally as a remedy for leprosy. In modern Europe, refined cade oil is taken internally as a vermifuge—three to five drops in a little water, followed by a weak purgative. It is applied sparingly at two- or three-day intervals to old wounds and ulcers to promote healing.[196] Cade oil has a long history of use in veterinary practice in the Old and New World for treating parasitic skin diseases. Its application to similar human afflictions, especially eczema, psoriasis and pruritic dermatoses, is more recent. It is usually utilized in the form of a 1 to 5% ointment, cream or paste. A 10% ointment or 4% shampoo may be prescribed for seborrheic dermatitis of the scalp.[423] The oil is also an ingredient in antiseptic soaps.[122, 233, 423]

Toxicity:

Cade oil is irritant and has no internal use in the United States.

Other uses:

The shrubby form of the prickly juniper is valued as a forest understory, preventing erosion on steep mountainsides and, because of its spininess, protecting forest tree seedlings from being devoured by grazing animals.[583]

In the perfume industry, the refined oil is prized in men's fragrances to which it gives a woody or smoky-leathery character; also in soap and other toiletries. It is also employed in food manufacturing to impart a smoky flavor to unsmoked fish and meat products.[41, 210] An oil distilled from the fruits is sometimes sold deceptively as juniper berry oil.[41]

10.

Figures 10 and 11. Prickly juniper (*Juniperus oxycedrus* L.), source of cade oil. Photographed in the National Arboretum, Washington, D.C.

11.

Ephedraceae

Figure 12. A bushy specimen of *Ephedra gerardiana* Wall., flourishing in a pot ordinarily kept inside a greenhouse at Longwood Gardens, Kennett Square, Pennsylvania.

Ephedra

Botanical name:

 Ephedra spp., principally *E. major* Host (*E. nebrodensis* Tineo); *E. gerardiana* Wall. (*E. vulgaris* Rich.); *E. intermedia* Schrenk & Meyer; *E. sinica* Stapf; and *E. equisetina* Bunge.

Other names:

 Joint fir, ma-huang (China).

Family:

 Ephedraceae, the Ephedra Family.

Description:

 Evergreen, erect or reclining shrubs, ranging from 6 in. (15 cm) to 6 ft. (1.8 m) in height; thick-stemmed, 3 to 5 in. (7.5 to 12.7 cm) at the base, with many slim, jointed branches and twigs having fine longitudinal ridges or ribs. Leaves are minute, scale-like, sheathing the nodes. Flowers are dioecious, borne in round or elongated catkins. Fruit (cone) is globular, usually red, occasionally yellow, to ¼ in. (6 mm) wide.

Origin and Distribution:

 E. major, a tall shrub, grows wild in the Mediterranean area of Spain, in Sicily, Afghanistan and Pakistan. *E. intermedia*, 1 to 2 ft. (30 to 60 cm) high, ranges from Inner Mongolia to Pakistan, where it occurs at lower altitudes than *E. major*. *E. gerardiana*, a dwarf species 6 in. to 2 ft. (15 to 60 cm) high, is native to the northwest Himalayas (at altitudes between 7,000 and 16,000 ft.–2,130 to 4,880 m) in northern India, West Pakistan, Tibet, Szechwan and Yunnan Provinces of China.[111] In Pakistan, it is less accessible than *E. major* and handicapped by more humid weather.[460] *E. sinica*, to 1½ ft. (45 cm), is found at the 5,000-ft. (1,515-m) level in northern China from Sinkian to Hopeh Province and north to Outer Mongolia. *E. equisetina*, from 2 to 6 ft. (.6 to 1.8 m) tall, is hollow-stemmed; it occurs at altitudes between 4,000 and 5,500 ft. (1,220 to 1,678 m) in Inner Mongolia. These species have been experimentally cultivated in Australia, Kenya, England and the United States. In this country, efforts were most successful in South Dakota but, because of high labor cost and other factors, the crop proved economically unfeasible. Commercial cultivation has been contemplated in India.[21]

Constituents:

The woody basal stems of *Ephedra* plants are very low in alkaloid content; roots and fruits are nearly alkaloid-free. The green branches and twigs of the cited Mediterranean and Asiatic species contain the alkaloids ephedrine ($C_{10}H_{15}NO$) and pseudoephedrine (isoephedrine) in varying amounts depending on the plant, the altitude at which it grows, and the time of year and weather.

Ephedrine was first isolated by G. Yamanashi, at the Osada Experimental Station, Japan, in 1885. Investigations by K. K. Chen and colleagues at Peking Union Medical College attracted worldwide attention to this drug in 1924.[236] It has continued to be the subject of much study.

E. major may contain over 2.5% total alkaloids, nearly 75% of the total being ephedrine. *E. intermedia* is low in ephedrine but fairly high in pseudoephedrine.[460] In *E. gerardiana*, total alkaloids vary from 0.8 to 1.4%, of which 50% is ephedrine.[117] *E. sinica* contains roughly 1.31% total alkaloids, 1.12% ephedrine; *E. equisetina*, 1.75% total, 1.58% ephedrine.[20] The pharmaceutical trade requires that dried ephedra contain no less than 1.25% ephedrine.

Ephedra has a strong pine odor and very astringent taste. Ephedrine and pseudoephedrine are very stable. A solution of ephedrine hydrochloride sealed for six years showed no oxidation nor loss of activity.[117]

Ephedrine acts much like epinephrine but has the advantage that it can be given orally as well as by injection. It is a mydriatic drug, stimulates the cardiac muscle, causes a marked rise in blood pressure and in pulmonary pressure; it is an effective dilator of the bronchioles, contracts the uterus,[20] and is a diuretic as well. Pseudoephedrine is weaker in action on the heart, is a more potent diuretic and has no effect on the uterus. The two alkaloids seem equal in their effect on voluntary and involuntary muscles. They are quickly absorbed from the gastrointestinal tract and pass through the liver unchanged.[117]

Propagation, Cultivation and Harvesting:

Ephedra plants can be grown from seed, layering or dividing the rootstocks. They thrive best where the rainfall is less than 20 in. (50 cm) annually. The fruits are best gathered by hand, dried, and then the seeds extracted. Seeds are planted in early spring, directly in the field at 2½ ft. (.75 m) spacing and ½ in. (1.25 cm) deep, in rows 2½ ft. (.75 m) apart. The seedlings must be watered and kept weed-free during the first year. Harvesting begins when the plants reach four years of age and takes place in autumn, during the blooming season, when alkaloid content is highest. Throughout the summer rains, alkaloid content declines, then it gradually increases until it is double that of springtime. Workers wielding hand sickles cut all stems that are less than ½ in. (1.25 cm) in diameter. These are dried in the sun for fifteen days to reduce volume by 50 to 60 percent. Drying by artificial heat for three hours at 120° F. (50° C.) has reduced alkaloid content from 1.22 to 0.27%. After drying, the stems are beaten with sticks to break off the joints and then screened to separate the unwanted joints from the internodes, which are then packed in bags or covered containers and stored in a dry atmosphere. Exposure to humidity during storage can result in complete loss of alkaloids.[20] The dried material was formerly exported in bales.

China was, until 1925, the only major supplier of ephedra. It was not until 1926 that native species were investigated and analyzed in India and found to be usable in place of the imported drug.[117] India began exporting ephedra from Baluchistan in 1928. That region is now part of West Pakistan, and the latter country is now the leading source of natural ephedra. Spain is a minor producer. Because of the variation in alkaloid content of the raw material, it has been found necessary to set up a factory at Quetta in the major growing area to process the plants and produce Ephedra Extract. The dry extract, containing 18 to 20% total alkaloids, serves as the source of the pure alkaloids.[20] In recent years, most of the ephedrine and pseudoephedrine used in the western world has been manufactured synthetically. In the 1950s, the price of synthetic ephedrine rose sufficiently to bring about an increased demand for the natural product, which some pharmaceutical companies still prefer. For one thing, synthetic ephedrine is not optically active. Also the re-entry of Chinese ephedra in the world market has lowered the price of the crude drug.

Medicinal Uses:

For at least 5,000 years, the Chinese have prized ephedra for the preparation of herbal teas and pills taken to cure colds, coughs, malarial and other fevers, headache and eruptions due to infection.[236, 469] The roots and joints reduced to powder and combined with oyster shells or other products are given to stop excessive sweating.[469] In present-day practice in the United States, ephedrine, orally or subcutaneously, is prescribed in cases of rhinitis, asthma, hay fever and emphysema. The relief is more enduring than that produced by epinephrine. After three or four days, patients may cease to respond. In such cases, dosage is suspended for a week or so and then resumed with good effect. Ephedrine salts in nasal sprays relieve congestion and swelling. Ephedrine is given subcutaneously to prevent hypotension during anesthesia. Orally, it has been used with success in treating certain forms of epilepsy, nocturnal enuresis, myasthenia gravis and urticaria accompanying angioneurotic edema. Pseudoephedrine, taken orally, is an effective nasal decongestant.[423]

The ten species of *Ephedra* native to North America are said to contain no useful alkaloids. They have been used mainly in folk-medicine decoctions taken primarily as remedies for venereal diseases. Some westerners in the past consumed such decoctions as daily beverages, and the plants acquired the local names of Desert Tea, Mexican Tea, Teamster's Tea, Mormon Tea and Whorehouse Tea. *E. viridis*, dried and packaged, is today widely sold by "health food" purveyors as "Squaw Tea" for beverage purposes.

Toxicity:

In large doses, ephedrine induces nervousness, headache, insomnia, dizziness, palpitations, flushing of the skin, tingling and numbing in the extremities, nausea, vomiting, and possibly brief spells of pain in the cardiac region. Some patients show a dermatological reaction.[117, 423]

Other uses:

The fruits of *E. gerardiana* and *E. intermedia* are eaten. Large woody knots develop on the rhizomes of the latter plant and are collected and useful as fuel in Tibet.[20] *E. intermedia* is employed in tanning sheep skins; the ashes of the burned plant are blended with powdered chewing tobacco in Pakistan.[460]

Bromeliaceae

Figure 13. Well-developed, unripe pineapples (*Ananas comosus* Merr.). The leafy crown and the slips (below the fruit) are used for propagation. (*See also* color illustration 1.)

Pineapple

Botanical name:

    Ananas comosus Merr. (*A. sativus* Schult.)

Other name:

    "Pine."

Family:

    Bromeliaceae, the Pineapple Family.

Description:

    A biennial, herbaceous plant, shallow-rooted, with a single stout, starchy stem 2 to 4 ft. (.6 to 1.2 m) high, supporting a rosette of stiff, recurved, strap-like, pointed leaves up to 40 in. (1 m) long, with spiny-toothed or nearly smooth margins. Flowers, borne in a (usually solitary, sometimes multiple) dense terminal head of 100 to 200, are blue-purple, six-parted, with 6 stamens and 1 style. The head develops into a compound fruit—round-oval to cylindrical, 6 to 10 in. (15 to 25 cm) long and 4 to 6 in. (10 to 15 cm) thick—which actually consists of the thickened, fleshy rachis, on the surface of which are embedded the abortive ovaries with their spiny bracts covering the deep-set floral remnants ("eyes") and fused together to form a hexagonally sectioned "rind," yellow to orange-red, dark-green or green-and-brown when the fruit is ripe. Mature fruit weight ranges from 1 to 16 lbs. (.45 to 7.2 kg). The fruit is topped by an erect tuft ("crown") of stiff, pointed leaves. New plants ("slips") emerge immediately below the fruit, and suckers develop in the leaf axils at the base of the main stem. Ratoons are shoots which spring up from below ground and have independent root systems. Internally, the ripe fruit is more or less fibrous, especially at the core, very juicy, nearly white to yellow, sub-acid and fragrant. Seeds (usually absent) are produced in cross-pollinated pineapples and utilized for breeding experiments.[129, 415, 458]

Origin and Distribution:

    It is believed that the pineapple originated in Brazil, though it was seen in cultivation in the West Indies by Christopher Columbus and his fellow voyagers in 1493. Spanish and Portuguese traders introduced the plant into the East Indies, India, China and Africa before the end of the 16th century. The fruit was reported to be abundant in the markets of

Bengal in 1800. The date of the pineapple's arrival in Hawaii is uncertain, but it was being grown on the Island of Hawaii in 1813, and the first cannery was established on Oahu in 1892. In 1860, Benjamin Baker of Key West, Florida, acquired slips from Cuba and raised pineapples on Plantation Key so successfully that pineapple culture quickly spread as far north as the Indian River. However, susceptibility to cold spells gradually reduced plantings in this state to dooryard status. The pineapple ranks among the major fruit crops of all tropical countries. Leading producers are Hawaii, Brazil, Malaysia, Formosa, Mexico, Thailand, South Africa and Australia. Important exporters of fresh fruit to Europe include the Ivory Coast, Kenya and the Azores, where pineapple culture in glasshouses is a major and long-established industry.[301] There is today a growing demand for the fresh and processed fruit and pineapple juice. Increasing labor costs are threatening to drastically reduce Hawaiian production. Six of the nine major Hawaiian processors (including Libby) have moved to other areas with more favorable conditions, and Dole has transferred 75 percent of its operation to the Philippines. In 1971, Dole had a total of 8,000 acres (3,237.5 ha) of pineapple fields on three islands of the Hawaiian group.[184] Philippine plantings in 1974 covered 59,127 acres (28,380 ha).[38]

Constituents:

Ripe pineapple contains a fair amount of calcium and thiamine but is low in phosphorus, iron and carotene. Ascorbic acid content varies with the cultivar from low to fair. Of the non-volatile acids, 87% is citric and 13% *l*-malic. In the final ten days before harvest, the starch in the stem and leaves is converted into sugar and transferred to the fruit, and there is no increase in sugar content after harvest. The fruit, unripe and ripe, the juice and the entire plant contain the proteolytic enzyme bromelain (or bromelin), first extracted in 1891.[57] The highest concentration occurs in the center of the lower portion of the mature stem.

Bromelain is completely precipitated by ammonium sulfate, methanol, isopropanol and acetone, the latter being the most suitable for large-scale extraction. Pineapple-stem bromelain contains several proteases which differ in "their susceptibility to oxidizing and reducing agents and in the pH value at which they hydrolyze their substrates most rapidly." It is rich in acid phosphatase, a fairly stable enzyme. When fresh, bromelain contains peroxidase, but this element disappears after three to six months of storage. Heinicke and Gortner present a detailed comparison of the enzymatic activities of bromelain, papain and ficin.[254]

Propagation, Cultivation and Harvesting:

The pineapple plant has been traditionally propagated by slips, suckers or crowns. When large quantities of planting material are required, two recently developed techniques may be resorted to: the division of crowns into about 20 leaf cuttings with axillary buds or the slicing of the stem into 15 to 24 cross-sections, each about 1 in. (25 mm) thick. The latter are dipped in a 1% solution of potassium permanganate before being placed in sand beds.[315] Slips and crowns may be stored for six months, upside-down in semi-shade. The plant tolerates a wide range of soil types, but it is important to fumigate the field against the rootknot nematodes and other soil pests and to pretreat with high-nitrogen fertilizer. Weed control may be accomplished with herbicides. Mulching with paper or plastic reduces weeding requirements and increases yield and quality.[191] Where labor is plentiful and inexpensive, most field operations are manual.[459] In highly developed countries, mechanization is necessary, and pretreatment, laying of mulch and planting activities are all performed by machine.

There are many named cultivars, though only a few are prominent. 'Smooth Cayenne' (with non-spiny leaves) is the most widely grown. 'Red Spanish' is favored in the Caribbean area, 'Abacaxi' in Brazil, 'Queen' in South Africa and Australia, 'Singapore Spanish' in Malaya.[458] 'Cabezona' of Puerto Rico and 'Sugar Loaf' of the Bahamas are large, very sweet and prized for local consumption.

Choice of cultivar, type of planting material and various other factors (including disease and pest control) affect rate of growth and fruit production. Blooming may take place fifteen months after planting, but it is becoming a common practice to induce flowering several months earlier by the application of ethylene, ethephon, calcium carbide (generating acetylene) or other chemicals.[10] The fruit generally ripens five to six months after flowering. There is extensive literature on fertilizer trials, variations in plant spacing to improve fruit size and quality and total yield. In Puerto Rico, a dense planting of 'Red Spanish' (set at 18,000/acre—roughly 43,000/ha) yielded 28.8 tons per acre (about 69 tons per hectare).

Movement through pineapple fields is difficult, and harvesting of the crop is laborious and increasingly unappealing to workers. It has been the custom to trim the fruits and cut them from the plant by hand and then place them in sacks or baskets for carrying to bins; or, they may be tossed to truckmen. Mechanical harvesting systems have been introduced which save much time and effort. These usually provide a conveyor boom extending from a tractor or truck across several rows. The first time through the field, the fruits are placed on the moving belt; the second time, the cut tops and slips are picked up for transfer to a planting machine. Some conveyors have "rubber finger" attachments which detach the fruit from the plant with crown attached. Crowns are later removed by a mechanical decrowner.[414] The crown-end of fruits to be sold or shipped fresh is then dipped in a fungicide to prevent spoilage. After mechanical sizing, the

14.

15.

Figure 14. Manual harvesting of pineapples in PRACO plantation, Vieques, Puerto Rico, in 1951.

Figure 15. Pineapples delivered to a canning factory in Puerto Rico in an era when labor was plentiful.

pineapples are packed in cartons which are loaded into refrigerated containers for ship, rail or air transport. Cannery operations are wholly mechanized, and the cores are processed for sirup to fill the cans.

Each main stem bears fruit but once. After harvest, slips and suckers are trimmed from the plants in the field for use in propagation and to encourage ratoon development. The second crop (18 to 23 months later) is produced by the ratoons. After that harvest, the plants are destroyed and the field cleared for replanting. Large producers have for a number of years disked the spent plants and plowed them under or have chopped them for silage. Some have turned from these practices to the production of dehydrated meal. In the process of creating this by-product (which is at the same time an efficient means of waste disposal), bromelain is extracted for pharmaceutical and other markets. The original patents for bromelain preparation were based on precipitation from pineapple juice, but the juice is too much in demand to be utilized for this purpose.[129, 184]

Medicinal uses:

Bromelain is usually combined with pancreatin (an animal enzyme) in the preparation of digestive aids.[184] As an anti-inflammatory agent, it is given after dental, gynecologic and general surgery and employed in treating sprains, contusions, hematomas, abscesses and ulcerations.[423] It has been injected in solution to dissolve mucus in human organs prior to x-rays.[129, 184] The pineapple has many uses in folk medicine. The juice is an active diuretic and, in quantity, promotes uterine contractions.[98] Dupaigne discusses bromelain and its uses—in dematology, stomatology, urology, gynecology, geriatry and in treating phlebitis and pulmonary edema—in a lengthy review with 304 bibliographic references.[163]

Other uses:

Bromelain is added to Knox® gelatin to render it more soluble, especially when liquefied for drinking to promote fingernail growth. It is also employed alone or in combination with papain and/or ficin as a meat tenderizer and for chill-proofing beer.[184, 552] In addition to the main food uses of the pineapple, it is also a source of vinegar and wine. Pineapple skins and other wastes from the canning factories are often converted into stockfeed. In Hawaii, the hexagonal sections of the fruit rind are dried and strung in leis. A very fine fiber extracted from the leaves is made into cloth in the Philippines, China and Formosa.[98]

Toxicity:

The unripe pineapple is poisonous and causes violent purgation. In the tropics, it is sometimes used unwisely as a vermifuge, abortifacient or treatment for venereal disease.[98]

Liliaceae

Aloe

Botanical and common names:

Of the more than 300 species of *Aloe*, only a few have been of importance in the pharmaceutical industry. The three leading species are *Aloe ferox* Mill., Cape aloe; *A. perryi* Baker, Socotrine or Zanzibar aloe (often confused with *A. succotrina* Lam. of the Cape Peninsula and adjacent mainland of South Africa); and *A. barbadensis* Mill. (syn. *A. vera* Tourn. ex Linn.), Mediterranean, Curaçao or Barbados aloe.

Family:

Liliaceae, the Lily Family.

Description:

Aloes are perennial, succulent plants with stout stems ranging from very short to 20 ft. (6 m) tall, bearing dense rosettes of fleshy, tapering, spiny-margined leaves to 2 ft. (60 cm) long and to 4 in. (10 cm) wide at the base. The leaf epidermis is smooth, thick, rubbery, waxy, grayish-green and clings firmly to the clear, jelly-like, mucilaginous pulp within. The flowers are borne in cylindrical, terminal racemes on central flower stalks 3 to 4 ft. (.91 to 1.2 m) tall, are pendent, tubular, yellow to orange-red and rich in nectar.

Origin and Distribution:

Aloe ferox is native to the Cape Province, Lesotho and Natal in South Africa [283]; *A. perryi* is native to the island of Socotra in the Indian Ocean and cultivated in East Africa and Arabia [475]; *A. barbadensis* is native to the Mediterranean region of southern Europe and North Africa and to the Canary Islands and commonly grown in Egypt, the Near East, Bermuda, the Bahamas, West Indies, tropical America and the tropics generally; also in southern Florida and Texas. In the Netherlands Antilles, this species has been cultivated commercially in Aruba and Bonaire, but the product was exported from Curaçao, hence the trade name, Curaçao aloe.

←

Figure 16. Mediterranean aloe (*Aloe barbadensis* Mill.) with spikes of yellow flowers and numerous offshoots around the mature plants. Photographed on the island of Curaçao, Netherlands Antilles.

It is no longer exploited in Barbados but to some extent in Haiti and Venezuela.

Constituents:

The very bitter, yellow latex which drains from the subepidermal, longitudinal cells of cut aloe leaves contains anthraquinone glycosides which are collectively termed "aloin." The aloin content of Cape aloe is about 10 percent; Socotra aloe, 25 to 28 percent; and Mediterranean aloe, around 30 percent. Aloin consists mainly of the pentosides *barbaloin, isobarbaloin* (in *A. barbadensis*), *β-barbaloin*, resins, saponins and other substances.[20] Barbaloin, on hydrolysis, yields a mixture in which *aloe-emodin* and *D-arabinose* have been identified.[423] *Aloe-emodin* content is highest in Curaçao aloe, which also possesses chrysophanic acid.[122]

Dried latex of Cape aloe contains aloinosides A and B,[584] (O-glycosides of barbaloin with an additional sugar).[122] Aloinoside B is an ll-*mono-a-L-rhamoside* of barbaloin.

The aloe resins are composed of resinotannols with cinnamic or *p-hydroxy-cinnamic* (*p-coumaric*) and other aromatic acids.[584]

Lorenzetti and coworkers reported that both the freshly drained "juice" [presumably the latex] and juice heated for 15 minutes at 80°F., inhibited the activity of *Staphylococcus aureus* 209, *Streptococcus pyogenes, Corynebacterium xerose* and *Salmonella paratyphi*.[347]

Many investigators have endeavored to establish the active principles of the mucilaginous pulp. It is 96% water. Among its polysaccharides, researchers have isolated glucose, mannose, traces of arabinose, galactose and xylose.[518] It is said to possess "biogenic stimulators"[188] and wound-healing hormone activity.[204]

Propagation, Culture and Harvesting:

Numerous offshoots, or suckers, develop around the base of the mature aloe plant. They may be separated and transplanted when 6 or 8 in. (15 or 17 cm) high and will have full-grown leaves in three years. Aloe plants will grow in sand or limestone but require perfect drainage. If fully exposed to the sun and not irrigated, plants of *Aloe barbadensis* will be somewhat stunted, with thin leaves generally tinged with pink, rich in bitter latex and strong in odor but with little mucilaginous pulp (often called aloe "gel"). If partially shaded, irrigated in dry seasons and generously fertilized, the plants reach maximum development, the leaves are lush and green, with less concentrated yellow latex and well filled out with pulp. Mature, fat leaves average 1 lb. (.45 kg) in weight.[388]

Aloe leaf harvesters must wear heavy gloves and protective clothing to avoid injury by the sharp spines. The latex, which turns purplish-red soon after exposure, makes indelible stains on clothing, containers and cement floors. To avoid undue leakage of latex, leaves are cut with the point of a knife, at the constricted, clasping base, close to the stem, and will self-seal and remain in good condition for two to three weeks without refrigeration. They can therefore be shipped long distances if wrapped individually in newspaper and well packed to avert damage.

Where aloe is cultivated solely for the latex, no effort is made in cutting to avoid leakage, the leaves being lopped off near the base with a short sickle. In South Africa, they are carried face up by the armload to a shallow pit in the ground which has been lined with a piece of hide or canvas. The leaves are stacked face down, base inward, around the perimeter of the pit until the stack is about 1 m high (approximately 200 leaves). The latex drips immediately and continuously for six or seven hours. The pit lining is then gathered up like a sack, and the latex is poured from it into a tin can for carrying to the vat, oil drum or kettle in which it is boiled and stirred for four hours until the water is evaporated.

The concentrated latex is then poured into 4-gal. (15-1 metal cans. As it cools it solidifies, and the semi-transparent finished product is then ready for export.[556]

Figure 17. "Bitter aloes"—a hardened block of the inspissated aloe latex, the powdered product ranging from yellow to khaki in color.

In the West Indies, the method is somewhat different, the cut leaf ends being placed so as to allow the latex to drip into a sloping trough and flow out the lower end into a container. Evaporation is achieved more slowly with less heat or by a vacuum process. The concentrated latex is accordingly opaque and referred to as "liver" aloe. The so-called Curaçao aloe has a characteristic iodoform odor.

South African production of evaporated aloe latex amounts to 500 metric tons annually.[466]

Medicinal uses:

In the tropics, aloe plants are popular in folk medicine. The peeled fresh "gel" is applied to inflamed eyes and on all kinds of skin inflammations, sores and burns.[225, 396] Its efficacy in healing x-ray burns was first attested in 1935 and confirmed by later experiments.[128, 353, 498] The pulp is widely taken internally to relieve sorethroat, ulcers and intestinal ailments. The potent fresh latex is taken in very small doses as a purgative.

In pharmaceutical practice, the main use of the dried and powdered latex ("bitter aloes") and of aloin extracted from it has been as a drastic purge and treatment for chronic constipation. Effectiveness is attributed to the irritant action on the colon. This use has declined, as both products tend to cause painful griping, though aloin less than the resinous latex. Belladonna, strychnine, cathartics or soap have frequently been given with aloe or aloin to reduce griping.[122, 423]

Aloe is mainly utilized today as an ingredient in Compound Tincture Benzoin (see under BENZOIN).

Toxicity:

Aloe is contraindicated in pregnancy and in individuals afflicted with hemorrhoids; also it is apt to cause kidney irritation.[423]

Other uses:

Many cosmetic manufacturers are incorporating the stabilized mucilaginous "gel" of the aloe leaf[482] into creams, lotions and ointments for softening, soothing or moisturizing the skin.[396]

199513

Figure 18. Autumn crocus (*Colchicum autumnale* L.). Reproduction from color painting by P. Turpin in *Flore Médicale* Vol. 3 (F. P. Chaumeton, Paris; 1816).

Autumn Crocus

Botanical name:

    Colchium autumnale L.

Other names:

    Meadow saffron, naked ladies.

Family:

    Liliaceae, the Lily Family.

Description:

    A perennial herb to 1 ft. (30 cm) tall, having a fleshy, conical or ovate corm 1¼ to 2 in. (3 to 5 cm) long when mature, with a dark-brown, membranous coat, an inner coat of lighter brown, and producing at the base numerous fibrous roots. Internally it is white, firm, with a milky juice, very bitter and acrid and with a radish-like odor. It is generally found 6 to 8 in. (15 to 20 cm) beneath the surface of the soil. From August to October, 1 to 6 flowers arise in succession, directly from the immature corm. They are tubular and enclosed at the base by a membranous sheath. The blooms are lavender to light-pink, with 6 flaring segments 1¼ to 1¾ in. (3 to 4.5 cm) long and 6 stamens tipped with oblong yellow anthers, and they wither and die down quickly. The ovary is at the bottom of the white, 2- to 8-inch (5- to 20-cm) tube within the corm. The tube is more than half below ground, and there is no part of the plant above ground during the winter. In April, there appear 3 to 4 dark-green, smooth, glossy, erect, oblong-lanceolate leaves 4¾ to 12 in. (12 to 30 cm) long and ⅝ to 1⅝ in. (1.5 to 4 cm) wide. In the center of the leaf cluster, there will be 1 to 3 oblong, pointed, three-lobed, three-valued capsules 1½ in. (40 mm) long and ¾ in. (20 mm) wide, which continue to develop and ripen and split open in June and July while the leaves are turning yellow and dying back. By this time, the corm has reached full growth, and new offshoot corms (1 or more) are developing at its base. The old corm gradually turns spongy and watery and decays by the following April.[63, 305] The numerous seeds are brown externally, white internally, ⅛ in. (3 mm) wide, minutely pitted on the surface, very hard, bitter and acrid. During drying they darken and exude and become coated with a sugary substance.[584]

Origin and Distribution:

The autumn crocus is native to grassy meadows and woods and riverbanks in southeastern Ireland, England, the Netherlands and Denmark and, at altitudes between 1,312 and 3,934 ft. (400 and 1,200 m) ranges east to Poland and south to Spain and central Italy and North Africa. It is sparingly naturalized in Scotland. It was formerly quite abundant and conspicuous as a European wild flower but is becoming scarce because of changing environmental conditions.[331] There are many cultivars grown as ornamentals throughout the north-temperate zone. Experimental plantings have been made in France and Indonesia with a view to commercial cultivation for drug production.

Constituents:

Besides the main active principle, colchicine (acetyltrimethylcolchicinic acid) ($C_{22}H_{25}NO_6$) present in amounts up to .6%,[122] the corm contains several other alkaloids: colchiceine, colchicerine, colchamine, 3-methylcolchicine, 2-demethylcolchicine, O-demethyl-N-methylcolchicine, N-deacetyl-N-methylcolchicine, N-formyl-N-deacetylcolchicine, and α- and β-lumicolchicine.[616, 617]

The seeds possess .6 to 1.2%[584] colchicine; also colchicerine, 2-demethylcolchicine, N-deacetyl-N-methylcolchicine, α- and β-lumicolchicine,[616, 617] plus a resin, fixed oil and sugars.[584]

Colchicine is not destroyed by drying, storage or boiling.[197] The drug is a pale-yellow powder which dissolves in water, alcohol and chloroform[122] but is only slightly soluble in ether or petroleum.[584]

Propagation, Culture and Harvesting:

Seeds must be planted as soon as they are ripe; they are covered with only ⅛ in. (3 mm) of soil, and the seedlings are transplanted when two years old.[233] However, the seedlings will not bloom and fruit for three to six years. For quick results, the young side corms may be taken up and transplanted, 3 in. (7.5 cm) deep, as soon as the leaves of the parent plant wither. The plant requires fine, moist, but well-drained soil, high in magnesium, with a neutral or slightly acid pH. For best growth, it needs a grass association; it will not flourish in open ground. A period of dormancy is essential. Under experimental cultivation, 50 percent of the plants have produced 2 or 3 capsules each, while wild plants may bear only 1.[331] Wild plants may produce only 2 axillary corms, but in cultivated hybrids there may be as many as 4.[494] The corms for pharmaceutical use must be taken up before the development of offsets.[584] The harvested seeds and sliced corms are dried slowly, without artificial heat.[457, 618] Seeds of autumn crocus are in limited supply and were priced at $40 per kg in 1967 with no guarantee of viability.[556] The commercial supply of seeds and corms comes mainly from Italy and Yugoslavia.[122]

Medicinal uses:

Well-known to the ancient Greeks and Romans,[513] the corm and seeds were formerly widely utilized as remedies for gout, rheumatism, arthritis, dropsy, gonorrhea and enlarged prostate, and they were sold fresh or dried for the preparation of tinctures and "wine."[305] The corm was included in the London Pharmacopoeia from 1618 to 1639, was eliminated but then was restored in 1788. In 1820, the seeds were introduced as being safer than the corm, and they were made official in England in 1824.[584]

In modern practice, only the extracted alkaloid, colchicine, is administered, usually orally in tablet form, rarely intravenously. It is a specific treatment for gouty arthritis and is also given when first symptoms appear in order to prevent impending attacks.[423, 481] The anti-inflammatory action has been attributed to the drug's interference with protoplasmic gelation.[366] Wright and Malawista observed colchicine's inhibition of intracellular mobilization and extracellular release of granular enzymes by phagocytizing leukocytes.[626] Because colchicine arrests cellular mitosis in metaphase, it was hoped that it might be of benefit in cancer therapy, but it is entirely too toxic for such use.[171] It is successfully employed in treating familial Mediterranean fever which is prevalent in Egypt.[252, 369, 448]

In the early 1960s there was an acute shortage of colchicine, and prices soared. In April, 1966, the price was $350 per oz. ($12.50 per g); in January, 1967, $566 oz. ($20 per g). Synthesis of colchicine was achieved in the laboratory of the Department of Chemistry, Harvard University, in 1965,[623] but it is not commercially feasible.

Toxicity:

All parts of the plant are poisonous. Horses generally avoid it.[457] Deer, cattle and young sheep have died from eating the leaves, flowers and seeds. The toxin is cumulative. Calves and children have been affected by the milk from cows that have ingested the plant.[197] People have died after eating the corms mistaken for onions and from overdoses of preparations from the corm or seeds administered as medication.[286] In 1960, in Strasbourg, the deaths of an elderly man and woman were attributed to ingestion of autumn crocus leaves prescribed by an herbalist.[528] Powdered seeds have been introduced into alcoholic drinks for homicidal purposes.[514] European children have toyed with the pods as rattles and have occasionally succumbed to the temptation to eat the seeds and suffered the consequences.

Swiss peasants used to hang the flower around a child's neck in the belief that it would prevent and even cure illness, but both children and adults have been fatally poisoned by consuming as few as three flowers.[457, 582]

Symptoms of poisoning include burning in the mouth and throat, extreme thirst, difficulty in swallowing, acute gastroenteritis, abdominal pain, nausea, violent vomiting and purging followed by bloody diarrhea, cessation of urine, weak pulse, chills, flatulence, gritting of teeth, pain in the extremities, loss of muscular power (in quadrupeds, beginning with the rear legs), sometimes paralysis. If treatment is unsuccessful, there will be graduate decline and death from respiratory failure after several days.[197, 412, 514] Autopsy shows kidney and vascular damage.[423]

The volatile emission from a fresh corm in the process of slicing irritates the nostrils and throat, and the fingertips holding the corm may become numb.[457]

Prolonged therapeutic use of colchicine may cause agranulocytosis, aplastic anemia, peripheral neuritis, azoospermia or oligospermia, and, sometimes, loss of hair.[137, 384, 423] Ingestion of as little as 7 mg of colchicine has proved fatal, though the usual lethal dose is regarded as 65 mg.[423]

Other uses:

Since 1937, colchicine has been extensively employed in plant breeding to double chromosomes and develop new strains of garden flowers or economic crops. Thus are achieved very large blooms or "super" fruits and vegetables with elevated chemical content and greater disease resistance or other desired features. Colchicine is applied also in animal studies, especially to explore embryonic growth and wound-healing.[171]

Related species:

Colchicum luteum Baker is a similar plant with yellow flowers that is native to Turkistan, Afghanistan and grassy slopes or margins of forests of the western Himalayas from Kashmir to Chamba, at elevations between 2,000 and 9,000 ft. (610 and 2,740 m) where the climate is temperate. A dry extract of the corms is esteemed and sold in native markets in Afghanistan for medical use. The air-dried corms have been found to contain 0.21 to 0.25% colchicine and the seeds, 0.41 to 0.43%.[117] The artificially dried corms are official in the Indian Pharmacopoeia, being regarded as carminative, laxative, aphrodisiac, alterative and aperient. Preparations are prescribed in cases of gout, rheumatism and liver and spleen complaints, and are applied externally to relieve pain and inflammation.[20] Extracts and tinctures of the seeds have the same uses. Corms and seeds are collected from wild plants and also from plantations. Seedlings are raised in covered beds until one year old and then set out in fields in rows 3 ft. (1 m) apart.[117] This species is grown as an ornamental in England, Europe and North America.

C. speciosum Stev., which has rose-purple flowers, yellow-spotted in the throat, is native to mountain regions of Asia Minor, Syria, Lebanon and Iran. The dried corms are sold as drugs in local bazaars.[266] There are numerous showy forms of this species grown as garden flowers in England and Europe.[115] Corms collected in Holland in September, 1964, by Doctor Robert Perdue were studied at the University of Virginia. They were found to contain colchicine, demecoline and 3-desmethylcolchicine. The latter alkaloid, in bioassay, showed "significant anti-leukemic activity against the L-1210 lymphoid leukemia in mice in the range of 1 to 5 mg/kg. and cytotoxicity against cells derived from human carcinoma of the nasopharynx."[325]

Figure 19. American hellebore (*Veratrum viride* Ait.). Courtesy Doctor Hans R. Schmidt, Ormond Beach, Florida. (*See also* color illustration 2.)

American Hellebore

Botanical name:

   Vertatrum viride Ait.

Other names:

   American false hellebore, false hellebore, green hellebore, white hellebore, American white hellebore, big hellebore, swamp hellebore, true veratrum, American veratrum, green veratrum, Indian poke, meadow poke, poke root (in New Hampshire), puppet root, Indian uncus, earth gall, crow poison, devil's bite, duckretter, tickle weed, itch weed, bugbane, bugwort, wolfsbane, bear corn.[114, 526]

Family:

   Liliaceae, the Lily Family.

Description:

   A perennial herb, 2 to 8 ft. (.6 to 2.4 m) high, having a fleshy, more or less conical, vertical rhizome, 1 to 3 in. (2.5 to 7.5 cm) long and ¾ to 1½ in. (2 to 4 cm) thick, dark-gray or brown externally, with a mass of slender, fibrous roots proceeding from its base. Internally, the rhizome is whitish, with wavy lines and dots in the center. The taste is initially sweetish, then bitter and finally acrid.[305] The stem of the plant is stout, cylindrical, solid, succulent, hairy, erect and leafy. Leaves, at first overlapping, later alternate; oval or elliptic, pointed, bright-green, pleated or fluted, downy, 6 to 12 in. (15 to 30 cm) long and 3 to 6 in. (7.5 to 15 cm) wide (upper leaves narrower than lower), the bases clasping the stem. Flowers, borne in an erect, open, branched, terminal cluster to 2 ft. (.6 m) high, are greenish-yellow, downy, 1 in. (2.5 cm) wide, six-parted (3 sepals, 3 petals), with 6 stamens. Fruit is a three-celled capsule to 1 in. (2.5 cm) long, splitting open when ripe, revealing flat, winged seeds.

Origin and Distribution:

   Native and abundant in swamps, wet meadows and woods, and on edges of streams, in Alaska and mountains of Idaho, Oregon and Washington eastward to Wisconsin and Quebec; southward through New England to Virginia and, in the Allegheny Mountains, to Tennessee and Georgia.[114]

Constituents:

The numerous alkaloids in the rhizome are placed in three groups. Group A, alkamines (esters of the steroidal bases) with organic acids, including germidine ($C_{34}H_{53}NO_{10}$) and germitrine ($C_{39}H_{61}NO_{12}$), most valued therapeutically; also cevadine, neogermitrine, neoprotoveratrine, protoveratrine and veratridine; Group B (glucosides of the alkamines), mainly pseudojervine ($C_{33}H_{49}NO_{8}$) and veratrosine ($C_{33}H_{49}NO_{8}$); Group C (alkamines), germine ($C_{27}H_{43}NO_{8}$), jervine ($C_{27}H_{39}NO_{3}$), rubijervine ($C_{27}H_{43}NO_{2}$) and veratramine ($C_{27}H_{39}NO_{2}$).[116, 117, 122] Ginzel has demonstrated muscle relaxation in cats by administration of protoveratine with phenyldiguanide and 5-hydroxytryptamine.[228]

Propagation, Culture and Harvesting:

American hellebore may be grown from seed or by division. It is cultivated as an ornamental in North America and abroad in moist, shady locations and requires rich soil.[115] For pharmaceutical use, the rhizome, with or without attached roots and with a tuft of leaf bases at the apex, is collected from wild plants in the fall. After cleaning, it is sliced lengthwise and dried thoroughly.

Medicinal uses:

American hellebore was employed for curative purposes by the Indians and pioneers.[584] Throughout the 18th and 19th centuries, it was valued as an analgesic in treating painful diseases, epilepsy, convulsions, pneumonia, peritonitis and as a cardiac sedative, and, in small doses, it was given to stimulate the appetite.[305] It was introduced into England around 1862.[584] There and in Europe, the powdered rhizome and tincture have been employed in treating respiratory afflictions, convulsions, mania, neuralgia and headache. Infusions have been recommended for gargling in cases of sorethroat and tonsillitis.[624] Because of the hazards associated with the drug, its use declined until 1950, when active investigation of its hypotensive action and improved chemical techniques opened up new and safer applications and it became widely utilized in the treatment of hypertension. It may be administered in the form of an alkaloidal mixture (of which there are various trademarked preparations)[122, 584] and is sometimes given in combination with alkaloids of *Rauvolfia serpentina*. The powdered rhizome (light-brown to pale olive-green) is also marketed as such for pharmaceutical use.[122] The ester alkaloids reduce both systolic and diastolic pressure, slow the heart rate and stimulate peripheral blood flow in the kidneys, liver and extremities. Given intravenously, they lower cerebral arterial resistance.[423] These veratrum drugs can be safely administered only in small doses. In chronic cases, tolerance develops, and, with increased dosage, there is the risk of emesis and other undesirable side-effects. Once again, enthusiasm for veratrum therapy is diminishing except in special emergencies such as hypertensive toxemia during pregnancy and the pulmonary edema which arises in acute hypertensive crises.[423]

Toxicity:

All parts of the plants, especially the rhizome, are highly toxic. The tops are grazed by elk, which are apparently unaffected but, because of their acridity, are usually avoided by other herbivores. Young animals, especially lambs, have died from leaf ingestion, and poultry have succumbed to eating leaves or seeds.[114]

Humans have been seriously poisoned by cooking and eating the leaves mistaken for those of marsh marigold (*Caltha palustris* L.) or believing them to be the new shoots of the true pokeweed (*Phytolacca americana* L.) which are edible in the early stage. Most poisonings have arise from overdoses of medical preparations.

The cut or powdered rhizome irritates the nasal passages and causes sneezing.[122]

Signs of internal poisoning include burning in the throat and esophagus, salivation, dilated pupils, impaired vision, abdominal pain, delayed nausea, retching, violent and prolonged vomiting, diarrhea, headache, cold sweats, dizziness, faintness, shallow respiration, muscular weakness, tremors, spasms, sometimes convulsions, irregular and, later, rapid pulse, partial loss of consciousness and, in extreme cases, paralysis and death from asphyxia.[122, 400]

Other uses:

The powdered rhizome or certain of the alkaloids are ingredients in some commercial insecticides.[632]

Veratrum Album

White Hellebore

Botanical name:
>Veratrum album L.

Other names:
>European white hellebore, langwort.

Family:
>Liliaceae, the Lily Family.

Description:
>A perennial herb to 3 or 4 ft. (.91 or 1.2 m) high, with a fleshy, oblong, partly horizontal rhizome, ½ in. (12.5 mm) thick, blackish or yellowish externally, whitish or light yellowish-gray within, bitter and acrid in taste and onion-like in odor, having numerous slim, grayish, fibrous roots growing from the base. The stem, erect, cylindrical and downy, bears 10 to 12 alternate, yellowish-green, ovate or oblong, fluted leaves, downy on the underside; the lower ones to 1 ft. (30 cm) in length and to 5 or 6 in. (12 to 15 cm) wide, their bases clasping the stem. Flowers, nearly sessile in a dense terminal raceme to 2 ft. (.6 m) high, are yellowish-white, tinged with green on the outside at the base, and composed of 6 spreading, toothed, somewhat wavy segments.[116, 305]

Origin and Distribution:
>White hellebore grows wild in moist meadows of central and southern Europe, mainly on the Alps and Pyrenees,[305] and in central Asia, Japan[266] and the Aleutian islands.

Constituents:
>The principal alkaloids of white hellebore are protoveratrine A ($C_{41}H_{63}NO_{14}$) and protoveratrine B ($C_{41}H_{63}NO_{15}$).[552] Also present are germerine, jervine, pseudojervine, veratrosine and a glucoside, veratramarine.[513]

Figure 20. White hellebore (*Veratrum album* L.). Reproduction from color plate in *Medical Botany: or, History of Plants in the Materia Medica of the London, Edinburgh, and Dublin Pharmacopoeias;* London; 1819 (author unknown).

Propagation, Culture and Harvesting:

White hellebore is multiplied by seed or division and is occasionally grown in England and Europe as an ornamental in shady borders or "wild" gardens.[115] The rhizome is collected from wild plants in September and October.[513]

Medicinal uses:

White hellebore was known and valued in ancient times and was administered in cases of cholera. Later, it was employed as a substitute for *Colchicum* in gout and, externally, in the form of an ointment, was applied to scabies and herpes and other skin afflictions. In Europe, it is still used in treating progressive muscular distrophy, rheumatism and arthritis and is also given as an emetic, purgative and diaphoretic.[513] However, American hellebore [q.v.] is preferred as less likely to cause gastric disturbance, and the main role of white hellebore in modern times has been in veterinary medicine.[233]

A tincture of the rhizome lowers blood pressure promptly.[513] In the United States, the crude drug, formerly in the Pharmacopoeia, is now utilized only as a source of protoveratrine A and B. Intravenously, they are nearly equal in activity, but, when given orally, the former is by far the more potent of the two; it may be used alone or the two alkaloids together, with or without alkaloids of *Rauvolfia serpentina* L., in the management of hypertension. There are various trademarked preparations of the active alkaloids on the market.[122]

Toxicity:

In early times, white hellebore served as an arrow and dagger poison, and it has been employed for homicidal purposes. A fatal dose for an adult human is said to be 1 or 2 g of the powdered rhizome.[518] The powder causes violent sneezing and running of the nose.[233, 555]

Other uses:

In England, the powdered rhizome is used to kill caterpillars, particularly those of the gooseberry sawfly.[115] It is an ingredient in some commercial insecticides in the United States.

Agavaceae

Figure 21. Sisal (*Agave sisalana* Perr.) with fast-growing, bamboo-like flower stalk rising from the center. A common sight in southern Florida.

Sisal

Botanical name:
>Agave sisalana Perr. (A. rigida var. sisalana Engelm.)

Other names:
>Sisal agave, sisal hemp, mescal, maguey, green agave.

Family:
>Agavaceae, the Agave Family.

Description:
>A shallow-rooted, succulent plant having copious watery sap, stemless at first, later with a thick stem reaching 3⅓ to 5 ft. (1 to 1½ m) in height and a rosette of spirally set, stiff, sword-shaped, flat or slightly concave leaves to 5 ft. (1½ m) long and 4 to 6 in. (10 to 15 cm) wide, tapering to a point and terminating in a sharp, brown needle to 1 in. (2.5 cm) long; margins are smooth or bear a few small teeth; the epidermis is leathery, dark-green, glossy, waxy. Flowers are tulip-shaped, yellow-green, ill-smelling, in clusters at the forked tips of the 30 to 40 horizontal branches extending from the stout green stalk which emerges from the center of the rosette and grows rapidly—2 to 2¾ in. (5 to 7 cm) per day—to a height of 15 to 23 ft. (4.5 to 7 m). Fruit (seed capsule) is oblong, 2¼ to 2½ in. (5.7 to 6.3 cm) long, .78 to .98 in. (20 to 25 mm) wide; seeds are ⅜ in. (10 mm) long. Few, if any, fruits form, but a great many—2,000 to 3,000—bulbils develop after the flowers are shed, fall to the ground, take root and form a colony of new plants. Blooming, which generally occurs in six to nine years (sometimes not before twenty years), causes the plant to die. By this time, the average plant will have produced 220 to 250 leaves. In addition to self-reproduction by bulbils, the plant sends out horizontal, creeping, white rhizomes from the tips of which arise offshoots, or suckers.[133, 288, 343, 349]

Origin and Distribution:
>Sisal is believed to be native to southern Mexico and it was introduced early into Central America. In 1836, Doctor Henry Perrine brought sisal from Yucatan to southern Florida as a potential fiber crop. From Florida, it was carried to the Bahamas, the West Indies and most

of the tropical regions of the world. Everywhere that the plant thrives it quickly becomes naturalized and multiplies freely. Sisal never succeeded as a crop in Florida, where it now exists only as a common escape from former sites of cultivation or is occasionally grown as an ornamental. The one-time sisal industry of the Bahamas declined to the local handicraft level many years ago. Today, sisal is cultivated on a large commercial scale in the state of Campeche, Mexico, in Brazil, Haiti, East Africa and, more recently, South Africa. Indonesia was formerly a major sisal-growing area, but acreage is now much less than before World War II.[343] There are only small plantations in Central America, Venezuela, Madagascar and India. In Hawaii and many other areas, it has been exploited for a time but later abandoned as impractical.[591] Henequen (*Agave fourcroydes* Lem.) is the similar fiber crop of Yucatan and Cuba.

Constituents:

Sisal leaves are composed of about 4% fiber, .75% cuticle (of which 5 to 17% is wax), 8% of other dry material, and 87% moisture.[605]

All parts of the plant contain steroidal sapogenins, principally hecogenin ($C_{27}H_{42}O_4$), tigogenin ($C_{27}H_{44}O_3$)[552] and neotigogenin. Present in small amounts are sisalagenin, gloriogenin, gentrogenin, delta-9-11-hecogenin, diosgenin and yamogenin. The content differs with the plant part and the stage of growth; thus, tigogenin is predominant in young leaves, bulbils and flowers. In mature and old leaves, there is usually no tigogenin, and hecogenin is dominant. Some investigators have found gitogenin to be the major component of the giant flower stalk; others, tigogenin. The rhizome contains both hecogenin and tigogenin, as do the roots but in much smaller amounts. Total sapogenin content is higher in the rhizome than in other portions.[88]

Propagation, Cultivation and Harvesting:

Sisal is a plant of semi-arid, somewhat humid tropical climates. It will not thrive in the very dry conditions suitable for henequen.[156] In Kenya, it does well up to an altitude of 1,828.8 ft. (6,000 m).[335] Best growth is attained in well-drained, open-textured soil. Either bulbils or suckers may be used for planting, which can take place any month of the year. Bulbils must be held in nursery beds for one to one and one-half years, but suckers are placed directly in the field after the site has been cleared and plowed. The plants are usually set out at the rate of 1,000 to 2,000 per acre (2,500 to 3,000 per hectare).[155] In a double-row system, 1,983 plants per acre (4,900 per hectare) allows a spacing of 3 ft. (.9 m) between plants and 13 ft. (4 m) between rows. In old fields, fertilizer is necessary to replace exhausted nutrients. Weeding is essential the first three years, but thereafter, soil disturbance is best avoided.

Harvesting commences three to four years after planting, when the plants are about 5 ft. (1½ m) high and have 120 to 125 leaves. All but the central 25 or 35 leaves are cut annually for seven or eight years.[288, 335] When flower stalks appear, they are quickly cut to a 4-ft. (1.2-m) level so that the leaves will develop to usable size instead of shriveling.[153] When 70 to 80 percent of the plants have sent up flower stalks, the plantation is plowed up.[288]

The few pests and diseases include the sisal beetle (*Scyphophorus interstitialis* Gyll.), black rot and yellow spot disease.[155] In East Africa, zebra leaf spot, bole rot and spike rot are all attributed to *Phytophora arecae*.

Young plants suffer depredation by monkeys, baboons, elephants, pigs, goats and other animals which relish the succulent base, or "heart."[335]

Harvested, tied in bundles and hurried to fiber extractors, leaves are decorticated with devices operated by hand or power. They must be processed within twenty-four hours or the fiber of the cut end will be discolored.[153] The extracted fiber is washed, dried in the sun or by artificial heat, brushed, graded, baled and delivered to handlers. Yield of fiber ranges from 2.5 to 4 percent; under good conditions, 1,760 lbs. (798.3 kg) per acre, 4,409.2 lbs. (2000 kg) per hectare, annually.[155] From 60 to 80 lbs. (27.2 to 36.2 kg) of dry fiber may be extracted from each 1,000 lbs. (454 kg) of leaves.[20] Total world production amounts to more than half a million tons per year.[343]

Chemical investigations with a view to commercial utilization of the pulp waste (12% of the leaf) were begun before 1910. That of henequen is a source of alcohol in Yucatan.[615] In Kenya, the waste is reduced to a concentrate, and the juice is extracted and fermented for one week. Then the sludge (which contains 80% of the hecogenin of the fresh leaves) is processed under steam pressure of 200 lbs. per sq. in., filtered and dehydrated. The resultant material (12% hecogenin plus other sapogenins) is exported to England, where it is utilized by Glaxo Laboratories as a source of cortisone.[584] In an extraction process developed at the Regional Research Laboratory, Bhubaneswar, India, the dry, dewaxed sisal waste is hydrolyzed by dilute HCl and the saponins converted to sapogenins. After filtering, washing and drying, extraction with petroleum ether yields pure hecogenin as a white crystalline product.[543] A pilot plant for the extraction of hecogenin and tigogenin has been established in northeastern Brazil.[395a] There is also an experimental steroid-recovery operation in Yucatan.

70   *Major Medicinal Plants*

Figure 22. Sisal flowers are succeeded by bulbils, which fall to the ground and take root, forming colonies of new plants.

Figure 23. Upper section of sisal flower stalk bearing an abundance of flowers and bulbils.

Medicinal uses:

See uses of sapogenins under MEXICAN YAMS.

Toxicity:

The raw sap of the plant is corrosive to metal and highly irritating to the eyes and the skin. It causes an instant, stinging red rash in gardeners who have occasion to cut any part of the plant and affects factory workers exposed to the sap and the wet fiber in the process of extraction.[598] Fresh sisal waste has molluscicidal activity, pulmonate snails being absent downstream from the sites where this residue is dumped into East African rivers.[424] The odor of sisal used in mattresses (generally in combination with some other material) causes allergic reactions in sensitive individuals.

Other uses:

Sisal fiber is valued mainly for rope, twine and marine cordage; also in sacks for coffee or sugar. About 85 percent goes into agricultural binder twine and balertwine. On the cottage industry level, sisal fiber is widely used for matting, rugs, coarse fabrics, hammocks, mops, brushes and sandals. Some is employed in upholstery and mattresses.

In Brazil, a factory has been erected in the state of Paraíba for the manufacture of cellulose from sisal which, after the age of three years, yields 13,227 lbs. (6,000 kg) annually as compared to 2,755.7 lbs. (1,250 kg.) from Paraná pine after twenty-five years.[395]

The central bud, or "heart," of sisal may be cooked and eaten. The fresh pulp waste is usable as stockfeed and can be fed to dairy cows at the rate of 30 lbs. (13.6 kg) per day. The dehydrated material is not as appealing to cattle.[200, 615]

A wax resembling carnauba can be recovered from sisal waste.[468]

As a mulch, soil conditioner or fertilizer, sisal pulp is beneficial to sisal plantations but may be detrimental to crops such as coffee which need acid conditions. Sisal extracts much calcium (as well as magnesium and potassium) from the calcareous soils in which it thrives, and these minerals are redeposited through the pulp.[413]

Dioscoreaceae

Mexican Yams

(1) *Botanical name:*

Dioscorea composita Hemsl. (*D. tepinapensis* Uline).

Other name:

Barbasco.

Family:

Dioscoreaceae, the Yam Family.

Description:

A climbing, stout-stemmed, non-hairy vine with white-fleshed, branched tubers, brown and corrugated externally, up to 4.9 ft. (1.5 m) long, 1.6 ft. (.5 m) thick and reaching 11 lbs. (5 kg) in weight. Leaves are alternate, long-petioled, ovate, somewhat leathery; in the juvenile stage not cleft at the base, in the adult stage deeply heart-shaped at the base, pointed at the apex, with 7 to 11 indented, longitudinal nerves and the leaf surface crinkled by numerous horizontal veins; the blade up to 9 in. (23 cm) wide. Flowers are small, sessile, purplish; male, in simple racemes 2 to 4 in. (5 to 10 cm) long arranged in a long panicle; female, on separate plants, borne in compound spikes. In var. *aggregata* Uline, the male flowers are in short, crowded clusters of 3 to 10 flowers. The fruit (seed capsule) is obovate to elliptic, .7 to 1.18 in. (18 to 30 mm) long and stiffly leathery. When mature, it turns brown and splits open on the upper side. Seeds are .11 to .19 in. (3 to 5 mm) wide and .03 in. (1 mm) thick, each centered in an oblong, membranous wing .59 in. (1.5 cm) long.[140, 255, 310, 377, 587]

Origin and Distribution:

This species is native to southern Mexico, particularly Oaxaca, Veracruz, San Pedro Tepinapa, Tabasco and Chiapas. It has been introduced into eastern and southern United States, Puerto Rico, Surinam and Costa Rica.

Figure 24. A diosgenin-yielding Mexican yam (*Dioscorea composita* Hemsl.)—female—in bloom.

(2) Botanical name:

Dioscorea floribunda Mar. & Gal.

Other names:

Alambrillo, corrimiento.

Family:

Dioscoreaceae, the Yam Family.

Description:

A twining, non-hairy vine with slender or stout stems and brown, branched, corrugated, yellow-fleshed tubers up to 2.7 ft. (.8 m) long and about 1 ft. (.3 m) thick; up to 6.6 lbs. (3 kg) in weight. Leaves are alternate, fairly thick, firm, with petioles to 2.75 in. (7 cm) long, triangular-ovate, shallowly or deeply cordate at the base, long-pointed at apex, nine-nerved. Flowers are dark-green; the male, tubular, nearly sessile, in spikes arranged in loose panicles; the female, on separate plants, bell-shaped, in short, solitary spikes. Capsule and seed characteristics are much like those of *D. composita*.[140, 141, 549]

Origin and Distribution:

D. floribunda is native to damp woods, thickets and oak forests up to an elevation of 4,917 ft. (1,500 m) in southern Mexico (Veracruz, Oaxaca, Tabasco) and Guatemala. It has been introduced into the eastern and southern United States, Puerto Rico and Costa Rica.[137]

Constituents:

These Mexican yams are rich in the glycoside saponin from which are derived, by partial synthesis, steroidal sapogenins, primarily diosgenin. ($C_{27}H_{42}O_3$).[552] This property was discovered by Japanese chemists in 1936 and converted to progesterone, an intermediate in cortisone production, in 1940.[58] This development started a worldwide search for plant precursors of cortisons. At first, interest centered on *Strophanthus* species, then *Agave sisalana* (source of hecogenin) and has, for the time being, focused on *Dioscorea* species. Of over 125 species tested,[371] *D. composita* and *D. floribunda* have proved most productive as commercial sources. Generally, *D. composita* averages 4 to 6% total sapogenins, *D. floribunda*, 6 to 8%.[377]

Propagation, Cultivation and Harvesting:

In southern Mexico, where the extraction of sapogenins from *Dioscorea* species originated and is being carried on by six companies,[420] most of the tubers are collected from wild plants of *D. composita*. In Guatemala, tubers of *D. floribunda* are gathered from the wild and delivered to the factory of the Compañia Agrícola Industrial Guatemalteca for processing.[501] Trial plantings have been established in Mexico, Puerto Rico and other suitable tropical regions, but extensive commercial cultivation will not be undertaken as long as there is an ample supply from the wild and until there is assurance that the natural sources of sapogenins will not be replaced by total synthesization.[370]

D. composita is most easily propagated by seed (except for some clones which can be vegetatively multiplied.) *D. floribunda* can be grown from seed, pieces of tuber or leaf- or stem-cuttings.[154, 376] Seeds remain viable for at least a year in dry storage; for three years if accompanied by a desiccant and refrigerated.[377] Seeds are planted, by hand or mechanically, ¼ to ½ in. (6 to 12.5 mm) deep and 1 to 1½ in. (2.5 to 3.8 cm) apart in nursery beds and will sprout in approximately three weeks. Support of some kind must be provided, or the young shoots will droop and die. Small sticks may be placed 6 in. (15 cm) apart, or a roof of chicken wire supported by sticks 4 to 5 in. (10 to 12.5 cm) high may be used.[377] When the plants are four to six months old, the tubers are lifted and, with vine cut off or merely reduced, set out in the field. *D. composita* needs 1½ to 2 ft. (.45 to .6 m) between plants and 4 ft. (1.2 m) between rows; *D. floribunda* plants can be spaced 1 ft. (.3 m) apart.[140, 141]

The soil, preferably sandy loam, should be well-drained [375] and plowed to a depth of about 8 in. (20 cm). The working in of organic material such as sugarcane filter presscake, which has been recommended, is unnecessary.[376] Small stakes are needed at first to support the vines. Later, when growth becomes more vigorous, these must be replaced by strong stakes at least 6 ft. (1.8 m) high. Bamboo may be used or posts of durable wood or cement, with wires strung between and with twine running down to each plant. Commercial fertilizer (9-10-5 formula) is applied at the rate of 1,000 lbs. (455 kg) per acre annually. Complete fertilization has shown a 30 percent increase in total yield.[140] Weeding (or black plastic mulch to prevent weed growth) is essential at the outset but is unnecessary after the vines have grown enough to shade out weeds. The plants grow faster in spring and continue to grow until the winter dry season when growth is temporarily suspended. *D. floribunda* at this time becomes dormant, the vine dying back.

The vines may be attacked by aphids and red spiders. *D. floribunda* is occasionally subject to green-banding virus (transmitted by aphids) which requires quick removal of affected plants. A fungus (*Helminthosporium*) has caused much leaf damage in Central America and is sometimes troublesome in nursery beds in Puerto Rico. Poor drainage or excessive rains may cause rot in old tubers. Snails, slugs and grubs often damage tubers and cause rotting. In light soils, roots and tubers may be infested with nematodes.[377]

The tubers of D. composita reach maximum size in four to five years. They have a vertical habit of growth and may extend to a depth of 2 or 3 ft. (.6 to .9 m) before spreading out. They are difficult to harvest, and usually a portion remains buried in the ground. In trials on Everglades peat soils of Florida, D. composita tubers stopped growing when they reached the high water table and had to be harvested in two and one-half years.[542] Tubers of D. floribunda are full grown in three to four years. They develop horizontally or on a slant and may not extend beyond 1 ft. (30 cm) and consequently are relatively easy to extricate.

Figure 25. Mexican yams in cultivation, the vines tied to bamboo stakes.

Harvesting takes place during the winter, inasmuch as the sapogenin content is known to decline in spring and increase during summer and fall.[141] Tubers of wild plants are dug up by machete, hoe or square shovel, an energetic laborer being able to take up 220 lbs. (100 kg) per day from a dense stand of D. floribunda.[501] In plantations, the vines and stakes are first eliminated, and then the tubers are plowed up, collected by hand,[373] rake or pitchfork and brushed to remove excess soil. There should be no delay in harvesting D. composita, otherwise new shoots may spring up. Finally, the tubers are washed and delivered immediately to sapogenin-extracting plants, or they are mechanically chopped into chips

Figures 26 and 27. A diosgenin-yielding Mexican yam (*Dioscorea composita* Hemsl.) in bloom: above, male; left, female.

or flakes and sun-dried to preserve them for later processing, locally. Only the derivatives of *diosgenin* are exported from Mexico.[370]

In an experimental planting, D. *composita*, over a period of four years, showed a steady increase in weight of tubers from 10 g fresh (3 g dry) with a .25% total sapogenin at 14 to 22 weeks, to 4,232 g fresh (1,259 g dry) with 4.37% total sapogenin at 194 to 202 weeks. D. *floribunda* increased from 9 g. fresh (3 g. dry) with .63% total sapogenin to 1,996 g. fresh (543 g. dry) with 5.39% total sapogenin.[141]

In Puerto Rico, four-year-old plants of D. *composita* have produced up to 14,387 lbs. (6,526 kg) of dry tubers, yielding 679 lbs. (308 kg) total sapogenin per acre; D. *floribunda* has produced up to 9,248 lbs. (4,195 kg) dry tubers, yielding 728.2 lbs. (331 kg) sapogenin.[375, 377]

In Florida trials on Everglades peat, D. *composita* produced in two and one-half years 42,301 lbs. (19,228.6 kg) dry tubers yielding 1,049 lbs. (476.8 kg) total sapogenin.[542]

Seed production is dependent on adequate pollination. While it is assumed that the flowers are normally insect-pollinated in the wild, there is little insect activity among the blooms in Puerto Rico, and pollen is carried only short distances by air. Female plants separated from males by 10 ft. (3 m) set little or no fruit. When male and female inflorescences are closely intertwined, fruit set is excellent.[374]

Medicinal uses:

The steroid drugs derived from diosgenin include oral contraceptives, anti-inflammatory compounds (topical hormones, systemic corticosteroids), androgens, estrogens, progestogens and other sex hormone combinations.[370]

Cortisone and hydrocortisone are prescribed in cases of rheumatoid arthritis, rheumatic fever, sciatica, certain allergies, Addison's disease, some skin diseases and contact dermatitis.

Androgens are given to counteract prostatic hypertrophy, male sexual insufficiency and psychosexual disturbances, and sometimes to women to offset excessive production of estrogens.

Estrogens are taken by women to regulate the menses, to control undesirable effects of the menopause, to overcome sterility caused by estrogen deficiency and also as therapy following ovariectomies. They may be administered in some skin disorders, high blood pressure, arterial spasms, migraine and to promote growth of hair through improving circulation in the scalp. Men may be given estrogens in treating cancer of the prostate.[466]

Toxicity:

The tubers of sapogenin-yielding species of *Dioscorea* are bitter and toxic and not consumable as human food as are several other species of this genus. In India, *D. deltoidea*, in very small amounts, has been administered as a diuretic, but it is known to be poisonous in larger doses.[116] The decoction is lethal to lice. Steroidal drugs taken over prolonged periods may produce serious side-effects.

Figure 28. Branched, corrugated tubers of *Dioscorea floribunda* Mar. & Gal. on the left; *D. composita* on the right.

Other uses:

Because of its saponin content, the tuber of *D. deltoidea* is traditionally utilized in the Western Himalayas to launder raw wool and woolen fabrics.[116]

Related species:

Several other species have received considerable attention, but most have characteristics that render them unsuitable for commercial plantings. Prominent among them are the following.

D. spiculiflora Hems., native to Mexico, is a virus-resistant species with small tubers possessing 8 to 14% of a mixture of sapogenins, but these are difficult to separate.[372] Dry tuber yield per acre is one-third that of *D. composita* and sapogenin yield per acre one-half that of *D. composita*.[542]

D. mexicana Guill., cabeza de negro, was the first to be exploited as a source of sapogenins. The vine grows for a period of twenty years, the tuber slowly becoming enormous—up to 100 lbs. (45.3 kg) in weight—but it is partly above ground and often infested with termites.[370]

Figure 29. Harvested tubers of *Dioscorea floribunda* being loaded for transport to laboratory.

Two South African species, *D. sylvatica* Ecklon and *D. elephantipes* Engl., develop huge tubers close to or above the surface of the soil in arid regions. They are extremely slow growers. Nevertheless, *D. sylvatica* is commercialized to some extent.[370, 372]

D. deltoidea Wall., of northern India, has slender, woody rhizomes containing 8 to 10% sapogenins. The plant is not a vigorous grower. Tubers of wild plants are gathered and processed in India. Experimental plantings in Kashmir and East Africa have demonstrated that cultivation of this species could be feasible if the demand should warrant it.

D. friedrichsthalii Knuth, of Central America, grows readily from seeds but is subject to virus diseases. It produces tubers of medium size—up to 8.8 lbs. (4 kg) in weight containing only 3% sapogenins.[371, 377]

Tubers of several native species are being collected and processed in China.[371]

Figure 30. A South African yam, *Dioscorea elephantipes* Engl., produces immense, curiously-formed tubers, often partly aboveground. Photographed at Longwood Gardens, Kennett Square, Pennsylvania.

Podophyllaceae

Figure 31. May apple (*Podophyllum peltatum* L.) in bud in a Maryland woods. (*See also* color illustration 3.)

May Apple

Botanical name:
   Podophyllum peltatum L.

Other names:
   Mandrake, wild mandrake, American mandrake, wild jalap, wild lemon, ground lemon, raccoon berry, hog apple, Indian apple, devil's apple, duck's foot, umbrella plant, vegetable calomel, vegetable mercury and yellowberry.

Family:
   Podophyllaceae, the May Apple Family; formerly included in Berberidaceae, the Barberry Family.

Description:
   An herbaceous plant with a perennial, creeping, fibrous, jointed, sometimes branching rhizome attaining a length of several feet and a thickness of ¼ in. (6 mm), brown on the outside and yellowish-white internally. When dried, it has a faint but unpleasant odor and sweetish, bitter, slightly acrid taste. From the rhizome there arises in spring 1 (or more) plants, each having a single smooth, fleshy stem up to 1 ft. (30 cm) in height, sheathed at the base and bearing at its apex 1 or two peltate leaves having petioles 3 to 6 in. (7.5 to 15 cm) long. Leaf blade, 5 to 13 in. (12.5 to 32.5 cm) wide, is nearly circular, divided into 5 to 9 triangular lobes, each toothed or cleft at the tip; dark-green and smooth on the upper surface, pale beneath. A single flower is borne (in May and June) at the axil of the petioles (on two-leaved plants), its peduncle 1 to 2 in. (2.5 to 5 cm) long and drooping. The corolla, ¾ in. to 2½ in. (2 to 6.5 cm) wide, has 6 to 9 white, smooth, waxy, concave petals and 9 to 20 stamens bearing yellow anthers. Fruit is green and disagreeably scented when immature, becoming light-yellow and brown-spotted and pleasingly fragrant when fully ripe (in September); oval, 1¼ to 2 in. (3 to 5 cm) long, contains mucilaginous pulp with a peculiar, subacid, faintly strawberry-like flavor and a dozen or so ovate, rough, dark-brown seeds. After fruiting, the leaves fall and the plant dies back.[305, 462, 526]

Origin and Distribution:
   The may apple occurs wild abundantly in moist, shady, deciduous woods and marshy meadows and ditches from southeastern Canada to mid-

Georgia, extreme northwestern Florida and westward to eastern Texas and southern Minnesota.[383] It has been introduced into England and Europe only as a specimen in drug plant collections in gardens or greenhouses.[115]

Constituents:

Active principles are present in the unripe fruit, the above-ground plant and especially the rhizome and roots. The latter contain 3.5 to 6% of a resinous mixture called podophyllum resin or podophyllin. Its primary constituents are lignan glycosides, including 20% of podophyllotoxin ($C_{22}H_{22}O_8$), a toxic principle yielding podophyllic acid ($C_{22}H_{24}O_9$) and picropodophyllin ($C_{22}H_{22}O_8$);[117, 552] also 10% of α-peltatin and 5% of β-peltatin. The purgative action is attributed mainly to the peltatins.[122] Among other properties reported in the rhizome are gum, starch, albumen, gallic acid, calcium oxalate, a fixed oil, a little volatile oil and the flavonol quercetin.[117, 305] Highest activity is found in rhizomes dug in the spring.[618] For the pharmaceutical trade, they must possess no less than 5% resin.[122]

Propagation, Culture and Harvesting:

In propagation experiments in Kentucky, there has been no success in germinating seeds.[322] The plant is readily grown by root division.[115] The demand for the dried rhizome has stimulated interest in cultivation, but, to date, the rhizomes are gathered from the wild primarily in Indiana, Kentucky, Tennessee, North Carolina and Virginia. Harvesting may take place in the fall or spring. The rhizomes with roots attached are washed, cut into pieces 4 to 8 in. (10 to 20 cm) long and dried before shipment.[618] Several hundred tons are consumed annually in the United States and abroad.[122]

Medicinal uses:

Cherokee Indians put drops of the juice of the fresh rhizome in the ear to relieve deafness.[618] The rhizome was valued by them and other American Indians as an emetic and potent though slow-acting purgative. It was soon adopted by the early settlers and became an official drug, at first in crude form and then, in 1864, the extracted, powdered podophyllin. It was employed as a vermifuge and in treating constipation, typhoid fever, cholera infantum, jaundice, dysentery, chronic hepatitis, scrofula, rheumatism, dysmenorrhea, amenorrhea, kidney and bladder and prostate problems, and was employed in the treatment of gonorrhea and syphilis in place of mercurials and without their deleterious effects.[305] An infusion of the rhizome is still prepared in some households as a laxative.[322]

In modern practice, podophyllum resin, or podophyllin, a yellow-brown powder, is used as a cathartic. For chronic constipation, it is combined with milder laxatives. Belladonna may be added to lessen the irritant effect.[423] Podophyllin resin with compound tincture of benzoin is an effective application on venereal warts (*Condyloma acuminatum*),[383] plantar warts and certain external papillomas of dogs, but it is so injurious to normal tissues that surrounding areas must be protected.[423] Podophyllum extract is an ingredient of various liver pills on the American market,[618] and ointments containing podophyllin are sometimes applied on hyperkeratotic and hypertrophic lesions. Podophyllin for a while appeared to be a potential remedy for skin cancer but has proved to be ineffective and hazardous.[423]

Toxicity:

The fresh plant is an irritant poison causing, if ingested, vomiting, diarrhea, dizziness, headache, bloating, lowered blood pressure and stupor. Young shoots were resorted to by some American Indians as a means of suicide.[618] The plant is bitter and usually avoided by grazing animals,[400] but pigs have died after eating the young shoots.[306] If cows eat the plant, the cathartic principle is transmitted to their milk.[229]

The freshly dried root is drastically purgative and emetic, but its potency is reduced by storing or roasting. Fatality has resulted from overdoses, one woman succumbing in thirty-one hours after ingesting 300 mg of podophyllum powder. Another died when podophyllin ointment was left too long (in excess of ½ to 1 hr.) on a venereal wart of the vulva.[423] Injected subcutaneously, .005 g of crystalline podophyllin is lethal to cats. Subcutaneous injection in dogs produces incoordination of the rear legs, progressing rapidly, accelerated respiration, lowered temperature, coma, clonic cramps and death.[117]

Handling the rhizome or exposure to the dust of the pulverized, dried product can produce severe skin irritation and eye inflammation.[229, 400]

The seeds are reportedly toxic, even in the ripe fruit.

Other uses:

The ripe fruit pulp is edible, enjoyed by some but considered nauseous by others. It has been made into ade, marmalade and jelly.[186]

Related species:

The Himalayan, or Indian, may apple (*P. hexandrum* Royle, syn. *P. emodi* Wall.) is native and abundant in Tibet, Afghanistan, Pakistan and northern India [584] at altitudes between 5,900 and 13,123 ft. (1,800 to 4,000 m). The plant resembles the American species, but the leaves are more deeply divided and bronze when young; the flower may be white or pink, and the fruit is scarlet.

In propagation trials, seeds with surrounding pulp were planted soon after removal from fresh fruits, and they germinated after a dormancy of nine to ten months.[383] The plant has been experimentally cultivated in high, moist situations, and rhizomes have been ready for harvesting in two to four years. Efforts are being made to encourage the pharmaceutical use of this domestic resource in place of the imported *P. peltatum*.[117]

The rhizome contains three times the resin of that of *P. peltatum*, and the resin possesses as much as 38% podophyllotoxin but no peltatin.[117, 584]

Primary use in India is as a purgative and in the treatment of skin diseases and tumorous growths.[277] Dombernowsky and coworkers have reported on the cytostatic activity of epipodophyllotoxin (derived from *P. hexandrum* by Sandoz Ltd.) in clinical trials with patients having malignant lymphomas and solid tumors.[159]

Menispermaceae

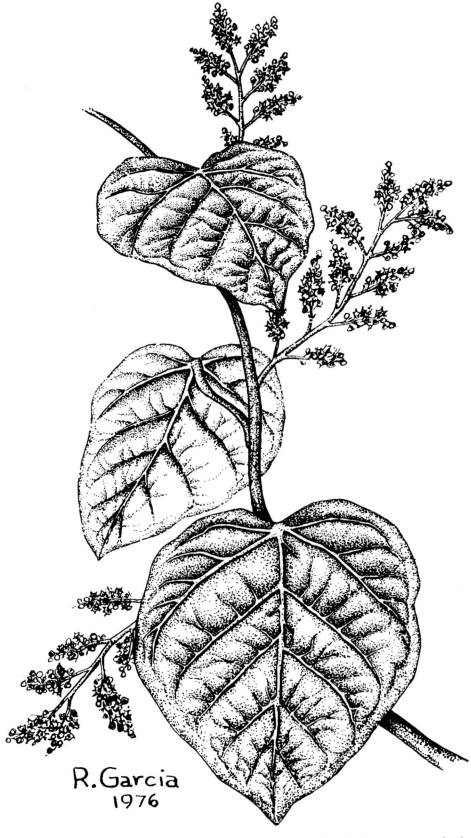

Figure 32. Levant berry (*Anamirta cocculus* Wight & Arn.), a climbing, twining shrub with thick, leathery leaves.

Levant Berry

Botanical name:

    Anamirta cocculus Wight & Arn. (*A. paniculata* Colebr., *Menispermum cocculus* L., *M. lacunosum* Lamk., *Cocculus lacunosus* DC., *C. suberosus* Wight & Arn.)

Other names:

    Crow killer, fish killer, fishberry, Indian berry, Indian coccles, Oriental berry, louseberry, poisonberry, hockle elderberry, the dried fruits in the drug trade often called "Cocculus fructus" or "Cocculus indicus."

Family:

    Menispermaceae, the Moonseed Family.

Description:

    A climbing, twining shrub with woody stems to 4 in. (10 cm) thick producing numerous aerial rootlets and having pale-gray, corky bark, cracked and fissured with age, and white, milky latex. Leaves are alternate, on stout petioles 2 to 6 in. (5 to 15 cm) in length; broad-ovate, cordate or truncate at the base, pointed at the apex, 4 to 9½ in. (10 to 24 cm) long, 3 to 6 in. (7.5 to 15 cm) wide; thick, leathery, glossy above, whitish beneath, with hairs in the axils of the 3 or, more often, 5 conspicuous main nerves. Flowers fragrant, in pendent, branched panicles 10 to 20 in. (25 to 50 cm) long; male and female on separate plants. Petals are absent; the yellowish-green perianth, ¼ in. (6 mm) broad, is composed of 6 sepals in two ranks; the female flowers have 3, or occasionally 4 or 5, ovaries; in male flowers, the staminal column bears numerous anthers at its rounded apex. Fruits, borne singly or in 2s or 3s in grape-like clusters, are white at first, turning dark-red when ripe; subglobose or kidney-shaped with the basal depression and the apex on the same side; ⅖ to ½ in. (10 to 12.5 mm) long; with a single, U-shaped seed. Dried fruit has a wrinkled, thin, brittle, nearly black outer shell and a thicker white inner shell enclosing the oily, bitter endosperm.[49, 148, 160, 266, 305, 584, 624]

Origin and Distribution:

    Native to eastern and southern India, Burma, Thailand, South Vietnam, Ceylon, Malaya and from the Philippines to New Guinea.[20, 93, 441]

Propagation, Culture and Harvesting:

The vine is easily propagated from seed or root cuttings but is cultivated only as a specimen in botanical gardens.[115] For pharmaceutical purposes, the fruits are gathered from the wild by village people and dried in the sun before export.[20]

Constituents:

The fruit pericarp is bitter; it contains the non-toxic alkaloids menispermine and paramenispermine.[20, 441]

The seeds contain 1.5% of the very bitter, crystalline principle, picrotoxin ($C_{30}H_{34}O_{13}$, m.p. 203-204°), which is soluble in alcohol and separable into picrotoxinin ($C_{15}H_{16}O_6$), a highly oxygenated sesquiterpene derivative,[584] and picrotin ($C_{15}H_{18}O_7$).[20] It is not altered by heat.[461] Also present are the tasteless cocculin or anamirtin ($C_{19}H_{28}O_{10}$) and 11 to 24% of a fixed oil.[20]

Analyses of the seed oil show fatty acids as palmitic, 6.1%; stearic, 47.5%; oleic, 43.3%; linoleic, 3.12%; trisaturated glycerides, 9.77%; disaturated mono-unsaturated, 41.55%; mono-saturated di-unsaturated, 48.78%, and tri-unsaturated, .0%. Sitosterol has been found in the unsaponifiable matter.[117]

Medicinal uses:

In India, the fresh leaves are inhaled as snuff to relieve malaria. The leaf juice and juice of the toxic rhizome of *Gloriosa superba* L. are used together to extract Guinea worms.[148] Juice of the fresh fruits is applied on scabies and foul ulcers.[160] Dried fruits in an oily solution, or the pulverized kernels blended with fat or castor oil, are much used as a remedy for ringworm and other parasitic skin afflictions and to destroy lice. Picrotoxin is incorporated into an ointment having similar uses, but all of these preparations are dangerous if the skin is broken or ulcerated.[148, 160] Picrotoxin, in doses of 1/200 to 1/100 gr is employed in the treatment of epilepsy; in doses of 1/120 to 1/60 gr to overcome night sweating in cases of phthisis. It is administered in paralysis of the pharynx and lower limbs; also to counteract morphine and chloral poisoning.[148] The fruit has been official in India, England, Denmark, Sweden, France, Spain, Mexico and Venezuela.[461]

In the United States Pharmacopoeia in 1890, picrotoxin was recommended for treating epilepsy and chorea, but the use has been abandoned. In the United States, it is now declared to be of no value in treating morphine poisoning and is employed only as a respiratory stimulant to offset the effects of barbituates and other central nervous system depressants [423] and sometimes to treat schizophrenic convulsions.[20]

It is given intravenously, and the reaction may not be evident for ten to 30 minutes, the time differing with the individual. Because the effect is transient, treatment is usually repeated in fifteen to thirty minutes. To avoid over-medication, doses are small at the outset and gradually increased until corneal and pupillary reflexes are restored.[423]

Toxicity:

Picrotoxin affects the central nervous system and is a gastric irritant. Overdoses cause salivation, vomiting, purging, restlessness, rapid, shallow respiration, slowing of the heart, sometimes palpitations, giddiness, tremors, impaired vision, delirium, stupor, loss of consciousness and, finally, repeated convulsions, mostly of the legs, just prior to death from asphyxia.[116, 423, 461] The lethal dose is said to be .05 gr per kilo of body weight.[264] Autopsy shows congestion of stomach, lungs and brain.[116]

Whole fruits or oven-dried, pulverized fruits have been used for killing crows, dogs and cattle or for stupefying fish.[20, 266] An oily extract also has been utilized for poisoning fish and game.[160] Malayan jungle tribes apply a paste of the seeds to their daggers and arrows.[116] Experimentally, an aqueous solution of the ground seeds killed a bean plant in twenty-four hours.[160] In India in 1928, it was reported that a decoction of the fruits caused serious illness in six persons, two dying within thirty minutes, the others surviving.[116]

Overdoses of picrotoxin are treated by injecting a barbituate intravenously.[423]

Other uses:

In the past, in England, Europe and in Bombay, a decoction of the fruits was often illegally and hazardously added to beer to render it more intoxicating.[305] A British publication entitled "Art of Brewing," advised adding 3 lbs. (1.36 kg) of the seeds to every "10 quarters" of malt.[160]

Rope is made by twisting the whole stems or their stripped bark.[93]

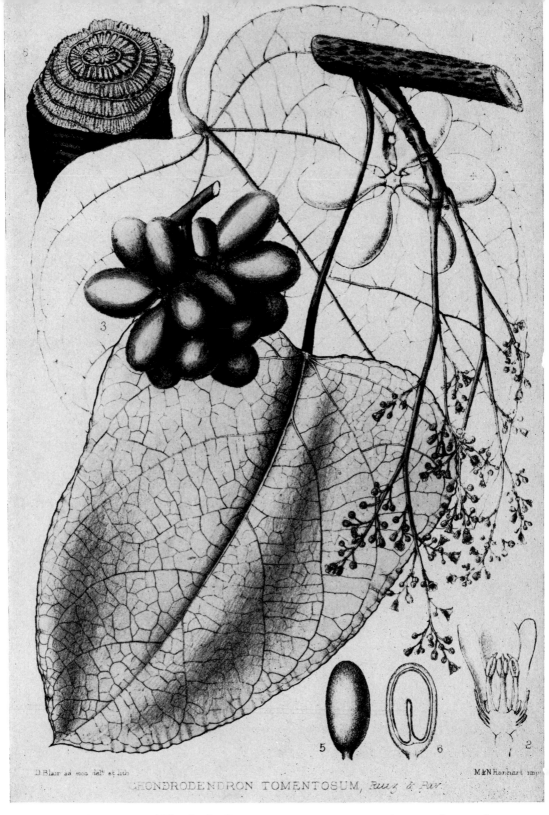

Figure 33. Pareira (*Chondrodendron tomentosum* Ruiz. & Pav.). Reproduction from color painting by David Blair in *Medicinal Plants,* R. Bentley and H. Trimen (J. & A. Churchill, London; 1880).

Pareira

Botanical name:

 Chondrodendron tomentosum Ruiz & Pav. (originally erroneously *Chondodendron*).

Other names:

 Pareira brava; wild grape.

Family:

 Menispermaceae, the Moonseed Family.

Description:

 A high-climbing vine with woody, twining, warty stems with minutely velvety gray bark. The rhizome, reaching several feet in length and up to 5 in. (12.5 cm) in thickness, is longitudinally corrugated and has transverse ridges; is dark-brown externally, orange-yellow within; in cross-section displays a waxy sheen and 3 or 4 concentric rings about ¼ in. (6 mm) wide. Leaves are broadly triangular, heart-shaped or nearly round, with blunt or pointed tip and downy petiole 3⅛ to 5½ in. (8 to 14 cm) long, smooth or slightly hairy on the upper surface, white woolly beneath; up to 4 or 7 in. (10 or 18 cm) long, sometimes with faintly scalloped margins. Male and female flowers borne on separate plants, the male with 9 to 16 sepals and 6 minute petals borne in axillary panicles 2½ in. (6.3 cm) or more in length, the female with 9 outer and 6 inner, larger, sepals which hang in panicles 4 to 10 in. (10 to 25 cm) long. Fruit, on thick stalks, is oblong-oval, about ¾ in. (2 cm) long, scarlet when unripe, black when ripe, fleshy, of acid flavor, and contains 6 flat, disk-like seeds.[333, 356, 476, 624]

Origin and Distribution:

 This vine grows wild in forests of southern Brazil, Peru, Colombia and Panama, up to altitudes of 2000 ft. (600 m).

Constituents:

 The roots and stems are intensely bitter. They contain the quaternary base *d*-tubocurarine and four crystalline tertiary bases, *d*-isochondodendrine, *d*-isochondodendrine dimethylether, *d*-chondocurine and a levorotatory alkaloid.[622]

The root contains the alkaloid chondodine; the stem, tomentocurine; the leaves, chondocurine, curine and tubocurarine.[617]

Other alkaloids reported in the plant include norcycleanine.[616]

Tubocurarine inhibits the enzyme acetylcholinesterase and thus has neuromuscular blocking action.[466]

Propagation, Cultivation and Harvesting:

The plant is not cultivated commercially. For pharmaceutical purposes, the stems and roots are gathered from the wild, dried, cut into short pieces and exported. Or, the plant material may be locally processed to produce a dark-brown, viscous extract which is sealed in tins for shipment.[584] The extracted, purified tubocurarine, or *d*-tubocurarine, chloride is a white, or near-white, bitter, odorless crystalline powder.[122]

Medicinal uses:

Brazilians regard the plant as an antidote for snakebite, applying the crushed leaves externally and drinking an infusion of the root in wine.[233] The bark of the stem has been employed as a febrifuge. In Europe, an infusion of the dried root (1 oz. in 1 pt. boiling water) was taken as a diuretic and remedy for kidney and bladder inflammation,[624] gonorrhea, leucorrhea, jaundice and rheumatism.[233] In modern practice abroad and in the United States, tubocurarine chloride is employed to induce paralysis of skeletal muscles for the purpose of reducing convulsions brought on by strychnine or tetanus and also those in mental patients during shock therapy, lessening the possibility of fracture.[122, 423]

It is injected as a muscle relaxant in abdominal surgery and frequently as a temporary relaxant in neurological cases and particularly in spastic cerebral palsy. It is administered before setting broken bones and dislocated joints; also during remedial therapy in cases of poliomyelitis. Tubocurarine is injected as an aid to diagnosis of myasthenia gravis, in which it emphasizes the symptoms. The natural product declined in use for a time and then in recent years has been again in demand, though it is frequently replaced by synthetic preparations.[423, 584]

Toxicity:

Overdoses or administration together with other drugs may cause respiratory paralysis which must be counteracted by injection of neostigmine methylsulfate.[423]

Other uses:

Extracts of this and other species of *Chondrodendron* and of *Strychnos toxifera* Rob. Schmob. are ingredients in the complex arrow poison, curare, of the South American Indians.[622]

Related species:

False pareira is the root of *Cissampelos pareira* L., a pantropic member of the same family. The roots contain isochondodendrine and other alkaloids with neuromuscular blocking action and have been exported in substitution for those of *Chondrodendron tomentosum*.[605]

Lauraceae

Figure 34. Camphor tree (*Cinnamomum camphora* Nees & Eberm.). Photographed in the main conservatory, Longwood Gardens, Kennett Square, Pennsylvania. The branches had been recently pruned.

Camphor

Botanical name:
> Cinnamomum camphora Nees & Eberm. (*Laurus camphora* L.; *Camphora officinarum* Nees; *Camphora camphora* Karst.)

Other names:
> Japanese camphor tree, camphor laurel, gum camphor.

Family:
> Lauraceae, the Laurel Family.

Description:
> A tree, usually to 40 ft. (12 m) high but occasionally attaining 90 to 100 ft. (27.4 to 30.4 m) in its homeland. It has a short trunk and dense crown, aromatic, dark-gray or dark-brown, rough, fissured bark; leaves are semi-evergreen, alternate, aromatic, long-stalked, ovate to oblong-lanceolate, pointed at both ends, 2¼ to 4¼ in. (5.5 to 11 cm) long and up to 2 in. (5 cm) wide, with 3 to 5 prominent nerves beginning a little above the base; leathery, dark-green, glossy above, whitish or glaucous beneath, drying to a dark red-brown. Flowers, in short axillary clusters, are sweetly fragrant, yellowish-white, .11 in. (3 mm) long, .20 in. (5 mm) wide. Fruit, seated on a brownish perianth tube, is round-oval, .39 to .47 in. (10 to 12 mm) wide, fleshy, with a single seed.[71, 134, 333, 549]

Origin and Distribution:
> Native to eastern China, Japan, Formosa and Tonkin. In 1676, the tree was planted in Holland and a few years later in other countries of Europe.[98] Before 1780, it was sparingly introduced into Java and later Malaya, the Philippines, eastern and southern Africa, Queensland, Central America, Cuba, Trinidad, central and northern Florida, Texas and California. It is common in gardens in Ceylon and India up to elevations of 7,000 ft. (2,133 m) but grows best between 4,000 and 6,000 ft. (1,220 to 1,828 m) where rainfall is between 45 and 145 in. (114 and 368 cm) per year. It is cultivated commercially in Formosa and to a limited degree in India[20] and in Georgian SSR.[628]

Constituents:
> Camphor is 2-bornanone[423]—a crystalline ketone containing 97% or more of $C_{10}H_{16}O$.[605] There are forms of *Cinnamomum camphora* (for-

merly believed to be distinct species) which contain no camphor, only camphor oil. The leaves of these forms smell of turpentine or eucalyptus. In others, camphor, which is produced by enzymatic action, is found in cells of yellow oil throughout the tree,[117] and the leaves of these forms are strongly camphor-scented.[20] In old trees, the camphor may be found concentrated in solid lumps.[98] In Japan, the sesquiterpenes have been studied in var. *linaloolifera*,[201] and one type of camphor tree having non-volatile sesquiterpenoids in the leaves is distinguished as "the sesquiterpene tree."[261]

Various environmental and other factors influence the amount of camphor in different parts of the tree. The crude oil from the wood contains camphor and safrole; from the leaves, camphor and no safrole; from the twigs, mainly cineol and less camphor.[98] There is more safrole in the roots than in other parts.[206] Camphor content is generally higher in the roots and base of the trunk than in the upper parts of the tree[98] though, in some strains, the camphor may be found mainly in the leaves.[93] Young leaves are richer in camphor than mature leaves. Leaves harvested after extended rainy or overcast periods are low in camphor, and those from shaded trees contain less than those from trees growing in the open.[20] Shade causes as much as 13.5 percent increase in safrole content.[201]

Camphor leaves in India have yielded 22.2% camphor, and camphor oil containing d-α-pinene, dipentene, cineol, terpineol and caryophyllene. Formosan leaves have yielded 44% camphor, and camphor oil containing 28% cineol, 0.4% aldehydes, pinene, camphene, phellandrene, limonene, dipentene and an unknown alcohol. Twigs distilled in India gave a low yield of a deep-brown oil containing 20% camphor, 26% d-α-pinene, 11% dipentene, 6% terpineol and 10% caryophyllene and cadinene. Indian camphor wood has yielded pale yellow oil consisting of 30% camphor, 16% d-α-pinene, as well as dipentene, terpineol, safrole, caryophyllene, cadinene and camphoracene.[20]

The seeds contain 42% of a fixed oil having a mild scent and consisting mainly of laurin.[98]

Propagation, Culture and Harvesting:

The camphor tree can be grown from seeds, cuttings, layers or root suckers.[117] Cuttings from trees rich in camphor, however, are difficult to root.[98] Propagation is usually by seed obtained from trees twenty to thirty years old. Seeds from earlier crops of fruits are infertile. Viability period is short, and germination rate is low. Fresh seeds, freed of pulp, should be air-dried for four to five days and, as soon as possible, planted in nursery beds of light, sandy loam enriched with manure. Germination usually takes about ninety days. When the seedlings are six months old, the tops and roots are trimmed and the plants are transferred to grass baskets filled with fertilized soil. Six months later, they are set out in the field in 2 × 2 × 2 ft. (.60 × .60 × .60 m) pits spaced 6 to 12 ft. (1.8 to 3.6 m) apart—605 to 1210 plants per acre.[20]

There are no major pests nor diseases requiring control, but the tree is subject to attack by thrips or red spiders and is also prone to chlorosis from manganese deficiency in some soils.[20, 398]

Trees fifty years of age and older yield the most camphor, and it was formerly the custom to fell seventy- to eighty-year-old trees, chop up the wood, young branches, twigs and roots and steam distill them for the crude oil, of which the yield is about 3%. In the late 19th century, the privately owned forests of camphor trees in the old provinces of Kiushiu, Shikoku, Iga, Suruga, Isé and Kishiu, Japan, were seriously depleted, though government-owned forests were still rich in camphor trees. At that time, replanting commenced, and also it was decided that trees twenty-five to thirty years old could be satisfactorily processed.[511] It was later realized that leaves could be harvested for camphor extraction without destroying the tree. Leaf-harvesting can begin when the trees are only five years old and can be performed three to four times a year, thus keeping the trees pruned to a height of 5 to 6 ft. (1.5-1.8 m).[20] A harvest of 50 lbs. (22.6 kg) of leaves yields about 1% of oil which is 10 to 15% camphor.[117]

Annual yields of camphor vary considerably. Indian plantations have averaged 60 lbs. (27.2 kg) camphor and 10 lbs. (4.5 kg) camphor oil per acre. The Hallakarai estate of 8 acres produces 500 lbs. (226.7 kg) of camphor and 150 lbs. (68 kg) of camphor oil annually.[20]

Prior to World War II, 80 percent of the 5,000-ton annual world supply of natural camphor came from Formosa,[117] where there are some 18 million trees. In post-war years, plantations on the southernmost island of Kyushu developed into another major source.[466] The oil is refined in Japan, Europe or the United States to obtain the camphor[117] and other important by-products including safrole, heliotropin, borneol, vanillin, terpin hydrate and terpineol.[466]

Synthetic camphor has been produced in Germany since 1910. A process, developed in the United States in 1930 and acquired by DuPont in 1931,[17] made possible the production of camphor from α-pinene derived from the turpentine of *Pinus pinaster* Ait. (*P. maritima* Poir.) and *P. palustris* Mill. and would be largely replacing the natural product were it not for the other compounds yielded by camphor oil.[466]

Medicinal uses:

Camphor production from *Cinnamomum camphora* was initiated in China and Japan before the 9th century AD.[98] During the Civil War,

35. 36.

Figure 35. The trunk of the camphor tree may be recognized by its deeply fluted, or fissured, dark-gray or brown bark.

Figure 36. A young camphor tree growing in the open on the grounds of the Morris Arboretum of the University of Pennsylvania, Philadelphia.

there was such a demand for camphor in the United States that this country contracted for the entire output of Formosa, which was then under the government of China, and it was even proposed that an effort be made to purchase Formosa to monopolize the camphor trade. Japan, instead, acquired Formosa in 1895.[17]

In India, various parts of the tree are employed as remedies for colds and chills.[71] In Asia and Europe, camphor is applied to sprains, inflammations, gout and rheumatic joints and taken internally to calm hysteria, abate convulsions and epileptic attacks; also as a carminative and as a respiratory and cardiac stimulant, and in some cases of diarrhea.[117, 277, 624]

In Cuba, camphor is employed as an antiseptic, antispasmodic, anaphrodisiac, antiasthmatic and anthelmintic. It is given in nervous and eruptive fevers. The leaves are crushed and steeped in alcohol, which is applied as a rub to relieve rheumatic pains.[485] In the southwestern United States, pioneers melted camphor and pork fat together or mixed camphor with whisky for the same purpose.[143] Camphor in olive oil is popular in Mexico and South America as an application on bruises, contusions and neuralgia, and camphor is much employed in treating rhinitis, asthma, pulmonary congestion, chronic bronchitis and emphysema.[368] In the past, it was a common and nearly universal practice to wear around the neck a little bag containing a lump of camphor in the belief that the emanation would cure colds and related ills.[624]

Today, in the United States, only 10 percent of the camphor supply is devoted to medicinal purposes—mainly as an antipruritic, as a rubefacient or counterirritant, or as an anaesthetic in lotions for external use to counteract pain and itching. It is often combined with cottonseed oil or with phenol and has been claimed to reduce phenol absorption, but camphor-phenol lotions have caused skin ulceration.[423]

Toxicity:

It is reported that the camphor fumes during the crude distilling process cause a continuous flow of tears from the eyes of workmen.[17] Camphor poisoning has occurred in children and, from large doses, in adults; the effects include a feeling of warmth, nausea and vomiting, headache, confusion, excitement or, occasionally, depression, restlessness, delirium, hallucinations, unconsciousness and convulsions; death may result from respiratory failure.[605]

Other uses:

The camphor tree has been planted as an ornamental shade tree and also for windbreaks and hedges.[619] While attractive, it has limitations in landscaping because of its broad spread, which may exceed 100 ft. (30.4 m), and its aggressive and massive surface roots which disrupt paving.

In China, camphor wood is greatly prized as it does not warp or split, is immune to insect attack and repels moths. It was formerly much used for sailors' storage chests, trunks, bookcases and panelling[333] but is today very expensive.

Powdered camphor is employed in embalming in Africa. Camphor is an ingredient in incense and plastics, and large quantities were formerly used in making celluloid. Much is consumed in the manufacture of explosives, disinfectants and insecticides. Camphor oil is utilized in perfumery, soaps, detergents and deodorants; also as a solvent in paints, varnishes and inks.[41] White camphor oil, free of safrole, contributes to the formulation of vanilla and peppermint flavors[17] and enters into the flavoring of certain soft drinks, bakery products and condiments.[210]

Papaveraceae

PAVOT.

Opium Poppy*

Botanical name:

>Papaver somniferum L.

Other names:

>White poppy, carnation poppy.

Family:

>Papaveraceae, the Poppy Family.

Description:

>An annual herb, 2 to 4 ft. (.6 to 1.2 m) tall, with a white, fusiform, branching, fibrous taproot and an erect, cylindrical, green stem, rarely branched, mostly smooth, slightly hairy near the apex. All parts contain white or gray latex. Leaves are alternate, clasping the stem, ovate- to linear-oblong, wavy, irregularly lobed and strongly toothed, 3 to 10 in. (7.5 to 25 cm) long. Flowers, borne singly at the tips of long, usually smooth stalks, are 3 to 4 in. (7.5 to 10 cm) wide, have 4 rounded, wavy petals, white, bluish-white with a purple splotch at the base, pale-lavender, purple, red or variegated, and a conspicuous, rounded green ovary capped with the five- to twelve-rayed stigma which is surrounded by a prominent fringe of yellow stamens—150 to 200 in five concentric circles.[145] Ornamental strains may have double or fringed petals. The fruit (seed capsule) is globular, 2 to 4 in. (5 to 10 cm) wide, crowned with the disk formed by the united stigmas, green when unripe with copious white, grayish, pale- or bright-pink latex. When mature, the capsule turns yellow, then it dries to brown and small apertures below the stigma open (in primitive types) releasing a great number of small, kidney-shaped seeds—white, yellowish, gray, lavender, blue, reddish-brown or black—which are rich in oil.[13, 63, 116, 145, 305, 345] In improved cultivars, there is virtually no opening of apertures and no loss of seed.[345]

---

* The literature on *Papaver somniferum* and its relatives is voluminous; see the *Annotated Bibliography on Opium and Oriental Poppies and Related Species* by Duke et al.[162]

←

Figure 37. The opium poppy (*Papaver somniferum* L.). Reproduction from a color painting by P. Turpin in *Flore Médicale* Vol. 5 (F. P. Chaumeton, Paris; 1818).

Origin and Distribution:

The plant is believed to be native to the western Mediterranean region of Europe and to have spread through the Balkan peninsula to Asia Minor.[318, 345] It was formerly much grown in Italy, Turkey, Greece and China. Turkey officially abandoned opium as a crop in June, 1971. As of 1973, there was no authorized Turkish production of opium.[36] At present, it is cultivated as a drug crop primarily in Egypt, Iran, the Soviet Union, India and Nepal. The latter country has announced its intention to prohibit opium growing.[36] Cultivation on a lesser scale is carried on in Yugoslavia, Bulgaria, Hungary, Austria, Czechoslovakia, Poland, Germany, the Netherlands,[329] Pakistan, China, Japan[33, 34] and Argentina.[487] In Thailand, it has been illegal since 1959, but there are illicit plantings in the hills of northern Thailand, the neighboring region of Laos and the Shan state of Burma,[33, 36, 505] and in these areas there are also factories for converting opium into morphine or heroin. Vietnam is burdened with over 100,000 opium addicts as well as several thousand enslaved to morphine or heroin.[36] In India, commercial cultivation was once widespread but is now largely restricted to Uttar Pradesh.[116] Pakistan is committed to progressive abolition.[36] Iran suspended opium culture in 1955 but resumed in 1969 for domestic consumption only.[584] By 1972, plantings in that country had increased to 54,562.5 acres (21,825 ha). In 1973, Iranian opium fields were reduced to a total of 5,660 acres (2,250 ha).[36] Afghanistan officially abandoned the cultivation, trade and use of opium in 1956,[33, 34] but the ban has not been thoroughly enforced (mainly because of the remoteness and near-inaccessibility of some growing areas), and opium produced, especially in the province of Badakhshan, is regularly smuggled out by caravan.[36, 317] In Russia, the crop is grown for seed in the Ukraine and for opium and alkaloids mostly in the Kirghiz region.[584] In central Europe, the plant is grown primarily for its seeds, with extraction of opium alkaloids a recently developed by-product industry.[263]

All legitimate opium culture is conducted under the supervision and regulation of the International Narcotics Control Board[232] and the control of the governments of the producing countries. In Macedonia, during periods of devaluation of currency, villagers have hoarded opium as a "gold reserve," especially as a dowry for daughters. This practice was outlawed by the Narcotics Control Act of 1938 and a stronger measure in 1950.[328]

As a garden flower, the opium poppy has been distributed throughout the temperate regions of the world. However, since the passage of the Opium Control Act in 1942, it can no longer be grown as an ornamental or for any other purpose in the United States.[306]

Constituents:

The latex of the immature fruit is the major source of alkaloids, the most important being morphine ($C_{17}N_{19}NO_3$), 5 to 15%; codeine, or methylmorphine ($C_{18}H_{21}NO_3$), .1 to 2%; thebaine, or dimethylmorphine ($C_{19}H_{21}NO_3$), .15 to .5%; noscapine, or narcotine ($C_{22}H_{23}NO_7$), 2 to 10%; narceine ($C_{23}H_{27}NO_8$), .1 to .4%; and papaverine ($C_{20}H_{21}NO_4$), .5 to 1%.[20, 423, 552] Other alkaloids include aporeine, codamine, cryptopine, gnoscopine, hydrocotarnine, lanthopine, laudanidine, laudanine, laudanosine, meconidine, narcotoline, neopine, oxynarcotine, papaveramine, porphyroxine, protopine,[617] reticuline,[616] and rhoeadine and xanthaline.[617]

Codeine, morphine, noscapine and narcotoline have been found in the leaves,[617] and codeine, morphine, noscapine, papaverine, rhoeadine and thebaine in the germinated seeds.[20]

Young roots contain .39% morphine, old roots, .13%.[20]

Morphine acts as an analgesic, narcotic, stimulant and euphoric. It is apt to lead to drug dependence.

Heroin, which results from the acetylation of morphine, is more potent, more addictive and prohibited.

Codeine is a milder analgesic and narcotic and induces little euphoria. It is less constipating than morphine and causes less respiratory depression. There is a growing demand for this alkaloid.[232] It may be prepared by methylation of morphine with phenyltrimethylammonium methoxide.[274]

Papaverine and noscapine are mainly muscle relaxants; they have a mild inhibitory effect on intestinal peristalsis.[423] Papaverine is non-addictive.[552] Commercial syntheses of papaverine are reviewed by Luk'yanov *et al.*[351]

Noscapine is slightly narcotic and accelerates respiration. When combined with morphine, it reduces the hazard of respiratory depression.

Narcotoline and oxynarcotine and hydrocotarnine are similar in action but weaker.[423]

Thebaine, laudanosine and laudanine are convulsants rather than narcotics. Thebaine is being increasingly utilized as a source of codeine.[232, 430]

Cryptopine depresses the higher nerve centers. Protopine in small doses is a sedative; in large doses it causes excitement and convulsions.[20]

Meconic acid, found only in opium and present in amounts varying from 3 to 5%, serves as a means of identifying this drug material.[122]

The presence of methyltransferase enzymes in poppy latex has been reported by Antoun and Roberts.[40] A review of *Papaver* alkaloids, old and new, is presented by Pfeifer.[443] Methods of quantitative determination and isolation are reported by Bechtel, by Gaevskii *et al.*, Kamedulski *et al.*,

Morice and Louarn, and Rondina et al.[66, 212, 292, 393, 487] Assay results for morphine are highest when the laboratory temperature is 68 to 69.8° F. (20-21° C.); results decrease as laboratory temperature increases.[465]

Propagation, Cultivation and Harvesting:

The opium poppy is grown from seeds which, in India, are mixed with soil or ashes and, in October and November, sown broadcast in well-tilled and fertilized fields at the rate of 3.5 to 5 kg per hectare.[20] In Bulgaria, broadcasting of seeds was supplanted by the use of a bottle with a goose quill in the cork, a method now largely superseded by mechanical seeders.[145]

In Bulgaria and Yugoslavia, there are two plantings each year, one in the fall which may be only partly successful and a more reliable one in the spring.[584] Fall planting is desirable where weather permits because the capsules are richer in morphine.[145] In Bulgaria, winter-hardy strains have been developed, as well as hybrids of subspecies *turcicum* and *eurasiaticum* which surpass their parent strains in morphine content.[452] However, tests over a five-year period have revealed higher morphine content in certain Russian and Hungarian strains than in any Bulgarian types.[553] In Hungary, a high-yielding hybrid has been achieved by crossing *P. somniferum* and *P. orientale*.[348] Variety selection studies based on genetic and environmental components are going forward in India with a view to increasing both the alkaloid content and yield of seed oil.[303]

Germination takes place in ten to twelve days.[145] Fertilization with NH_4NO_3 at planting time contributes to morphine production.[183] Irrigation also is essential at the outset. When the seedlings are 2 to 3 in. (5 to 7.5 cm) high, they are thinned out, allowing a distance of 8 to 10 in. (20 to 25 cm) between the remaining plants. The crop is very sensitive to cold, heavy rain and strong, drying winds, and is attacked by bacterial and fungus diseases, a virus disease, cutworms, beetles, weevils, caterpillars and other pests.[20, 317, 525] Temperature and rainfall at the time of capsule maturity strongly affect morphine content.[367]

Flowering commences in seventy-five to eighty days, and the petals are shed in twenty-four to seventy-two hours. Thereafter, the capsules enlarge and, in eight to fifteen days, are lanced to cause the latex to exude. This operation is performed with a single- or 3- to six-bladed knife, or spiked instrument, in the afternoon, the incisions being shallow and made either vertically or, most often, horizontally, partly encircling the capsule. The workers walk backwards to avoid brushing the latex from the capsules just incised.[328] The following morning, the darkened, coagulated latex is scraped off and stored in metal or earthen pots (perforated at the bottom or placed at a tilt) and turned over every ten days or so to facilitate draining off of a dark-colored fluid. Thereafter, the opium is placed out on flat plates to dry in the sun. Lancing may be repeated from three to ten times until the latex ceases to flow. The various collections are kept separated because the first lancing yields more morphine and noscapine than subsequent flows.[20] Multiple lancing has been practiced mostly in the production of opium of low alkaloid content for smoking.[328] Each plant bears from 5 to 8 capsules. The terminal or oldest capsule has a higher percentage of morphine than those that are laterally borne.[20] When latex collection is finished, the plants are left standing for twenty to thirty days. Then the dry capsules are havested for their seeds.[529]

In Bulgaria, Yugoslavia and Russia, the modern practice is to make only one lancing and collection of latex. Thereafter, the entire aboveground plant with its capsules is harvested for bulk extraction of alkaloids. Mechanical harvesting techniques are being developed, and considerable attention is being given to the breeding of high-yielding strains, some producing 110 to 176 lbs. (50 to 80 kg) of opium per hectare. Average yields of 66 to 78 lbs. (30 to 35 kg) per hectare are obtained from Russian plantings thinned to 121,000 to 130,000 plants per hectare. Lower plant density —70,000 to 80,000 per hectare—has yielded 59 lbs. (26.9 kg) of opium per hectare. Each Soviet farm is planted to three types of poppy to lengthen the harvesting period of fifteen to twenty days. All Russian production and processing is managed by the All-Union Office for the Production, Preparation and Processing of Medicinal Plants.[525]

Turkish and Yugoslavian opium was formerly scraped off onto poppy leaves, molded into bricks of various shapes and sizes, stamped, dried and then dusted with earth and put in sacks or tins with dry *Rumex* fruits which adhered to the surface and provided a non-sticky coating.[584] Since 1958, a more refined product is exported from these countries, usually coated with powdered, dried poppy leaves.

In India, after inspection of the sun-dried product for impurities and adulterants, the lumps are classified as to quality and packed in double cloth or polyethylene sacks, or, if of low consistency, in pottery jars. The substance is plastic at first, becoming brittle with age.

Beginning in mid-April, the entire output is delivered to the Government Opium and Alkaloid Works in Ghazipur for processing and export. The Government Opium Factory in Madhya Pradesh serves as a warehouse for stock. After chemical tests are made, the opium is graded according to morphine content and consistency, the five grades being as follows: A (M. S. 12%), B_1 (M. S. 11 to 12%), B_2 (M. S. 10 to 11%), B_3 (M. S. 8 to 10%), and C, inferior (impure). Morphine is quickly lost if the opium is exposed to the air at 60° F., but there is little loss in sealed containers at 98 to 100°. In moist atmospheres, the latex may become covered with mold without affecting its morphine content.[20]

The Ghazipur factory processes opium into three forms: *Excise*, sun-dried to 90° consistency, cut into cubes, wrapped and marked individually and packed in compartmented chests of various capacities; *Export Opium*, prepared as 1 kg cakes or 5 to 10 kg slabs; *Indian Medicinal Opium, cake and powder*, various grades blended, sun-dried to 90° consistency, formed into .5 to 1 kg slabs. For powdering, the opium is heated to raise the consistency to 100°, then pulverized by machine, sieved and packed and sealed in metal cans.[20] Russian opium is kneaded mechanically, and metabisulphite is added to prevent fermentation.[584]

Solid opium is a light khaki color externally and dark-red-brown internally, aromatic and bitter. Powdered opium is light-brown or yellow-brown.[423] Unadulterated opium is highly inflammable and burns with a clear flame.[160] An improved and simplified process for extracting alkaloids from opium was developed in Canada and described in 1957.[259]

While most of the opium and opium alkaloids of commerce are obtained from the latex, in central Europe the alkaloids are extracted from "poppy straw"[329]—4 to 12 in. (10 to 30 cm) of the upper stalk together with the crushed capsules[524, 525] as a by-product of the crops grown for the production of poppy seed and poppy seed oil.[584] The empty, dry capsules, particularly, first used for morphine extraction in Hungary in 1931, are, because of improved techniques, becoming more and more important as a source of morphine, competing successfully with the more costly opium.[263, 453] The process is wholly mechanized.[525] The upper 4 in. (10 cm) of the fruit stalk may contain as much morphine as the dry capsule but usually contains 50 to 70 percent less.[328] In 1958, over 14,000 metric tons (14 million kg) of capsules and straw yielded 10,000 lbs. (22,003 kg) of morphine.[329]

It has been demonstrated that harvesting of dry capsules thirty-six days after flowering results in a 25 percent increased yield of morphine per acre or hectare compared with the yield from immature capsules.[581] In 1964, 32.2 percent of the world's morphine was extracted from "poppy straw" (including dry capsules).[33]

Medicinal uses:

High esteem for the narcotic properties of the opium poppy among the ancients is evidenced by the many representations of the capsules discovered by archeologists in eastern Mediterranean ruins. Likenesses of the capsules appear as designs on coins, jewelry, vases, plates and sculpture. Jars, pendants, pinheads, beads and balls of stone, clay, metal or ivory were shaped to resemble the capsules. On some such objects there are grooves clearly suggesting incisions for the extraction of opium.[318]

In the 3rd century BC and earlier, opium poppy capsules were valued by the Arabs, Romans and Greeks for the preparation of sedative potions.[117] Children were given poppy capsules to chew as pacifiers.[167] Opium was unknown in China before 763 AD. Arab traders introduced the capsules to the Chinese for medicinal use in the 13th century. Later, Chinese writers extolled the capsules as magical in curing dysentery.[117]

One hundred years ago, poppy-capsule decoctions, sirups and extracts were widely prescribed by physicians for internal use, to calm the nerves, alleviate pain and to induce relaxation.[305] Opium preparations were in casual household use.[167] In England, opium was a familiar ingredient in a number of "soothing" patent medicines.[160]

Today, in India, a warm decoction of the dry, ripe capsules is applied as a sedative on inflammations and painful swellings.[116] A hot infusion is even advocated as a beverage comparable to coffee or tea.[529]

The extracted latex has a long history of use in Mohammedan medicine, internally and externally as a sedative and internally as a treatment for diarrhea, dysentery, coughs and asthma. Since 1914, great efforts have been made in India to produce high-quality medicinal opium for domestic purposes and export.[117]

About 350,000 lbs. (158,757 kg) of medicinal opium reaches the United States annually, mainly from India.[122] Opium has been commonly administered as a narcotic, sedative, anodyne, antispasmodic, hypnotic and sudorific. Powdered opium and ipecac with lactose constituted the Dover's powder popular in the past as a sedative and sudorific at the onset of colds and influenza and in cases of muscular rheumatism.[423] The major role of opium in medical practice, however, has been its gastrointestinal effect, inducing spasms which delay the emptying of the stomach and bowels. At the present time, the form in which it is chiefly employed is Paregoric (the tincture of camphorated opium), given to control severe diarrhea and dysentery,[423, 552] and sometimes prescribed as a mild anodyne to control coughing and nausea. Paregoric represents 12% of the Brown mixture which is taken in 5 to 15 ml doses as an expectorant.

Various mixtures of opium alkaloids in virtually their natural proportions are preferred for administration by injection.[423]

Morphine was first isolated by Derosne in 1803 as "salt of opium." Two years later, Sertürner published an account of his isolation of morphine and gave it this name. Despite its hazards, it has been universally regarded as a great boon to mankind for nearly 200 years. Heroin, adopted into therapeutic use in 1898 in the erroneous belief that it had a less depressing effect on the respiration than morphine and that it could be used to overcome morphine addiction, proved more dangerous and had to be abandoned.[167]

Morphine sulfate is official in the United States Pharmacopoeia and is given orally or by injection to relieve extreme pain of myocardial infarction, inoperable cancer and other severe afflictions; it is also often prescribed in typhoid fever, internal hemorrhages and traumatic shock. It may be combined with atropine in renal and intestinal colic and coronary thrombosis. It is hazardous to infants, aged persons and pregnant women and contraindicated in epilepsy, tetanus, toxic psychoses, emphysema, bronchitis and certain other conditions.[423]

Apomorphine hydrochloride is utilized as an emetic. Hydromorphone hydrochloride is given orally or in rectal suppositories as an analgesic and narcotic and causes less nausea, vomiting and constipation than morphine.[552]

Codeine, discovered in 1832, is derived from opium or from morphine. Both codeine phosphate and sulfate are employed orally or subcutaneously to relieve pain, sometimes with aspirin. Codeine is most often prescribed to relieve persistent coughs.

Papaverine was isolated by Merck in 1849 and may be recovered from opium residue after extraction of morphine and codeine, or it may be prepared synthetically. It is generally resorted to for relief of gastric or intestinal spasms or asthmatic attacks and may be given orally, by injection or in rectal suppositories. This alkaloid has received much acclaim for relieving pain in coronary artery disease but has proved inferior to aminophylline unless given in large doses or in moderate doses over a period of several weeks. It is valued for reducing vasospasms in cases of arterial embolism. It is contraindicated in complete heart block or depression of cardiac muscle.[423]

Noscapine and noscapine hydrochloride are given orally in 15 mg (or up to 30 mg of the hydrochloride) doses as antitussives.

Cotarnine and cotarnine chloride were formerly employed as hemostats, especially in uterine hemorrhage, but are no longer in use.[552]

There is much veterinary use of morphine sulfate and hydrochloride as analgesics and antispasmodics, mainly for dogs, the drugs being unreliable in larger animals and there being danger of excitation in horses.[552] Apomorphine hydrochloride serves in veterinary practice as an emetic for dogs and an expectorant for horses, cattle, sheep, goats and pigs.[552]

Toxicity:

Workers incising poppy capsules in the fields complain of a kind of "narcosis" from the fumes. If there is not enough wind to carry the fumes away, the laborers may wear masks to prevent excessive inhalation; they also keep their homes closed to shut the fumes out.[317] Because of the danger of addiction, opium and morphine are not recommended in modern medical practice where other agents will suffice.[423] There is little evidence of ill effects from very light doses (½ to 1 grain per day), but habitual use develops the craving for the drug and larger amounts are required to give satisfaction. Opium produces mental confusion, thirst, languor and somnolence, deep but slow respiration, flushed, moist skin and contracted pupils becoming at length almost invisible. Among aftereffects may be dermatitis, headache, constipation and retention of urine. People taking over 10 grains daily become sallow, emaciated, have constricted pupils, low globulin level, lowered vitality, dry skin and tongue, foul breath and susceptibility to respiratory affections; at length there is a hardening of the blood vessels, loss of appetite, impaired digestion and liver function and chronic constipation. The addict becomes indifferent to and unable to perform work and shows moral decline; his euphoria is temporary and is followed by depression, melancholy and occasionally mania.[116] Abstinence gives rise to intense nervousness, digestive disturbances, abdominal pain, diarrhea, headache, insomnia and collapse.

Fatal doses, varying greatly with the individual, may be as low as 300 mg of opium, though addicts tolerate 2 g of morphine taken in a four-hour period. Death from circulatory and respiratory failure is preceded by coldness and clamminess of skin, weak, rapid pulse, pulmonary edema and cyanosis.[423] Pupil dilation occurs just before death.[116] Injected morphine sulfate has produced strong hyperglycemia, shivering, pupil dilation and excitation in cats.[182]

Overdoses of codeine may bring on excitement, narcosis, convulsions, nausea and vomiting, rapid pulse and contraction of pupils, but very seldom does fatality occur.[423] Codeine is reportedly a source of contact dermatitis.[20]

Papaverine, intravenously in doses of 65 to 100 mg, must be administered very slowly, otherwise it can bring on apnea and death.[423]

In Iran, prior to 1955, opium was linked "to the premature death of 100,000, the suicide of about 5,000 and the abandonment of 50,000 children each year."[28] The number of addicts was estimated as between 1,500,000 and 2,000,000 in a population of 19,000,000.[625]

In India, opium and morphine are common means of suicide. According to Chopra et al., nearly 40 percent of fatal poisonings reported to the chemical examiners are attributed to opium. Accidental deaths of infants sometimes result from overdoses when opium is given them to make them sleep.[116]

Chronic poisoning by opium and morphine occurs frequently in India.

In the 17th and 18th centuries, potenates indulged in a beverage made from the poppy capsules, wine and hemp.[117] Opium smoking has been a widespread custom throughout the Orient. In 1891, India initiated efforts to ban or limit opium smoking and, by 1954, the vice was totally prohibited in seven states, prohibited, except for registered smokers (of which there were none recorded) in nine states, and limited to 2,519 registered smokers in twelve states.[25] Not prohibited, and the usual mode of taking the drug in that country, is by mouth ("eating opium").[116] The novice will at first experience irritation of the lips and tongue and even blisters in the mouth.[305] As this habit tends to upset the stomach, many people are adopting the western way of injecting opium alkaloids.

Opium addiction was rampant in China in the latter half of the 18th century, fostered mainly by the British East India Company which furnished illicit opium to the Chinese in exchange for tea. In the Opium Wars of 1840 and 1855, China tried unsuccessfully to stop the smuggling of opium into the country. In 1935, the Republic of China launched an elaborate program of control and treatment designed to finally wipe out opium abuse.

In 1969, Hungarian scientists reported the death of a three-year-old child twenty-four hours after ingestion of unripe poppy capsules.[314]

The plant has rarely caused poisoning in grazing animals and then only when consumed in quantity. The seed residue, or presscake, after oil extraction has been much used as feed for cattle and poultry. Presscake evidently contaminated by alkaloid-containing particles of the capsules was fed to dairy cattle in France during World War II, causing few deaths but many cases of gastroenteritis, nervous excitement, loss of appetite, colic, progressive emaciation and cessation of milk flow.[197]

Other uses:

Seeds of the opium poppy (often called maw seed), harvested from the intact ripe capsules or from the lanced capsules after the gathering of the latex, are widely sold as birdseed and consumed as human food, especially in or on bakery products and in confectionery. The seeds also are an important source of oil, much like olive oil, valued for culinary purposes and as an illuminant and used also in soap and as a drying oil in paints and varnishes. Further, the seeds and seed meal are a commercial source of lecithin.[20] In India, Iran and elsewhere, the young, tender plants (presumably those removed in thinning out the plantation as well as those deliberately grown as potherbs) are cooked and eaten as greens or in soup.[20, 529]

Great Scarlet Poppy

Botanical name:

Papaver bracteatum Lindl. (*P. orientale* var. *bracteatum* Ledeb.)

Other name: [None shown in the *Flora of Iran*]

Family:

Papaveraceae, the Poppy Family.

Description:

A perennial herb reaching 20 to 50 in. (50 to 125 cm) in height, the unbranched stems (as many as 15) covered with whitish hairs. Lower leaves are 8 to 18 in. (20 to 45 cm) long; upper leaves smaller; more or less deeply cut into lanceolate or oblong, toothed segments. Blooms and borne singly, 7 to 18 on each plant; buds are erect, oval to oblong; mature flowers are 4 to 7 in. (10 to 18 cm) in diameter, have 4 to 6 broadly obovate petals, blood-red, often with a dark-purple splotch at the base. The stigma is ten- to eighteen-rayed. The blooms closely resemble those of the Oriental poppy (*P. orientale* L.) of flower gardens but are distinguished by having 3 to 8 leaf-like or calyx-like bracts, two or more persisting. Further, the petal color is long-lasting, while in *P. orientale* it quickly fades. The capsule is round-oval, to 1½ in. (4 cm) long and to 1¼ in (3 cm) wide, smooth, topped by the stigmatic disk.[46, 115, 179, 226]

Origin and Distribution:

This species is native to the Russian Caucasus and northern and western Iran (Alborz mountains and Kurdistan). It is common in dry, stony regions at elevations between 4,000 and 10,000 ft. (1,500 to 3,000 m).[46, 226] It has been widely grown as an ornamental. Introduced into

---

Figure 38. Physically, the great scarlet poppy (*Papaver bracteatum* Lindl.) closely resembles another common species in Iran and Turkey, *P. pseudo-orientale* Medw. Both species have floral bracts which persist below the fruit. Distinguishing features are the short, appressed calyx bristles on the buds of the former and the deep-red of its flowers. Photographed in the Botanischer Garten, Munich, West Germany. (*See also* color illustrations 4 and 5.)

England from Siberia in 1817, its hybridization with *P. orientale* resulted in many named crosses with flowers of various colors.[123]

Since 1963, certain strains found wild in Iran have attracted worldwide attention to *P. bracteatum* as a potential source of thebaine,[46, 521] from which codeine can be synthesized as well as naloxone and other needed narcotic antagonists.[112, 179] This species does not contain opium, and morphine can be derived from thebaine only with complex chemical procedures; it is therefore hoped that substitution of this species for *P. somniferum* would greatly reduce the international opium abuse problem. The United Nations and a dozen countries, including the United Kingdom and the United States, have devoted much effort to the investigation of *P. bracteatum* since 1972. It is being experimentally grown in England. The United States Department of Agriculture and a few pharmaceutical companies have established test plantings in Maryland, Illinois, Arizona, Oregon and Washington. Uppermost is the controversial issue as to whether this crop should be grown commercially in the United States in view of this country's efforts to discourage narcotics production in other countries.

Constituents:

The major alkaloids of *P. bracteatum* are thebaine ($C_{19}H_{21}NO_3$) and alpinigenine ($C_{22}H_{27}NO_6$). Fresh capsule latex may contain .125% thebaine; the dried, .25 to .26%. Thebaine content of the dried roots ranges from .50 to 1.3%; of the stems, .09 to .22%.[46, 180] None or only a little occurs in the leaves.[46] In some strains, thebaine concentration in shoots of young plants has been found to range from 233 to 808 ppm; of older plants, 437 to 1,606 ppm.[125] Alpinigenine is found in all above-ground parts of the plant and may constitute 18% of the total alkaloids in the capsule latex.[180] Because of the predominancy of thebaine in *P. bracteatum*, this species can be identified in the field in the absence of flowers by use of a testing kit. The latex is tested with a solution of sulfuric acid containing molybdate. A "slight red brilliant" color reaction distinguishes the latex of this plant from that of related species.[599]

Propagation, Cultivation and Harvesting:

In comparison with *P. somniferum*, this is a slower-producing crop. In trial plots in England, plants have been raised from seed obtained from Iran and then multiplied by root division and seed. Seedlings commenced flowering the second year after transplanting to the field. Plants grown from root cuttings flowered at the end of the first year.[180] In Iran, the plants bloom in June; the petals fall in July.[521] Thebaine content in the capsules is highest three to four weeks after the flowers open, but seeds require another three to four weeks to ripen. It seems advisable to harvest the plant tops at seed maturity in summer (to acquire seeds for replanting) and to dig the roots in the fall. Experimental harvests indicate that the combined yield of thebaine may range from 15 lbs. per acre (15 kg. per ha) to 40 to 70 lbs. per acre (40 to 70 kg. per ha), depending mainly on the strain of plant grown.[125, 180, 181]

Medicinal uses:

Thebaine is in demand mainly as a source of codeine, widely employed as an analgesic and cough remedy. It is also a source of other analgesics, ethenotetrahydrothebaines, and of naloxone, a narcotic antagonist.[180] Naloxone is a life-saving drug administered to infants born of heroin addicts. Etorphine, or M99, one of the "Bentley Compounds" derived from thebaine is much used to sedate large wild animals for scientific purposes.

Other uses:

The seeds of *P. bracteatum* are eaten in Iran and can be utilized as a source of culinary and drying oil, as are those of *P. somniferum*.[179]

Related species:

Various *Papaver* species were examined for alkaloid content in Japan. Morphine was found only in *P. setigerum* DC. and *P. somniferum*. The morphine content of *P. setigerum* was low, as was the level of codeine; noscapine was very low, but thebaine and papaverine levels were high. A hybrid between the two species was nearly twice as high in morphine as *P. setigerum* and codeine content almost four times as high.[43]

Hamamelidaceae

Figure 39. The sweet gum (*Liquidambar styraciflua* L.), common from the eastern United States to Central America. Photographed on Johns Island, South Carolina.

Sweet Gum

Botanical name:
　　Liquidambar styraciflua L.

Other names:
　　Red gum, bilsted, alligator tree.

Family:
　　Hamamelidaceae, the Witch-Hazel Family.

Description:
　　A tree to 150 ft. (45.7 m) in height with a tall, straight trunk to 5 ft. (1.5 m) in diameter. With age, the tree secretes a fragrant balsam in pockets under the bark. Leaves are deciduous, alternate, star-shaped, three- to seven-pointed, 3 to 7 in. (7.5 to 17.7 cm) wide, minutely toothed; they turn red in the fall. Flowers are dioecious, the male tiny, yellow-green, massed in rounded heads on an erect spike to 3 in. (7.5 cm) long; female are larger, pale-green, in one large round head to 1½ in. (3.8 cm) wide, dangling below the male spike. Fruit (in autumn) a round, hard, spiny capsule, usually containing 2 glossy, brown or black, oblong, winged seeds ½ in (1.25 cm) long.[61, 474]

Origin and Distribution:
　　The sweet gum grows wild from Connecticut west to Oklahoma and Texas and south to Central Florida; also from southern Mexico to Nicaragua. It is widely planted as a shade tree, especially in parks. The balsam occurs in abundance only in Central American trees, and Honduras is the chief source of supply.[41]

Constituents:
　　The balsam, called storax, or American storax, is liquid at first but hardens on exposure to the air. It has an agreeable odor and a bitter taste, contains 20 to 30% water and is 82 to 87% soluble in alcohol. When purified, it is a dark-brown, semi-solid, viscous substance composed mainly (30 to 50%) of an alcoholic resin (storesin), together with 5 to 15% free cinnamic acid, 5 to 10% styracin (cinnamyl cinnamate), about 10% phenylpropyl cinnamate and small quantities of ethyl cinnamate, benzyl cin-

namate and styrene (phenylethylene). It also contains an aromatic liquid, which has been called styrocamphene, and traces of vanillin. Oil of Storax, obtained by steam distillation, possesses, in addition to styrene and cinnamates, free phenylpropyl and cinnamyl alcohols and cinnamic acid.[20]

Propagation, Cultivation and Harvesting:

The sweet gum flourishes in moist to very wet, rich soil. It may be grown from seed, but germination is slow and may take as long as two years. Air layers root readily. However, there are no commercial plantations. Storax is harvested from the wild trees which form extensive stands in Guatemala and Honduras. Indians are employed to tap the pockets of balsam, indicated by excrescences on the trunk. The yellow resin flows out readily into containers (formerly troughs of banana leaves wrapped around the tree).[544] It is exported, unrefined, in 5 gal. (20 l) tins or 55 gal. (220 l) used oil drums.[41] The yield ranges from 44 to 220 lbs. (20 to 100 kg) per tree.[466]

Medicinal uses:

In the past century, the bark decoction was taken to halt diarrhea and dysentery, and it is employed even today in rural areas. The leaves, rich in tannin, are chewed to curb diarrhea and relieve sorethroat.[397] In Latin America, the gum is popular in folk medicine as a sudorific, pectoral and diuretic; it is taken as a remedy for gonorrhea and is also applied on sores and wounds.[378] It was formerly widely prepared as an ointment for use on hemorrhoids, ringworm of the scalp and other parasitic skin afflictions.[305, 423] In the United States today, its only official use is as an ingredient in Compound Tincture Benzoin.[423]

Other uses:

The sweet gum is valued also for its red-brown timber, sometimes called satinwalnut, used for making furniture, doors, veneer, boxes and various small articles. The gum has been chewed as a masticatory and to clean the teeth.[549] In Mexico, Indians use it as incense and for flavoring tobacco.[544] Resinoids derived from storax are important in the manufacturing of perfumes.[41] In the food industry, storax extract is used to flavor soft drinks, candy, desserts and chewing gum.[210]

Figure 40. The Oriental sweet gum (*Liquidambar orientalis* Mill.) has deeply lobed leaves with irregular secondary lobes. This branch is from a tree in the Secrest Arboretum, Ohio Agricultural Research and Development Center, Wooster, Ohio.

Oriental Sweet Gum

Botanical name:
    Liquidambar orientalis Mill.

Other name:
    Turkish sweet gum.

Family:
    Hamamelidaceae, the Witch-Hazel Family.

Description:
    A bushy tree, to 25 or 40 ft. (7.6 to 12.1 m), with deciduous leaves, usually deeply five-lobed, with irregular secondary lobes; 2½ to 3 in. (6.4 to 7.5 cm) wide, coarsely toothed. Flowers are monoecious, yellow, in spherical clusters. The fruit, a hard, spiny ball, is about 1 in. (2.5 cm) in diameter.

Origin and Distribution:
    This tree is native to Asia Minor and occurs in solid stands in the southwestern part of Asiatic Turkey (mainly near Budrum, Melasso, Moughla and Marmorizza).[233] It is of very limited distribution; it has been introduced into Palestine and into botanical gardens of Europe and the British Isles. Its commercial cultivation has been proposed in India.[20]

Constituents:
    The balsamic resin, called Levant, Asiatic, or Liquid storax, is grayish to gray-brown, semi-liquid, semi-opaque, aromatic, with a burning taste. It consists mainly (50%) of two resin alcohols, alpha and beta storesin (partly free and partly combined with cinnamic acid); also 10 to 20% storesin cinnamate, 5 to 10% styracin (cinnamyl cinnamate), 10% liquid phenylpropyl cinnamate, .5 to 1% volatile oil, 2 to 5% free cinnamic acid and a trace of vanillin.[122] On standing, Levant storax deposits a dark-brown oleoresin.

Propagation, Cultivation and Harvesting:
    The Oriental sweet gum grows in dry situations; in Palestine, it thrives on warm hillsides. It can be propagated by seed or air-layering

but is slow-growing. The wild trees furnish the storax for the Old World market and part of that which is utilized in the United States. The balsam is not found in pockets as in the American species but is a pathologic product secreted when the tree is abused. In the summer, workers bruise the bark by pounding or puncturing, which stimulates the secretion of resin in the sapwood and bark. By autumn, the inner bark will be saturated. The outer bark is peeled away, and the inner bark is scraped off, repeatedly, in chips which are heaped in pits, later gathered into sacks and squeezed in a wooden press. Then hot water is poured over the sacks, and they are pressed again and most of the resin extracted. Or the bark may be first boiled in kettles, the exuded resin skimmed off the water, and then the boiled bark is put in sacks and pressed to release more resin. The semi-liquid storax, after being "washed" in boiling water, is packed in barrels or cans for shipment. It retains about 25% water, which rises to the surface on standing. When the water is poured off and the storax purified, the latter has the consistency of thick honey.[233]

Medicinal uses:

In Europe and Asia, Levant storax is employed as an expectorant in asthma, bronchitis and various pulmonary complaints. Incorporated into an ointment, it is commonly used in treating skin diseases. Its only pharmaceutical use at present in the United States is in Compound Tincture Benzoin.

Other uses:

The spent bark after storax extraction is pulverized and used as a fumigant.[20]

Related Species:

L. formosana Hance, the Fragrant Maple of southern and central China, grows to a height of 75 to 100 ft. (25 to 30 m), has usually three-lobed leaves, rarely five-lobed. It yields Chinese storax containing 16% cinnamic acid, cinnamyl alcohol, borneol, a resin alcohol which may be storesin and 1.8 to 8% volatile components.[20] It is valued in Oriental medicine as a hemostat and astringent. The wood is employed for tea chests and idols. Because the branches wave and the leaves flutter in the wind, the Chinese believe the tree to be inhabited by spirits. For this reason and for its colorful autumn foliage, it was planted by an early emperor around the Imperial Palace at Peking.[566]

Leguminosae

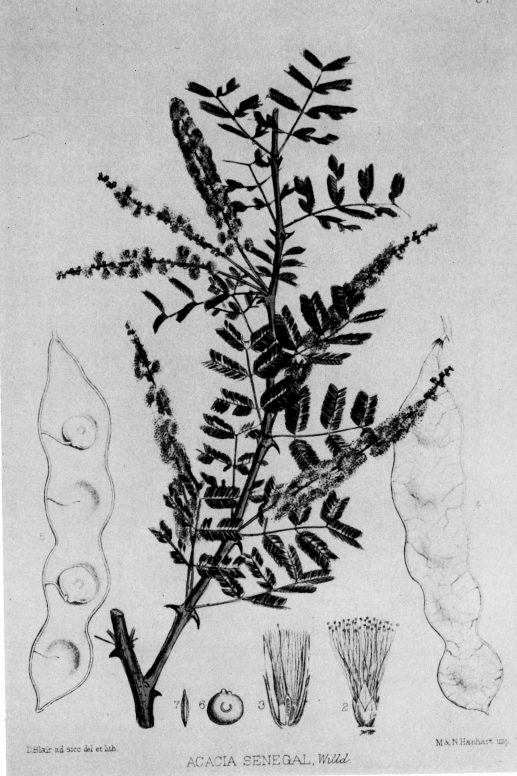

Figure 41. Gum acacia (*Acacia senegal* Willd.). Reproduction from a color painting by David Blair in *Medicinal Plants*, R. Bentley and H. Trimen (J. & A. Churchill, London; 1880).

Gum Acacia

Botanical name:

    Acacia senegal Willd. (syn. *A. verek* Guill. & Perr.)

Other names:

    Gum arabic tree, Sudan gum arabic, Somali gum, yellow thorn (var. Geelhaak).[124]

Family:

    Leguminosae, the Pea Family.

Description:

    A tropical shrub or flat-topped tree to 30 ft. (9.14 m) high but usually less than 15 ft. (4.5 m), with short trunk and pale bark, fissured and flaking with age. It often forms thickets. A translucent gum accumulates in the cambium and exudes from cracks or wounds. The flexuous branches bear sharp thorns, to ¼ in. (6 mm) long, grouped in 3s on the enlarged nodes, the central thorn being strongly recurved. Leaves are deciduous, bipinnately compound, to 1½ in. (4 cm) long, with 3 to 6 pairs of pinnae, 8 to 18 pairs of gray-green, oblong, leathery leaflets ¼ in. (6 mm) long. Flowerbuds are red. Flowers, usually emerging before the new leaves, are ivory-white in dense spikes, solitary or in 2s or 3s, slender, 2 to 4 in. (5 to 10 cm) in length. Fruit is a flat, papery seedpod, brown, pubescent, oblong, ¾ to 1 in. (20 to 25 mm) wide and 1½ to 5 in. (4 to 12.5 cm) long, containing 1 to 6 greenish-brown, disk-like seeds. Pod does not split open. The tree is said to produce seed at five-year intervals.[16, 144, 148, 169, 275]

Origin and Distribution:

    Acacia senegal occurs wild on sand dunes in Senegal and Mauritania and in dry, rocky land throughout tropical Africa, east to Ethiopia and south to Bechuanaland. A dwarf form, called yellow thorn because the peeling bark leaves yellow patches on the trunk, is an aggressive invader of overgrazed land in South Africa. *A. senegal*, despite its name, is indigenous also to the Western Desert of Iraq and the arid hills of northwestern India. It has been sparingly introduced into other parts of the world, including the West Indies. Currently it is being advocated for revegetation of the drought-plagued Sahel region.[222]

Product and its Properties:

The gum, known by various names as picked turkey gum, white sennar gum and kordofan gum, as well as senegal gum, hardens on exposure into globular or angular "tears" up to 1½ in. (4 cm) wide, white, yellowish or pinkish in color.[20] It is very brittle, nearly scentless, of bland taste and is composed mainly of calcium, magnesium and potassium salts of a polysaccharide (arabic acid, or arabin). It yields, on hydrolysis with dilute acid, l-arabinose, l-rhamnose and 3-d-galactoside-1-arabinose. The residue consists of galactose and uronic acid.[15, 122] When hydrolyzed with dilute sulphuric acid, it yields l-rhamnopyranose, d-galactopyranose, l-arabofuranose and the aldobionic acid 6-β-d-glucuronosido-d-galactose.[584] It possesses an oxidase-type enzyme which is destroyed by heating at 100° C.[423] and 12 to 15% of water.[122]

There is considerable variation in the quality of the gum, depending on whether it is a natural flow caused by extreme drought, obtained by tapping or induced by boring of beetles which attack sites of injury. The natural exudate is preferred in some markets. The gum from wild trees is usually darker than that from cultivated trees.[20] A sweet, dark-colored gum called "Hennawi" has lower rhamnose content and low uronic acid residue.[15] The gum is soluble in water; insoluble in alcohol,[20] ether and oils.[423]

Propagation, Cultivation and Harvesting:

Though most of the gum on the market is collected from wild trees, the tree is also extensively cultivated, especially around Kordofan in the Sudan, the source of 85 percent of the world supply of gum arabic.[168] Solid strands have been planted as windbreaks in the region between the Blue Nile and White Nile rivers.[431] The tree is grown from seed. Planted 3 per site and with good germination, 2.2 lbs. of seeds will yield about 4,000 plants. *Acacia senegal* regenerates spontaneously on abandoned cultivated land in the Sudan. During the first three or four years of growth, many seedlings are eaten or uprooted by grazing gazelles, pigs and goats.[221]

Gum harvesting starts when the seedlings are between four and eighteen years old and may be continued for two decades. The yield is highest from unthrifty trees on impoverished soil. Cultural improvement reduces the excretion of gum. Extreme drought causes cracks in the bark from which the gum exudes. Deliberate tapping is done during February and March or August and September by making horizontal axe cuts and stripping away the bark above and below the incision, baring an area of cambium 2 to 3 in. (50 to 76 mm) wide and 2 to 3 ft. (.6 to .91 m) in length.[584] In twenty to thirty days, the exuded gum will be crystallized and ready for collection, which may be done with long-handled nets to avoid scratches by thorns.[431] Harvesting is repeated once a week, the "tears" being accumulated in leather bags and then transferred to 100 lb. (45.3 kg) sacks and delivered to market centers where government-sponsored auctions are held. The collected gum must be freed of sand and debris, cleaned, sorted, and some is sun-cured (bleached) for three to four months. This causes many fine cracks on the surface, rendering it semi-opaque, and it reduces the weight by 20 percent.[584]

Figure 42. Gum acacia tree photographed in Northern Kordofan by Doctor Abraham D. Krikorian, State University of New York, Stony Brook, New York.

Figure 43. Gum exuding from *Acacia senegal*. Photographed by Doctor Abraham D. Krikorian, State University of New York, Stony Brook, New York.

Annual yields of 200 to 2,800 g are obtained from young trees (average, 900 g) and 40 to 6,700 g (average 2,000 g) from old trees.[423]

Medicinal uses:

Gum arabic (probably originally collected from *A. arabica* Willd., now replaced by *A. senegal*) was a valued article of commerce among the early Egyptians. It is locally utilized as a remedy for gonorrhea, dysentery, burns, sore nipples, inflammations and nodular leprosy.[117, 605]

In modern pharmacy, it is commonly employed as a demulcent in preparations designed to treat diarrhea, dysentery, coughs, throat irritation and fevers.[624] It serves as an emulsifying agent and gives viscosity to powdered drug materials; is used as a binding agent in making pills and tablets and particularly cough drops and lozenges.

Because of its enzyme, the gum is not suitable for use in products having readily oxidizable ingredients. For example, it reduces the vitamin A content of cod liver oil by 54 percent within three weeks. It is incompatible with aminopyrine, morphine, vanillin, phenol, thymol, α- and β-naphthol, guiacol, cresols, creosol, eugenol, apomorphine, eserine, epinephrine, isobarbaloin, gallic acid and tannin; also with strongly alcoholic liquids, solutions of ferric chloride and lead subacetate and strong solutions of sodium borate.[423]

It was formerly given intravenously to counteract low blood pressure after hemorrhages and surgery and to treat edema associated with nephrosis, but such practices caused kidney and liver damage and allergic reactions, and have been abandoned.[423]

Other uses:

Over 11,000 tons of gum arabic are imported into the United States annually, primarily for use in the food industry. Much is utilized in the manufacture of confectionery and in various bakery products to give body and texture and uniform distribution of flavor. The gum is an important material in lithography and enters into water colors and pigments for fabric-printing and ceramics. It is employed in adhesives, especially for postage stamps and envelope flaps and for coating one side of wrapping paper; it is also used in wax polishes and was formerly used in textile sizing. The dark heartwood of old trees of *A. senegal* has been made into weaver's shuttles.[20] Tool handles are made of young trunks and old roots. Strong fiber from the surface roots is valued for rope-making and fishnets.[146] The young foliage is grazed by camels and goats.[20]

Related species:

In the French Sudan, gum is collected from the closely related *A. laeta* R. Br. as well.[221, 222] The talha or suakim gum arabic tree (*A. seyal* Del.; *see* color illustration 6) is abundant from northern Nigeria to Egypt and western Sudan.[146] It yields a darker, more astringent gum which represents about 10 percent of the gum arabic in world trade.[168]

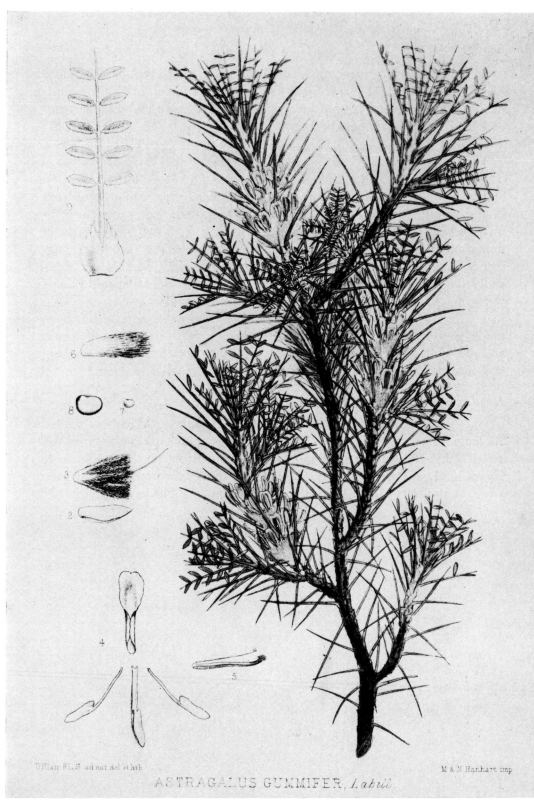

Figure 44. Tragacanth (*Astragalus gummifer* Labill.). Reproduction from a color painting by David Blair in *Medicinal Plants*, R. Bentley and H. Trimen (J. & A. Churchill, London; 1880).

Tragacanth

Botanical name:

    Astragalus gummifer Labill.

Other names:

    Gum tragacanth, Syrian tragacanth, Anatolian tragacanth, gum dragon, green dragon, goat's thorn.

Family:

    Leguminosae, the Pea Family.

Description:

    A perennial, bushy shrub, 1 to 2 ft. (.3 to .6 m) high, with large taproot, smooth branches and downy twigs bearing stiff, yellow spines ¾ to 2 in. (2 to 5 cm) long. The woody parts and the taproot excrete a gummy substance when injured. Leaves are glabrous, pinnately compound, with 4 to 6 pairs of ellipitic-oblong, acute leaflets to 3/16 in. (5 mm.) long, which are shed in the dry season, leaving the tough, sharp rachis. Massed near the tips of the branchlets, these remnants have the aspect of a formidable array of thorns. The flowers, borne in 2s and 3s, are white-petaled; the tubular calyx is hairy.[454]

Origin and distribution:

    This shrub grows wild in dry regions, usually between altitudes of 4,000 to 10,000 ft. (1,219 to 3,040 m) in Greece, Corsica, Asiatic Turkey, Syria, Lebanon and Iran, but not in Iraq.

Constituents:

    Tragacanth gum has a high molecular weight (840,000). It consists primarily of bassorin (60 to 70%), tragacanthin (8 to 10%) and a small amount of starch. Bassorin is referred to as a methylated acidic polysaccharide.[610] It contains about 5.38% methoxyl.[584] It is insoluble, but swells, in water. Tragacanthin consists of uronic acid and arabinose and is water soluble. On hydrolysis, tragacanthin yields d-galacturonic acid, d-galactose, 1-fucose (6-deoxy-1-galactose), d-xylose and 1-arabinose.[610] The best grades of tragacanth are the lowest in tragacanthin.[584] In Turkey, A. *gummifer* reportedly contains no 1-fucose.[65] In cold water, the gum forms a gelatinous mass, only 8 or 10 percent dissolving[233] after one or two days.[269]

Propagation, Cultivation and Harvesting:

Tragacanth is not cultivated as a crop. The gum is collected from wild plants. It exudes in round or conical "tears" where the plant has been injured by goats or other grazing animals but is commonly induced to flow by tapping, which is begun when a plant is two years old and may be repeated in alternate years for about seven years. The best gum is derived from the roots. Soil is scraped away for a depth of 2 in. (5 cm), and the blade of a knife is plunged into the root; a wooden wedge is then inserted and left in the cut for twelve to twenty-four hours to keep it open. The gum begins immediately to exude, protruding as much as ¾ in. (2 cm) in the first half-hour.[584] From vertical knife incisions, it emerges in ribbon-like strips which are hard enough to collect in two days in the absence of rain. If the root is punctured with a slender pick, the exuding gum is vermiform and called "vermicelli." Saw cuts on branches yield a form of gum called "arrehbor." Plants have been burned after tapping to stimulate gum flow, but this practice discolors the gum and may kill the bush.

The gum, brought to collection centers, is purchased by dealers and graded by form, size and color. Flat, white ribbons rate highest in quality, next, curly ribbons and, in descending order, ivory-white or pinkish ribbons, light-colored flakes, amber flakes and reddish-brown flakes. The best grades come from the cooler climates.

Medicinal uses:

Tragacanth is occasionally employed as a demulcent in treating coughs and diarrhea.[624] It is seldom used externally, but the mucilage (6% of the gum in water and glycerine preserved with benzoic acid) is sometimes applied to burns.[233] In modern pharmaceutical practice, tragacanth is especially valued in spermicidal jellies and creams, which immobilize spermatozoa on contact. It is important as a suspending agent for insoluble powders and mixtures containing resinous tinctures[610] and much employed as a binding agent in pills, tablets and lozenges.[584] It serves as an emulsifier in cod liver oil and facilitates the absorption of steroid glycosides and fat-soluble vitamins.[610] Tragacanth produces a coarse emulsion and is therefore often combined with gum arabic for better results.[269]

In 1965, it was reported by workers at the Chester Beatty Research Institute, London, that gum tragacanth exerts a strong inhibiting action on cancer cells. In 1972, Indian investigators concluded that "depending upon the nature, structure and change of the bare tumour surface and also based upon the polymer concentration and the ionic strength of the gum solution applied to the malignant region, the rigid portion of the adsorbed biopolymer in direct contact with the interface may assume a special and highly specific conformation which results in an antimalignant effect." [588]

In 1971, the United States imported about 800 tons (720,000 kg) of tragacanth for pharmaceutical and other uses. [610]

Other uses:

Much use is made of tragacanth in the food industry—in salad dressings, condiments, dessert toppings, pie fillings, confectionery and as a stabilizer in ice cream and sherbets. It is an ingredient in toothpaste, hand lotions and other cosmetics; also in furniture, auto and leather polishes; it is an emulsifier in insect repellants and a binder in ceramics and is utilized in fabric printing and as a stiffener for textiles. [122, 610]

Related species:

The tragacanth exported to Great Britain, Russia and India is derived mainly from *A. kurdicus*, *A. brachycalyx*, *A. verus*, *A. leiocladus*, *A. adscendens*, *A. strobiliferus* and other species growing in southern Kurdistan, Iran and neighboring regions. [610] In Turkey, tragacanth is collected from *A. gummifer* and *A. kurdicus* in southern and eastern Anatolia but only from *A. microcephalus* in western and central parts of that region. [65, 217]

Figure 45. Indian senna (*Cassia angustifolia* Vahl). Reproduction from a color painting by David Blair, in *Medicinal Plants*, R. Bentley and H. Trimen (J. & A. Churchill, London; 1880).

Indian Senna

Botanical name:
    Cassia angustifolia Vahl (*C. lanceolata* Royle).

Other names:
    Tinnevelly senna, Mecca senna.

Family:
    Leguminosae, the Pea Family.

Description:
    An annual, herbaceous subshrub, 2 to 2½ ft. (.6 to .75 m) high, with compound leaves to 6 in. (15 cm) long, having 4 to 8 pairs of lanceolate, nearly sessile leaflets 2 to 2⅓ in. (5 to 6 cm) long and ⅕ to ½ in. (5 to 12.5 mm) wide; bluish-green, smooth on the upper surface, slightly downy beneath. Flowers are yellow and borne in erect axillary and terminal racemes. Seedpod is dark-brown, oblong, 1½ to 2¾ in. (3.75 to 7 cm) long and ¾ in. (2 cm wide), thin and flat, containing 5 to 7, obovate, dark-brown, fairly smooth seeds.[20, 160]

Origin and Distribution:
    This plant is native to Somaliland and southern Arabia, also the Punjab and Sind areas of northwestern India. It is cultivated commercially in northwest Pakistan, southern India (Tinnevelly, Madura, Trichonopoly and Mysore districts)[20] and more recently in Jammu.[2] It has been grown also in California.[432]

Constituents:
    The plant possesses .33% β-sitosterol.[605] The active constituents are mostly anthraquinone derivatives. The leaves possess two crystalline glycosides, sennoside A and B,[91] and a third more potent glycoside having two additional glucose molecules. Among other properties are sennosides C and D (glycosides of heterodianthrones involving rhein and aloe-emodin), aloe-emodin-dianthrone-diglucoside, rhein-anthrone-8-glucoside, rhein-8-glucoside, rhein-8-diglucoside, aloe-emodin-8-glucoside, aloe-emodin-anthrone-diglucoside and possibly rhein-I-glucoside. During fruit development, the aloe-emodin and rhein content drastically declines.[584] The

leaves also contain the flavonols kaempferol (1:3:4'trihydroxyflavone), kaempferin, and isorhamnetin, mannitol, sodium potassium tartarate, salicylic acid, chrysophanic acid,[20] calcium oxalate, resin[584] and saponins.[68] The British Pharmacopoeia requires that senna leaves contain no less than 2.5% sennoside B. The sennoside content of the pods ranges from 1.2 to 2.5%.[584] They are less resinous than the leaves.[117] Green pods are more active than ripe pods. The cathartic properties of senna are not lost during five years in storage[20] but are destroyed by boiling,[305] unless in a sealed vessel.[233]

Propagation, Cultivation and Harvesting:

This species is usually grown on dry land, sometimes on wet land as a succession to rice soon after the rice is harvested. It is grown from scarified seeds sown in rows 1 ft. (30 cm) apart in March-April or in August,[2] at the rate of 15 lbs. (6.8 kg) seed per acre. The plants require sunshine and little rain and are given only slight irrigation. After the first three to five months, when flowering commences, the racemes are cut off in the early stage to promote lateral branching. The mature leaves are stripped off by hand in two harvestings one month apart. Fresh leaves contain 75% moisture and are spread in the shade to dry for a week to ten days (becoming yellowish-green) and then are graded and baled for shipment, being carefully packed to avoid leaf breakage. After the second harvesting, the plants are left to bloom and fruit. Thereafter, a third picking of leaves may take place while the pods are collected, thrashed to separate the seeds and packed in cartons.[437]

Dry-land fields yield 300 lbs. (136 kg) dried leaves and 75 to 150 lbs. (34 to 68 kg) empty pods, while irrigated land produces 750 to 1,250 lbs. (340 to 556.5 kg) dried leaves and 150 lbs. (68 kg) of pods.[20] Senna, being a nitrogen-fixing plant, enriches the land for the benefit of subsequent crops.[2] Most of the senna cultivated in India has been exported to Germany and France. In 1969-70, 270 tons (270,000 kg) were imported into the United States; 11 tons (23,245 kg) in early 1972.[437]

Medicinal uses:

Locally, an infusion of the leaves is given as a laxative. An infusion of the pods in cold water is milder and slower in action and taken several nights in succession to regularize evacuation. A paste made of the powdered leaves and vinegar is applied to skin diseases and eruptions.[148]

Extracts and powdered forms of both leaves and pods are extensively employed in formulating commercial laxative preparations.[122] A trial compound of pure sennosides A and B was effective in 84.7 percent of functional obstipation cases treated in Finland.[228] The action is particularly on the lower bowel, where it increases peristaltic movements by its effect on the intestinal wall.[423, 555] If taken by nursing mothers, the laxative agent is transmitted to the infant.[277]

Toxicity:

Senna causes griping unless combined with saline laxatives or ginger, cloves, coriander or other spices. It cannot be prescribed in cases of colitis or spastic constipation.[423]

In sensitized individuals, the leaves may cause severe and painful dermatitis on contact, and this reaction is attributed to the saponin content.[68]

SÉNÉ.

Alexandrian Senna

Botanical name:
>Cassia senna L. (C. acutifolia Del.)

Other names:
>Aden senna, Nubian senna.

Family:
>Leguminosae, the Pea Family.

Description:
>An annual, semi-herbaceous subshrub to 3⅓ ft. (1 m) high, with pale, somewhat zigzag branches. Leaves are alternate, compound, with 3 to 7 pairs of leaflets shorter and narrower than those of *C. angustifolia* (q.v.) being 1¼ to 1-9/16 in. (3.2 to 4 cm) long and 3/16 to ⅜ in. (.5 to 1 cm) wide; elliptic to lanceolate, acute at the tip, glaucous, with more hairs present on the underside than on those of *C. angustifolia*.[584] Flowers are bright-yellow in erect racemes. Seedpod is greenish-brown, 1½ to 2½ in. (3.75 to 6.4 cm) long, ¾ to 1 in. (19 to 25 mm) wide, flat, elliptic-oblong, tending to be more curved than the pod of *C. angustifolia*. The 6 to 7 seeds, when magnified, show a distinctly continuous raised reticulum in contrast to the discontinuous wavy ridges on the seeds of *C. angustifolia*.[181, 584, 624]

Origin and Distribution:
>This plant is native and abundant in the northern and central Sudan[16] and Sinai.[454] It is cultivated along the Upper Nile in the Kordofan and Sennar regions[2] and to some extent in India.[117]

Constituents:
>Investigators have found no significant differences in the chemical properties of *C. senna* and *C. angustifolia*.[181] The sennoside content of the pods ranges from 2.5 to 4.5%.[584]

←

Figure 46. Alexandrian senna (*Cassia senna* L.). Reproduction from a color painting by P. Turpin in *Flore Médicale* Vol. 6 (F. P. Chaumeton, Paris; 1818).

Propagation, Cultivation and Harvesting:

Alexandrian senna is derived mostly from the wild plants. The tops, bearing leaves and pods, are cut in April and in September to about 6 in. (15 cm) above the ground and dried on rocks in the full sun, the leaves becoming a grayish-green or olive color. Thereafter, the material is sieved to separate the pods and the larger stems from the leaves and smaller stems which are then winnowed, a process which moves the leaves (leaflets) to the surface. The leaves are sorted into grades: whole, mixed whole and half leaves, and siftings. The leaves are rather loosely packed in sacks, as are the pods after removal of stems, and sent to Port Sudan for export.[584]

Medicinal uses:

In Africa, the dried, pulverized leaves are applied to wounds and burns. An infusion of the tops (leaves, flowers and pods together) is taken as a purge to allay fever.[146] In Iran, the imported leaves are mixed with rose petals and tamarind pulp for purgative dosages.[266] In the pharmaceutical industry, leaves (leaflets) and pods of *C. senna* serve the same purposes as those of *C. angustifolia*.[426]

Related species:

C. obovata Collad. (*C. obtusa* Roxb. *C. italica* Lam., *C. senna* Lam. *Senna italica* Mill.), SENEGAL, TRIPOLI, ITALIAN, JAMAICA or PORT ROYAL SENNA (*see* color illustration 7). This species is native to the northern Sudan and dry regions of India and Ceylon. It has been cultivated throughout southern Europe and in the West Indies and is now encountered as a naturalized alien in the Caribbean islands and northern South America. The plant is low-growing, about 18 to 20 in. (.5 m) high, and the pod is strongly curved. The leaves and pods occur as adulterants in Indian and Alexandrian senna and have local uses in folk medicine. They are reportedly more drastic than the products of either of the foregoing species.[117]

Figure 47. Licorice (*Glycyrrhiza glabra* L.). Photographed in the Beal-Garfield Botanic Garden, Michigan State University, East Lansing, Michigan.

Licorice

Botanical name:
> *Glycyrrhiza glabra* L.

Other names:
> Liquorice, sweet wood.

Family:
> Leguminosae, the Pea Family.

Description:
> A perennial, temperate-zone herb or subshrub, 3 to 7 ft. (1 to 2 m) high, with a long, cylindrical, branched, flexible, burrowing rootstock and horizontal, creeping, underground stems (stolons) attaining 5 to 6 ft. (1.5 to 1.8 m) in length, having buds which send up stems in the second year.[233] Leaves are alternate, pinnate, with 9 to 17 ovate, yellow-green leaflets, 1 to 2 in. (2.5 to 5 cm) long, viscid on the underside. Flowers, in erect, axillary, long-stalked spikes 4 to 6 in. (10 to 15 cm) long, are pea-like, lavender to purple, ½ in. 1.25 cm) long. Seedpod is maroon, to 1¼ in. (3 cm) in length, oblong, pointed, flattened, and contains 2 to 4 kidney-shaped seeds.[51]
>
> *G. glabra* var. *typica* Regel & Herd., SPANISH LICORICE, of Spain and Sicily, has smooth seedpods. Its rootstock and stolons are ¼ to ¾ in. (6 to 18 mm) thick, dark-maroon in color, longitudinally wrinkled and sweeter than other types.
>
> *G. glabra* var. *glandulifera* Waldst. & Kit., RUSSIAN LICORICE, of southern Russia is a viscid-hairy plant with spiny seedpods.[233] It has no stolons but multiple spindle-shaped roots, 2 in. (5 cm) wide, with purplish, scaly bark, which are moderately sweet and somewhat bitter, also more or less acrid.
>
> *G. glabra* var. *violacea* Boiss., PERSIAN LICORICE, has very thick roots; ANATOLIAN or TURKISH LICORICE has rootstocks up to 3⅛ in. (8 cm) thick.[584]

Origin and Distribution:
> Licorice is native from southern Europe to Pakistan and northern India. It has a long history of cultivation in England, Belgium, France,

Germany, Spain, Italy, Greece, Turkey, Russia, Egypt, Syria, Iraq and, in recent years, has been planted commercially in northern India.[20] It has been grown experimentally in the United States.

Constituents:

The chief property of licorice is the saponin-like glycoside glycyrrhizin (glycyrrhizic acid), $C_{42}H_{62}O_{16}$, occurring in amounts varying from 5 to 20%. Spanish Licorice averages 6 to 8%; Russian, 10 to 14%. Glycyrrhizin is fifty times sweeter than sucrose, but the addition of 1 lb. (.45 kg) glycyrrhizin will double the sweetness of 100 lbs. (45.3 kg) of sugar.[132]

On hydrolysis, glycyrrhizin yields two molecules of glucuronic acid and one molecule of glycyrrhetinic acid (glycyrrhetic acid), the latter being a pentacyclic terpene, $C_{30}H_{46}O_4$, somewhat resembling in action the steroid desoxycorticosterone.[423] A new triterpenoid, 28-hydroxyglycyrrhetic acid, has been isolated in Egypt.[173, 175] Canadian scientists have observed typical steroid and triterpenoid reactions in single cell suspension cultures.[627]

Licorice contains also 20% starch, 3.8% glucose, 2.4 to 6.5% sucrose, 2 to 4% asparagine, 0.8% fat, resins, mannitol, bitter principles (glycyramarin)[117, 122, 584] and 22,23-dihydrostigmasterol.[122] An isoflavan derivative, licoricidin, $C_{25}H_{32}O_5$ (3′,6-diisopentenyl-2′,4′,5-trihydroxy-7-methoxyisoflavan) was isolated in 1968 in Japan.[522]

Other properties reported are the coumarin derivatives herniarin and umbelliferone;[122] also the flavonones liquiritin, liquiritigenin, isoliquiritin and isoliquiritigenin (responsible for the yellow color), and the glycosides liquiritoside and isoliquiritoside, which, together, are credited with licorice's anti-ulcer activity. The isoflavonoid formononetin was found to represent .01% of commercial licorice root by investigators in Cairo in 1971.[174] Two new flavonoids, rhamno-liquiritin and rhamno-iso-liquiritin, were announced in 1971 by Belgian investigators.[589] Extracts of the above-ground parts of the licorice plant have produced estrogenic activity in experimental animals and the phenolic principle glycestrone has been shown to have 1/533 of the potency of estrone.[313, 402]

Propagation, Culture and Harvesting:

The plant may be grown from seed, but the usual method of multiplication is by division of the crown or by root cuttings, the settings in the prepared, pre-fertilized field being 2 ft. (.6 m) apart, with 3 to 4 ft. (.91 to 1.2 m) between the rows. The plant in the natural state is found in deep, moist soil, as on river banks subject to flooding, but where the late summer is hot and dry.[233] Under cultivation, irrigation is necessary at the outset. Thereafter, the crop requires little other attention apart from the control of weeds.[20] Growth is slow during the first two years.[233] Glycyrrhizin content increases as the plant develops and may not vary with the season.[336] When the plants are three to four years old and before they have borne fruit,[122] they are taken up in the autumn and the tops are cut off and set aside for compost, being high in nitrogen. Remnants of roots in the soil generate a new crop which is augmented by the planting of root cuttings.

The harvested roots and runners are cut into sections. The roots of the Russian type and some of the others are peeled at this point before shade-drying, which may take four to six months[122] and reduces the moisture content from 50 to 10%.[20] The average yield of roots and runners is 4 to 5 tons per acre (10 to 12.5 tons per hectare).[233] A new spectrophotometric method for determining the active principles in licorice has been demonstrated in Madrid.[421]

Medicinal uses:

Glycyrrhizin increases extracellular fluid and plasma volume, induces sodium retention and loss of potassium.[423] Licorice has been rationed to desert troops to prevent extreme thirst on low water intake. Decoctions of the peeled, dried root were formerly given to allay coughs, catarrh, bronchitis, sorethroat, laryngitis, urinary irritation and the pain associated with diarrhea.[305] In modern pharmaceutical practice, licorice is utilized mainly in the form of extract, a dark-brown paste obtained by crushing or shredding the fresh roots, decocting under low steam pressure and evaporating. It is formed into sticks or rolls for marketing. The glycyrrhizin content of the extract varies from 1.2 to 24%.[20] Licorice powder may be the pulverized extract or the finely ground dried roots (brownish yellow if unpeeled; pale-yellow if peeled). Ammoniated glycyrrhizin, in the form of glossy brown flakes, is the result of precipitating glycyrrhic acid from the extract, dissolving in ammonia and spray-drying.[20, 132]

Licorice is a common ingredient in cough sirups and cough drops[117] for its flavor as well as its demulcent, mildly expectorant and anti-inflammatory effects. It is customarily added to bitter laxative preparations (of senna, aloe, cascara sagrada and other drugs) to improve the flavor and because it sensitizes the intestines and thus potentiates the action.[466] A thick sirup of the powdered extract and water, or commercially prepared N-glycyrrhetinoyl amino acid derivates,[258] may be prescribed in cases of gastric and duodenal ulcers. Though relief is achieved, some patients develop edema and hypertension or cardiac asthma on account of the increased venous and systolic arterial pressure and pulse pressure. Licorice or glycyrrhetinic acid may be beneficial in Addison's disease[423] and rheumatoid arthritis,[584] and in the treatment of dermatitis.[122] In India, licorice powder mixed with fat and honey is applied to cuts and wounds.[20]

The United States imports more than 27,000 tons of licorice root and 150 tons of licorice extract annually.[122]

Toxicity:

In addition to the above-mentioned adverse effects experienced by some patients as a result of glycyrrhizin ulcer therapy, excessive consumption of licorice candy (as much as ¼ lb. or 113.3 g daily) has induced abnormal heart action through potassium depletion and incipient kidney failure.[86, 480] Licorice should be avoided by persons with cardiac problems, hypertension, kidney complaints and those who are overweight or having difficult pregnancies.[117]

Other uses:

Dried pieces of licorice root up to 6 in. (15 cm) long are popular chewsticks in Italy, Spain and parts of the West Indies. In the Netherlands Antilles, they are called "palu dushi" (sweet stick). Large quantities of licorice are employed in chewing gum and in confectionery. Glycyrrhizin potentiates the flavor of cocoa, replacing 25 percent of cocoa in manufactured products.[132] Licorice is increasingly utilized in soft drinks, liqueurs, ice cream, puddings, bakery products,[210] soy sauce and soybean-protein meat substitutes.[132] It is added to beer to enhance the "head"[122] and aroma,[20] and to porter and stout to provide more body and a darker hue.[233] It also enters into mouthwashes, breath "purifiers" and toothpaste,[132] and has been used for flavoring tobacco.[411]

The pulp residue from the preparation of ammoniated glycyrrhizin is extracted with a solution of dilute caustic soda; the product is an important foam stabilizer in fire extinguishers and is also used as a wetting, spreading and adhesive material in insecticides. The spent pulp is usable as culture media for food yeasts or mushrooms; it may serve as fertilizer or mulch [132] and can be made into composition board or insulation panels.[20]

Related species:

G. uralensis Fisch., MANCHURIAN LICORICE, is found in Turkistan, Mongolia and Siberia.[233] It has light chocolate-brown roots, the corky exterior peeling readily. It is of good glycyrrhizin content but possesses little sugar and is distinctly pungent in flavor.[584]

G. echinata L., native from Hungary to Korea has somewhat bitter roots. The pods are small and ovoid with long spines.[233]

Figure 48. Balsam of Tolu (*Myroxylon balsamum* Harms.) Reproduction from color painting by David Blair in *Medicinal Plants*, R. Bentley and H. Trimen (J. & A. Churchill, London; 1880).

Balsam of Tolu

Botanical name:

 Myroxylon balsamum Harms. (*M. balsamum* var. *genuinum*, *M. toluiferum* HBK).

Other names:

 Tolu balsam, quinoquino.

Family:

 Leguminosae, the Pea Family.

Description:

 A handsome tree, 40 to 100 ft. (12 to 3.4 m) tall with a spreading crown, the rough bark, when injured, excreting a fragrant, pungent, yellow-brown resin. Leaves are evergreen, odd-pinnate, having 7 to 13 glossy leaflets 3 to 4 in. (7.5 to 10 cm) long and to 2 in. (5 cm) wide, rounded at the base, long-pointed at the apex. Flowers, in simple axillary racemes to 6 in. (15 cm) long, the rachis slightly brown-woolly; the five petals are yellowish, the standard .47 in. (12 mm) wide. Fruit (seedpod), 3 to 5 in. (8 to 12.5 cm) long with a wing 1 in. (2.5 cm) wide, the curved apex containing two balsam pits and a single, kidney-shaped seed, .47 to .55 in. (12 to 14 mm) long, covered with a thin, dry testa.[357]

Origin and Distribution:

 Native to Argentina, Paraguay, Brazil, Venezuela, Bolivia, Peru and Colombia. Introduced into Singapore in 1876 and 1882[98] and has been grown successfully in India, Ceylon and Sumatra as well as in some islands of the West Indies. Seeds were taken from Ceylon to Hawaii in 1918, and several specimens were planted on the University grounds.[483]

Constituents:

 The balsam, solid at room temperature, plastic in the warmth of the hand and when chewed, is nearly insoluble in water, soluble in 90% alcohol, ether and chloroform. It contains 75 to 80% resinous material—cinnamic and benzoic acid esters of a complex alcohol, tolu-resinotannol ($C_{17}H_{18}O_5$). Free cinnamic acid level is about 12 to 15% and free benzoic acid, 8%.[584] Datta *et al.* describe a spectrophotometric method of estimating the cinnamic acid content.[149] The balsam has an acid value of 97 to 160; ester value of 47 to 95; saponification value, 170 to 224; balsamic acids, 35 to 50%.[20] It contains benzyl benzoate, benzyl cinnamate and traces of vanillin and yields 1.5 to 7% volatile oil by steam distillation.

Balsam of Tolu can be distinguished from balsam of Peru by a water-sulphuric acid test, the tolu balsam becoming a gray mass instead of showing a vivid violet color. When heated with sulphuric acid, it becomes bright-red.[233] An ethanol extract of tolu balsam has antibiotic activity against *Mycobacterium tuberculosis*.[20]

Propagation, Cultivation and Harvesting:

The seeds germinate readily, and the tree is sometimes grown for timber, especially in Argentina and Brazil. In various tropical or subtropical regions, it is planted as an ornamental in parks and gardens. However, the resin is harvested from wild trees. The method of tapping is like that used in collecting rubber latex from *Hevea* trees. V-shaped incisions are made in the trunk, and the oleoresin drips into calabash (or gourd) cups. It is accumulated in rawhide bags and finally exported in five-gallon cans. It hardens and turns brittle with age; it becomes pourable at 60° C.[41]

Yield averages 17½ to 22 lbs. (8 to 10 kg) per tree.[395] The balsam has been exported mostly from Colombia and derived its name from the small town of Tolu near Cartagena.[269]

Medicinal uses:

Balsam of Tolu has been commonly formed into lozenges for the relief of coughs and throat irritation, and was formerly widely employed in cough sirups and pill coatings. Also, it was heated and the vapor inhaled as a treatment for respiratory ailments.[423] Today, in the United States, it is utilized mainly as an ingredient in tincture benzoin.[190]

Toxicity:

Allergists have found that balsam of Tolu cross reacts with tincture benzoin and with balsam of Peru.[190]

Other uses:

The extract of balsam of Tolu is utilized by the food industry in soft drinks, ice cream, candy, baked goods and chewing gum. The gum enters into soft drinks, ice cream, candy, baked goods and sirup.[210]

Various products of tolu balsam, such as neutralized extracts, absolute and the distilled oil, are much used in the formulation of perfumes and long-lasting scents for soaps.[41]

The wood of the tree, red-brown or purplish when fresh, becomes redder as it cures and resembles mahogany.[98] Being fragrant, hard, heavy and strong, it is valued in cabinetwork, even though it is hard to work and difficult to stain.[470] Much is fashioned into furniture and into railway ties.[395]

Figure 49. Balsam of Peru (*Myroxylon balsamum* var. *pereirae* Harms.). Reproduction from color painting by David Blair in *Medicinal Plants*, R. Bentley and H. Trimen (J. & A. Churchill, London; 1880).

Balsam of Peru

Botanical name:

    Myroxylon balsamum var. *pereirae* Marms. (*M. pereirae* Klotzsch, *M. peruiferum* Millsp., *Toluifera pereirae* Baill.)

Other names:

    Peruvian balsam, black balsam, Indian balsam.

Family:

    Leguminosae, the Pea Family.

Description:

    A slow-growing tree, usually up to 60 ft. (18.28 m) occasionally to 115 ft. (35 m) tall, with a straight, smooth trunk attaining a diameter of 1½ to 3 ft. (.45 to .9 m) at the base. Bark is pale gray and when wounded exudes an aromatic, dark-brown resin. Leaves are evergreen, odd-pinnate, having 7 to 11 thin, glossy, ovate to oblong leaflets, pointed at the apex, 1¾ to 3¼ in. (4.5 to 8 cm) long, ¾ to 1½ in. (2 to 4 cm) wide, with conspicuous translucent oil glands. Flowers, borne in axillary racemes 4 to 8 in. (10 to 20 cm) long, are very fragrant, white, five-petaled, the standard two-lipped and ⅜ in. (1 cm) long, the calyx white-hairy. Fruit, a yellow pod 2¾ to 3½ in. (7 to 9 cm) long, is narrow at the base and ¾ to 1 in. (20 to 25 mm) wide at the apex. There are 1 or 2 kidney-shaped, yellow seeds .59 to .70 in. (15 to 18 mm) long, having an agreeable odor.[378, 438, 548, 550]

Origin and Distribution:

    The tree is native to moist or dry lowland forests, up to an elevation of 1,968 ft. (600 m) from Veracruz, in southern Mexico, to Panama. It is most abundant on the Pacific side of El Salvador, an area famed as the "Balsam Coast."[269] It is cultivated to some extent in Central America, tropical South America and the Old World tropics including West Africa, India and Ceylon. In Ceylon, it has become naturalized and is said to be common in the forests of Kandy.[14]

Constituents:

    The balsam is a viscous, translucent, flammable liquid not hardening on exposure, bitter and acrid in taste but with a pleasant vanilla-like odor. It consists of 25 to 30% resinous material and 60 to 65% essential oil; it is nearly insoluble in water and dissolves readily in 95% alcohol, chloroform and glacial acetic acid but only partly in ether and petroleum ether.[210] It has a specific gravity (at 26° C.) of 1.152 to 1.170, an acid value of 56.2 to 83 and cinnamein content of 47 to 58.7%.[602] The oil extracted with petroleum ether or benzene contains mainly benzoic acid, benzyl benzoate and benzyl cinnamate and small amounts of d-nerolidol, farnesol, cinnamyl cinnamate and vanillin.[8, 9, 20] The seeds yield a "balsam" composed of 67.68% resin, 14.83% wax, 11.93% acid resin, .39% coumarin, .40 tannin, 4.64% water and residue.[379] Oil distilled from the wood is 68 to 70% nerolidol.[41]

Propagation, Cultivation and Harvesting:

The tree is easily propagated by seeds or cuttings but is cultivated only as a shade for coffee plantations or as an ornamental tree. The oleoresin is derived from natural stands, mainly in El Salvador and Belize. The trees must be at least twenty-five years old and do not naturally exude much resin. Tapping is begun at the base and progresses up the trunk and the larger branches. A crude method of stimulating the flow is by beating the bark severely at several places on the trunk, waiting about six days and then stripping the loosened bark from these areas. The balsam begins to exude in about three weeks and is absorbed by cloths placed over the exposed surfaces. Later, these regions are scorched with a torch and the inner bark peeled away to further promote the flow. As a third operation, the areas may be scraped. A week or ten days after each operation, the cloths are collected, soaked in hot water and the resin squeezed out. It separates naturally from the water, is accumulated and boiled to evaporate excess moisture.[379] The tree does not recover from such rough treatment for eight years or more. Therefore, an improved method was developed in Surinam which requires only the removal of narrow strips of bark at wide intervals and the use of a hot iron in place of a torch. The iron may be applied three times, but the wounds heal in six months. Bark that has been stripped off is soaked in boiling-hot water, and the resin that exudes is salvaged also.[210, 609] Techniques for obtaining maximum yields of balsam are being advocated in Brazil.[7]

Each tree may yield 3 to 5 lbs. (1.36 to 2.26 kg) of balsam annually. The product is exported in 60 lb. (27 kg) drums. In the 16th century, the balsam was carried to Peru by the Spaniards and shipped from Callao to Spain, hence the trade name. In 1948, the annual production of El Salvador was reported to be 115,000 lbs. (52,163 kg),[470] nearly one-half being exported to the United States.[550]

Medicinal uses:

Indians and peasants in southern Mexico and Central America take a decoction of the leaves of the tree as a diuretic and vermifuge. The fruits were formerly official in the Farmacopea Mexicana as having antispasmodic and stimulant properties.[378] The dried fruits are still sold in native markets of Guatemala for use in relieving itch.[550] In rural Cuba, the fruits are steeped in alcohol, which is then used as a rub to relieve headache and rheumatism.[485] Mexicans and Central Americans have long used the resin as a remedy for asthma, catarrh, rheumatism, and gonorrhea, and to heal cuts and wounds.

The balsam is bactericidal. It has been widely utilized by physicians in treating syphilitic sores and other ulcerous complaints, and it is today officially employed (in the form of a tincture or blended with castor oil as an ointment) as an antiseptic, fungicide and parasiticide, particularly in cases of scabies, ringworm, pediculosis, granulations, superficial ulcerations, wounds, bed sores, diaper rash and chilblains.[423, 602, 624] It is currently an ingredient of dental cements and of suppositories intended to relieve the discomfort of hemorrhoids and anal pruritus.[423]

Internally, it was formerly given in cases of bronchitis and laryngitis, also amenorrhea, dysmenorrhea,[378] diarrhea, dysentery and leucorrhea.[624] At the present time, it has no internal use in the United States.[423]

Toxicity:

Balsam of Peru has skin irritating and sensitizing activity and has caused dermatitis in chemists, pharmacists, dentists and in other persons in contact with the balsam or products containing it. It may cross react with various essential oils.[190]

Other uses:

In older times, the balsam was so prized in its homeland that it at times served as a medium of exchange. It was adopted for ceremonial use and for incense by the Catholic Church in Rome, and in 1562 and 1571, there were Papal edicts forbidding the abuse or destruction of the trees.[379]

Balsam of Peru is employed in china painting and some oil paints.[190] The balsam or its distilled oil are much used in the creation of heavy, Oriental-type perfumes and for scenting soaps, cosmetics and hair tonics. Because of its long-lasting fragrance, it is valued as a fixative in blends. The essential oil distilled from the chopped wood of the tree has some utility in the perfume industry but is considered inferior to the balsam.[41]

In the Cobán region of Guatemala, the seeds are put into *aguardiente* as flavoring.[550] Both the balsam and its essential oil are employed sparingly by food manufacturers as flavoring in soft drinks, ice cream, candy and baked goods; the balsam is also used in gelatins, puddings, sirups and chewing gum.[210]

The dark-red wood of trees which die after excessive tapping is employed in furniture-making; also as indoor trim and for railway crossties.[470]

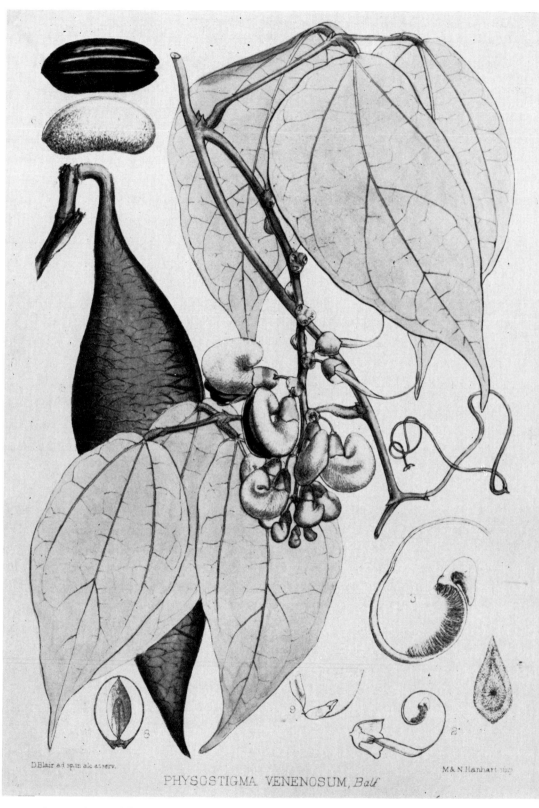

Figure 50. Calabar bean (*Physostigma venenosum* Balf.). Reproduction from color painting by David Blair in *Medicinal Plants*, R. Bentley and H. Trimen (J. & A. Churchill, London; 1880).

Calabar Bean

Botanical name:
    Physostigma venenosum Balf.

Other name:
    Ordeal bean.

Family:
    Leguminosae, the Pea Family.

Description:
    A woody vine, ascending 20 to 50 ft. (6 to 15 m) or more, with smooth, twining branches and trifoliolate leaves to 6 in. (15 cm) long and 4 in. (10 cm) wide, the leaflets rounded at the base, pointed at the apex, the central leaflet ovate and symmetrical, the side leaflets oblique on the inner margin. Flowers, borne in axillary, pendent clusters, are purple and kidney-shaped and, when they fall, leave protruding nodes on the zigzag axis. Fruit is a yellow-brown pod to 6 in. (15 cm) long and 2 in. (5 cm) wide, with prominent horizontal veins; it splits open when ripe, revealing 2 or 3 dark-brown or maroon, oblong or slightly curved seeds, 1 to 1¼ in. (2.5 to 3 cm) in length, ⅜ to ⅝ (1 to 1.5 cm) wide, and 1½ in. (4 cm) thick, hard, fairly smooth but wrinkled near the indented hilum which lies between 2 longitudinal ridges on the convex side and extends around one end.[117, 275, 584, 624]

Origin and Distribution:
    Native to West Tropical Africa (Sierra Leone to the Belgian Congo); usually found along the banks of streams. It has been introduced into Brazil and has been recommended for trial in coastal areas of South India and in the Nilgiris.[20]

Constituents:
    The seeds contain several alkaloids, principally physostigmine (or eserine) ($C_{15}H_{21}N_3O_2$), present in amounts varying from .04 to .3%, which acts as a strong sedative on the spinal cord; also eseramine; eseridine (geneserine) ($C_{15}H_{21}N_3O_3$), which is a purgative; calabarine, which has a stimulating effect on the spinal cord; and physovenine, which is

strongly myotic.[275] Isophysostigmine and N-8-norphysostigmine have been identified as well.[624] Boiling and burning are said to destroy or inactivate these alkaloids.[275] The starch content of the seeds is about 48%, and albuminoids constitute 23%.[122, 275, 477, 584]

Propagation, Culture and Harvesting:

Cultivation of this vine was prohibited by British authorities in colonial times because of the native practice of using the seed as an ordeal poison. In fact, local rulers had long required the destruction of all vines except those under their own control in order to monopolize the supply of the notorious bean. Eradication was discontinued with the outlawing of ordeal trials in the mid-19th century.[477] Seeds for pharmaceutical use are collected from wild plants.

Medicinal uses:

The seeds are commonly sold in local African markets for use in small doses as a remedy for edema and, in combination with other drugs, in treating articular rheumatism. They are also crushed and applied on skin diseases.[275] In Europe, a liquid extract has been administered sparingly as a sedative in cases of extreme nervous irritation.[624] Physostigmine resembles muscarine in action.[117] It is employed as an antidote for strychnine, atropine,[423] nicotine and curare.[419] The seed extract is given by injection in cases of acute tetanus, has been administered to overcome flatulence and atony of the urinary bladder and intestines after abdominal operations, and aids in avoiding adhesions.[20, 275, 423] For these latter purposes, synthetic neostigmine may be preferred.[419]

The principal use of physostigmine is in ophthalmology to achieve extreme and long-lasting contraction of the pupil, sometimes to counteract the effect of atropine, cocaine or homatropine.[275] Alternated with atropine, it serves to break adhesions between the iris and the lens.[423] In cases of glaucoma, physostigmine may be applied as an ointment to lower intraocular pressure.[275] Physostigmine salicylate, given intravenously, has, in recent years, been successfully employed to reverse intoxications by tricyclic antidepressants, antihistamines and certain sedatives—imipramine hydrochloride,[99] amitriptyline hydrochloride,[540] methylphenidate,[280] butyrophenone droperidol and diazepam.[492] Physostigmine salicylate is official in the British and International Pharmacopoeias. It may be given orally, subcutaneously or intramuscularly as an antispasmodic in rheumatoid arthritis, fibrositis and bursitis.[20] In ophthalmia, the solution is applied to the conjunctiva. Physostigmine sulfate is less stable and is hygroscopic. It is administered to the conjunctiva in solution, often combined with antiseptic agents.[423]

In veterinary medicine, physostigmine is given for colic in horses,[20] and rumen impaction in cattle.[552]

Toxicity:

Physostigmine, like acetylcholine, inhibits cholinesterase.[419] It paralyzes the central nervous system, affecting both smooth and striated muscles and causing extreme contraction of the entire digestive tract, also the bladder, spleen, uterus and bronchial muscle.[116]

In African ordeal trials of accused offenders or in testing of warriors, the subject was required to drink an infusion of the seed or chew the seed thoroughly and was also given an enema of a solution of the powdered seed.[275] If he experienced nothing worse than nausea, vomiting and headache, he was pronounced innocent of the presumed offense or a fit and capable warrior. Positive signs of poisoning, beginning within a few minutes of ingestion, are thirst, severe abdominal cramps, flushing and swelling of the face, protrusion of the eyes, trembling,[477] lacrimation, perspiration,[116] salivation and foaming at the mouth, purgation,[477] constricted pupils; respiration at first rapid and deep, later slow and labored; [116] progressive paralysis beginning in the lower extremities, inability to stand; and cardiac and respiratory failure in about one-half hour. As many as thirty-five seeds have been ingested without fatal results, probably because large doses induced effective emesis, removing the poison from the system.[477] On the other hand, twelve grains of the powdered seeds, given experimentally, caused acute illness.[116]

Other uses:

The powdered seeds are used to kill mice and, mixed with palm oil, are smeared on the head to destroy lice, which may take several days.[275] The leaves yield a black dye valued for staining wood. Split stems of the vine serve as matting on which to dry cacao.[20]

Erythroxylaceae

Figure 51. Coca foliage and flowers. Photographed at Longwood Gardens, Kennett Square, Pennsylvania, July, 1976. (*See also* color illustration 8.)

Coca

Botanical name:

  Erythroxylum coca Lam., including var. *coca* (*E. bolivianum* Burck), var. *novogranatense* (*E. novogranatense* Morris) and var. *Spruceanum* (*E. truxillense* Rusby).

Other names:

  Coca bush, cocaine plant, cocaine tree.

Family:

  Erythroxylaceae, the Coca Family.

Description:

  A small, bushy tree to 18 ft. (5.4 m), decreasing to a shrub to 6 ft. (1.8 m) tall as it ascends to higher altitudes; branches are slender, with red-brown bark roughened by leaf scars. Leaves are evergreen, alternate, mostly near the ends of the branches, oval to elliptic, pointed at both ends or sometimes rounded at the apex; 1½ to 4 in. (3.75 to 10 cm) long, ¾ to 1¾ in. (2 to 4.5 cm) wide, thin, characterized by two nerve-like lines paralleling the midrib, bright-green, slightly glossy above, dull and paler beneath; flowers fragrant, ⅜ in. (9 mm) wide, five-petaled, ten-stamened, in groups of 6 to 20 in axils of absent leaves of previous season, white or yellowish; fruit oblong-ovoid, ⅓ to ½ in. (8 to 12.5 mm) long, scarlet with ripe, with thin flesh and a single seed.[18, 84, 358, 461]

Origin and Distribution:

  Coca is native to uplands of Colombia, Ecuador, Peru, Bolivia and western Brazil. Specimens are grown in botanical gardens throughout the tropics and subtropics, but large-scale cultivation takes place mainly in Peru and Bolivia between 1,500 and 6,000 ft. (457 to 1,828 m). There are less extensive plantations in southern and northern Colombia.[230] Until recent years, coca was grown with irrigation on the arid coast of northwestern Venezuela.[432] Commercial cultivation flourished formerly in Ceylon but was discontinued at the request of the British government many years ago.[116] Experimental culture began in 1958 in Japan.[389] Trial plantings in India were generally low in alkaloids, and the crop has not been considered a feasible one for that country.[116]

Constituents:

The alkaloid content of coca leaves ranges from .7 to 2.5%, the main alkaloids (derivatives of *ecgonine*) being *cocaine* ($C_{17}H_{21}NO_4$), *cinnamylcocaine* and *α-* and *β-truxilline*, the proportion of each differing with the variety and the stage of growth. Leaves produced in Bolivia and Peru are high in *cocaine* (75% of total alkaloids) mainly because they are fully mature when plucked, while cinnamylcocaine is the primary constituent of the immature leaves harvested in Java (largely from var. *Spruceanum*) with a higher level of total alkaloids (up to 2.4%).[116] Among other coca alkaloids are the *tropines* (*tropacocaine* and *valerine*) and the *hygrine* group (*hygroline* and *cuscohygrine*). Also present are four crystalline glycosides, *cocatannic acid*, a coloring substance, *coca citrin*,[117, 384] and an essential oil composed of methyl salicylate, acetone and methanol.[210] *Cocaine* occurs in the bark of the tree and in the seeds as well.[117] *Cocaine* has been prepared wholly synthetically, but it is commercially more practical to produce it from the total ecgonines obtained from the dried leaves[116] or coca paste.[580]

Propagation, Cultivation, Harvesting:

Coca is sometimes propagated by cuttings 18 in. (45 cm) long, planted 6 in. (15 cm) deep.[507] However, inasmuch as the plant presents little seedling variation,[102] coca is usually grown from seeds. Viability is short-lived. In Japan, 80 to 100 percent germination was achieved with fresh seeds in a greenhouse temperature of 68 to 86° F. (20 to 30° C.).[389] If seeds are soaked for five days before planting, shoots will emerge in 1½ weeks.[484] Sowing is done in December and January in small, shaded nurseries. When the seedlings reach 2 ft. (.6 m) in height, they are transplanted to carefully prepared forest clearings in valleys or terraces on mountain sides, called "cocales" at a density of 7,000[18] to 28,000[484] plants per acre. In humid situations, a virus disease called "estalla" transmitted by insects, may completely destroy a plantation. The best quality leaves are from the relatively dry terraces. In Bolivia, the native tree, *Inga luschnathiana* Benth., is interplanted for shade but with ample allowance for ventilation. The plants are kept pruned to a height of no more than 5 or 6 ft. (1.5 to 1.8 m) to facilitate harvesting.

One or two years after transplanting, leaf-harvesting begins; the plants could survive for forty or fifty years, but since they become less productive with age, they are usually replaced in twenty years.[102, 484]

Women and children, or, in some Indian tribes, only men, pluck the leaves by hand at least three times a year—in April, June and November—storing them in the pocket of an apron or poncho. Loads are taken to collection areas where the leaves are spread out to sun-dry for eight to ten hours on hard ground or a patio paved with flat stones, clay bricks or cement. Next, they are heaped in piles to sweat for three days and then sun-dried again for an hour.[84] Thereafter, the leaves are compacted in wooden presses (formerly by treading) and sacked or wrapped in coarse cloth, making bales securely bound with dry banana leaves for transport by llama, burro, mule or truck. Each bale weighs 25 lbs. (11.3 kg), and two bales are bound together into a bundle also wrapped and tied with banana leaves. Sometimes charcoal is added to keep the leaves from becoming brittle.

The yield is about 4 oz. (113.39 g) of leaves per plant at each picking. It requires four or five pickers to collect 25 lbs. (11.3 kg) each day. The total yield from 1,350 plants each year is roughly 1,000 lbs. (453.1 kg) of fresh leaves, 500 lbs. (230.8 kg) after drying.[18]

From the Yungas de la Paz region of Bolivia, caravans in the past have carried out as much as 35 million pounds (15,895 metric tons) of dried leaves to the La Paz market.[18, 102] In 1961, estimates of Bolivian production ranged from 3,000 to 12,000 tons, and the Peruvian crop was believed to be 10,000 tons.[33] The amount of coca grown in Peru is controlled by the Coca Monopoly. Sometimes acreage is reduced to maintain price levels.[27] Colombian production in 1961 was reported to be about

Figure 52. Coca (*Erythroxylum coca* Lam.). Photographed in the Botanical Garden of the University of Havana, Cuba, June, 1958.

143,650 kg.[69] The greater part of the supply is domestically consumed. It is estimated that over 8,000,000 people chew the coca leaf.[18, 102] There is much illicit traffic in coca from Bolivia to Argentina.[31]

The leaves which formerly were common in world drug trade were identified by type, the large-leaved Bolivian or Huanaco coca and the smaller Peruvian or Truxillo coca produced in Peru and Java. There is a strong tea-vanilla odor when the bales are opened. To reduce bulk, Java now pulverizes the leaves for export.[624] The alkaloid content declines in storage and is lost in about seven months. Since 1890, the leaves have not been exported from Peru but have been processed in factories in that country. The total alkaloids are extracted with dilute sulphuric acid or by organic solvents and sent abroad in paste form.[580]

Medicinal uses:

As a South American folk medicine, an infusion of 5 g of coca leaves is taken as a sedative for the nerves, a stomachic (in cases of dyspepsia and other gastric ailments), pectoral, a remedy for asthma and as a sudorific.[440] Infants with colic have been given warm milk in which coca leaves have been steeped, and the leaves have been chewed or smoked in the form of cigarettes to relieve asthma, colds and catarrh.[461]

A Spanish physician, Doctor Nicolas Monardes, brought coca to European attention in 1569. A full account of the plant and its use was given by Eduard Poeppig of Leipzig in 1836.[579] Later, in France, an infusion of the leaves was popularized as a stimulating beverage.[18] Cocaine was first isolated by Niemann in Germany in 1858 and was introduced as a local anesthetic by Doctor Carl Koller in 1884. Coca leaves were formerly included in the United States Pharmacopoeia, being employed as a stimulant for the central nervous system.[423] Today, cocaine and, more often, cocaine hydrochloride, and various less toxic preparations derived from them, are administered under strict governmental control and used topically only. They are no longer given by injection because of their toxicity. These drugs reduce the sense of pain and contact by paralyzing the peripheral ends of sensory nerves. At the same time, they stimulate the muscular coats of blood vessels.[423] They are incapable of penetrating unbroken skin[116] but are rapidly absorbed by mucous membranes. When applied to mucous membranes, cocaine produces insensibility in five to ten minutes, and the effect lasts about twenty minutes. It was for a long time the leading anesthetic in eye surgery but required saline irrigation of the eye during use to avoid serious damage to the cornea; it has been replaced by safer synthetics. It is still utilized in nasal and oral operations and cases of inoperable cancer. Cocaine solutions may be resorted to in earache, and cocaine is an ingredient of ointments and suppositories for relief of neuralgia, urticaria and hemorrhoids.[423]

Toxicity:

Applied to the eye, cocaine has caused clouding, pitting and ulceration of the cornea. Taken internally, it stimulates the brain, spinal cord and medulla, increases muscular activity, masks the sense of fatigue and constricts the arteries, reducing body heat loss. In toxic doses, it causes delirium, hallucinations and mania, increased pulse rate succeeded by decrease; there may be exaggerated reflexes, tremors and mild or violent convulsions. In extreme cases, death results from respiratory and circulatory failure.[423]

Cocaine abuse is most widely by injection or by means of snuff. In India, the snuff is taken orally with betel leaf, and some prostitutes use a cocaine solution as a pre-intercourse douche.[117] Repeated use brings on psychic addiction, mental and moral decline, impotence, pallor, insomnia, melancholy, loss of appetite, emaciation; frequently the addict is afflicted with intolerable sensations of itching and of vermin ("cocaine bugs") creeping under or over the skin, or imagines sand or pebbles under the skin.

Orally, the immediate effect of cocaine is a tingling of the tongue and throat, increased by the addition of lime, throat constriction and complete loss of oral sensation and an illusion that the tongue is no longer present, dizziness, palpitations, body warmth, and gradual development of excitement, thirst, constipation and deafness. Large doses produce nausea, vomiting, cramps and delusions of persecution. For man, a convulsive dose is usually .2 g (3 grains); a fatal dose, 1.2 g (18 grains), though death has resulted from .02 g (⅓ grain).[116]

The chewing of coca leaves was originally a practice limited to the Inca highpriests and chieftains, athletes, message-runners and other favored subjects, but the habit spread to the lowest classes when the regimes and strictures were broken down by the Spanish conquerors. The latter, and the exploiters who followed them, encouraged coca-chewing to temporarily enhance the physical activity and endurance of laborers and to enable them to ignore hunger, fatigue, cold and other hardships.[243] Recent observations of fourteen coca-chewing Quecha Indians exposed to cold for two hours indicate mild vasoconstriction and somewhat retarded loss of body heat.[247]

Lime, obtained by burning snail shells, corn stalks, quinoa (*Chenopodium quinoa* Willd.),[579] *Cecropia* or *Pourouma* leaves, or by pulverizing limestone, is carried in a small gourd shell and is mixed with or chewed with the whole or toasted and finely pulverized leaves to extract the maximum amount of alkaloid or render it more fully absorbed.[243] Some Colombian Indians flavor the lime with an aromatic resin.[516] The natives carry their coca supply in an ornamented pouch (*chuspa*) and have calculated distances according to the amount of leaves taken to sustain the walker on the journey.

There has been much controversy as to the long-range effects on the leaf chewers, who take a new quid three or four times a day,[18] the total amounting to a daily intake of about 50 g of leaf—an ingestion of approximately 350 mg of basic cocaine or 235 mg of ecgonine.[392] Habitual chewing causes an enlargement of the cheek inside which the quid is maintained.

The United Nations Commission on Narcotic Drugs first met in 1946, and coca-leaf chewing by South American natives was one of the problems under consideration. After elaborate investigation, it was unanimously declared by the members of the Commission nine years later that coca-leaf chewing was harmful, and it was resolved to recommend to the governments of the countries concerned that there be a gradual but determined effort to reduce production, local use and illicit trade, that wild plants be destroyed and that cultivation for legal use be strictly controlled.[33] In 1968, Peru reported to the Commission that coca plantings were limited to 16,000 hectares and no new licenses were being issued.[34]

In 1967, scientists at McGill University reported results of sophisticated tests and statistical analyses of coca chewers and non-chewers on a large sugar plantation in northern Argentina. This study revealed in the long-term coca chewers a psychological deficit derived from brain damage, despite outward similarities in the general attitude of chewers and non-chewers.[410]

Two years after these tests, workers from the same plantation were studied to determine the effect of abstinence and subsequent learning ability in those who had chewed coca for at least ten years. The findings were clear that long-term coca users who discontinue the habit "can learn through training but cannot catch up with corresponding non-users" in overall performance. While they exhibited normal manual dexterity in simple tasks, they were clearly handicapped in problems requiring abstract thought.[403]

Another carefully controlled study made by epidemiologists of the Johns Hopkins University, partly supported by the United States Army, was based on users and non-users in a Bolivian village. The coca chewers exhibited malnutrition (largely because coca enabled them to ignore hunger), anemia, poor hygiene, increased susceptibility to hookworm infections, twice as much hepatomegaly and twice as much loss of time from work due to illness as non-chewers. They displayed inferior performance at work as compared to non-chewers, contradicting the chewers' euphoric illusion of superior performance.[96]

Poeppig related that a coca leaf infusion, taken at night, caused restlessness, discomfort and sleeplessness. Mantegazza, in an essay published in 1859, reported that a two-dram (¼ oz.) infusion increased the pulse rate fourfold; 3 drams induced fever, hot skin, seeing of flashes, headache, dizziness, roaring in the ears and excitement, a transient sense of strength and agility. A total of 18 drams infused and consumed in one day produced delirium and hallucinations. The strong odor emanating from a stack of dried coca leaves has caused severe headache in persons resting nearby.[579]

Other uses:

Coca leaves which have lost most of their alkaloid content because of improper drying or storing or exposure to dampness are commonly used for fertilizer.[27]

Decocainized extracts of coca leaves are utilized by food and beverage manufacturers to lend flavor and aroma to soft drinks, liqueurs, ice cream and candy.[210]

Related species:

Aynilian and coworkers have reported on the cocaine content of *E. coca* and seven other tropical American species, as assayed from herbarium specimens. In two species, alkaloids were undetectable; in all others except *E. coca*, they found only traces of cocaine.[47]

Rutaceae

D Blair F.L.S. ad nat del et lith. PILOCARPUS PENNATIFOLIUS, *Lem.* M&N Hanhart imp.

Jaborandi

Botanical and common names:

Pilocarpus jaborandi Holmes, Pernambuco jaborandi; *P. microphyllus* Stapf., Maranham jaborandi; *P. pennatifolius* Lem., Paraguay jaborandi; and *P. trachylophus* Holmes, Ceará jaborandi.

Other names:

Alfavaca (in Pernambuco).

Family:

Rutaceae, the Rue Family.

Description:

Shrubs or small trees, from 5 to 10 ft. (1.5 to 3 m) high, with alternate, simple, trifoliate or pinnate leaves borne mainly at the branch tips. The leaflets are elliptic or oblanceolate, notched at the tip, dull-green or grayish and, when held against the light, are seen to be dotted with translucent oil glands on the underside. Those of *P. microphyllus* are ¾ to 2 in. (2 to 5 cm) long and ⅜ to 1⅛ in. (1 to 3 cm) wide, with midvein prominent on the upper surfaces; those of *P. jaborandi* from 1⅝ to 6¼ in. (4 to 16 cm) long, ¾ to 1⅝ in. (2 to 4 cm) wide, with midvein prominent on the underside; those of *P. pennatifolius* up to 9 in. (23 cm) long and without prominent veins. Leaflets of *P. trachylophus* are smaller than those of *P. jaborandi*, are hairy on both surfaces but especially below. Flowers are numerous in axillary or terminal racemes or spikes reaching 1 or 1½ ft. (30 to 45 cm) in length. They are white or purple, star-like with 4 to 5 triangular, leathery petals and 4 to 5 stamens; gland-dotted and aromatic when crushed. The fruit consists of 1 to 5 nearly separate, ridged, one-seeded carpels, splitting midway into two halves.[115, 205, 358, 584]

Origin and Distribution:

Pilocarpus species are very abundant from northeastern Brazil westward to the Rio Acre region.[368] *P. pennatifolius* is common also in

---

Figure 53. Paraguay jaborandi (*Pilocarpus pennatifolius* Lem.). Reproduction from color painting by David Blair in *Medicinal Plants*, R. Bentley and H. Trimen (J. & A. Churchill, London; 1880).

Paraguay, and specimens of this species are grown under glass in Great Britain.[115] Repeated efforts to cultivate *Pilocarpus* species in Malaya have been unsuccessful.[98]

Constituents:

The leaves contain an essential oil of strong odor and pungent flavor [205] consisting of methyl nonyl ketone and various hydrocarbons.[116] They possess up to 72% of total alkaloids, mainly pilocarpine ($C_{11}H_{16}N_2O_2$) —0.15 to 1.97%, pilocarpidine ($C_{10}H_{14}N_2O_2$), pilosine ($C_{16}H_{18}N_2O_3$), isopilosine and isopilocarpine ($C_{11}H_{16}N_2O_2$),[552, 617] most of the latter believed to be formed during the extraction process.[584] *P. pennatifolius* contains, in addition, jaborandine ($C_{18}H_{28}N_2O_2$), which has been isolated from the leaves, flowers and fruit.[617]

Pilocarpine stimulates the secretions of the respiratory tract,[423] the salivary, lachrymal, gastric and other glands,[116] weakens the heart action, accelerates the pulse rate, increases intestinal peristalsis [423] and promotes uterine contractions.[264]

Propagation, Cultivation and Harvesting:

P. pennatifolius is easily propagated by cuttings.[115] There are no commercial plantations. The leaves are harvested from wild plants and exported mainly from the state of Maranhao in northeastern Brazil.[395]

Figure 54. Jaborandi (*Pilocarpus jaborandi* Holmes) in northeastern Brazil. Courtesy Merck Maranhão Produtos Vegetais, S. A.

Alkaloid content declines in storage, 50 percent being lost in a period of twelve months.[122]

Medicinal uses:

In Brazil, the leaf decoction is taken internally as a sudorific and dieuretic and externally applied to the scalp to prevent baldness.[205] In Europe, an infusion of powdered, dried leaves is recommended as a stimulant and expectorant in diabetes and asthma.[624] The leaf extract has been prescribed in cases of dropsy, pleurisy, rheumatism and Bright's disease.[544] In the United States, it was formerly administered in doses of 5 mg, orally or subcutaneously, in nephritis and post-operative urinary retention.[423]

The leaves are no longer official in the United States nor in England. Now, pilocarpine nitrate or hydrochloride are utilized almost solely as cholinergics. The drug may be applied to the conjunctiva one to six times daily in an ointment or gelatin preparation but usually in a 0.5 to 4% solution. The miotic effect begins within fifteen minutes, reaches a peak in one-half to one hour, and wears off in less than twenty-four hours. In the treatment of early glaucoma, pilocarpine stimulates drainage and reduces pressure. It is also alternated with a mydriatic agent to break adhesions between the lens and iris and may be applied to counteract the effects of atropine. Pilocarpine may still be given in cases of saliva deficiency.[423]

Toxicity:

Overdoses of pilocarpine cause excessive salivation and heavy perspiration; sometimes nausea, vomiting and diarrhea and central nervous system paralysis. In weak patients, small doses may result in mental confusion, disturbed vision, bronchial edema and heart failure.[116] It is counteracted by atropine. Arctander suggests that the essential oil may be a skin irritant.[41]

Other uses:

The essential oil, which must be distilled from the leaves before extraction of the alkaloids, is not available in quantity and has only minor use in perfume formulation.[41]

Euphorbiaceae

Figure 55. Castor bean (*Ricinus communis* L.). A common strain in southern Florida with bluish-green immature fruits and small, mottled-brown seeds. (*See also* color illustration 9.)

Castor Bean

Botanical name:

    Ricinus communis L.

Other names:

    Castor oil plant, oil nut, palma cristi; ricino, higuereta, higuerilla and a multitude of other local names.

Family:

    Euphorbiaceae, the Spurge Family.

Description:

    A vigorous, perennial, erect, branched herd. Dwarf forms may rise no higher than 5 ft. (1.5 m) but most specimens become tree-like with stout, fibrous roots and thick, soft-woody trunk, reaching a height of 20 to 30 ft. (6 to 9.1 m). Stems and branches may be flushed with deep-red or maroon. Leaves are alternate, long-stalked, peltate, medallion-like, thick but soft, to 1 or 2 ft. (30 to 60 cm) in width, with 5 to 9 deep lobes which are coarsely toothed and pointed at the apex; young leaves are purple-bronze and silky, mature leaves gray-green or dark-purplish-red. Flowers, borne in small clusters on terminal spikes 6 in. to 2 ft. (15 to 60 cm) long, are petalless; female flowers uppermost, bristly, greenish with bright-red stigmas; male flowers below have green or red, four- to five-parted calyx and ball-like tuft of yellowish stamens. Fruits (seed capsules) are usually bur-like—in some forms, spineless—three-lobed, $\frac{5}{8}$ to 1 in. (16 to 25 mm) long; when immature, green, bluish, or brilliant-red, turning brown when mature and dry; three-celled, dehiscing with force and scattering three oval, flattened, glossy seeds $\frac{5}{16}$ to $\frac{5}{8}$ in. (8 to 16 mm) long, with a prominent white caruncle at the pointed end. The seed coat varies in color from all-black, red or white to white mottled with yellow-brown, pale-gray with black or red blotches, or light-brown mottled with dark-brown. The flavor of the raw seeds is disarmingly nut-like and agreeable.[146, 415, 549, 595]

Origin and Distribution:

    Native to the Old World tropics, most likely Africa. Seeds found in Egyptian tombs are believed to date back 4,000 years. The plant was apparently introduced to the Chinese between 600 and 900 AD.[98] It has been widely distributed and has become naturalized in nearly all subtropical and tropical countries. In low, moist land it multiplies rapidly and forms extensive thickets. Throughout many regions, including southern Florida, it is a pestiferous weed. In temperate climates, it is often grown as a summer annual.

Commercially, the castor bean is cultivated on the greatest scale in Brazil (35 to 45% of the world crop), India (15%) and Russia. Other important producing areas are Thailand, the United States (mainly Texas), Mexico, Peru, Paraguay, Rumania, Yugoslavia, Tanzania, South Africa and Ethiopia.[35, 172]

Constituents:

Castor seeds contain the toxic glycoprotein *ricin* (which is a blood coagulant resembling the *abrin* of *Abrus precatorius* L.;[607] also the much less toxic alkaloid *ricinine* ($C_8H_8O_2N_2$), the enzyme *lipase* and a potent allergen called CBA. The seed consists of 20% coat and 80% kernel. Unshelled seeds contain 40 to 53% of a fixed oil; decorticated kernels, 58 to 66%. Small seeds are proportionately richer in oil than larger seeds.[117] The toxic properties of the seed are not extracted with the oil. However, the oil contains the purgative principle *ricinolein* (the glyceride of ricinoleic acid) and lesser amounts of the glycerides of oleic, isoricinoleic, linoleic, stearic and dehydroxystearic acids.[423, 584]

Ricin is separable into highly toxic ricin D, acidic ricin and basic ricin.[208, 209, 248] Injected subcutaneously, ricin is highly inflammatory but, at the same time, causes mobilization of corticosteroids from the adrenals which reduces the inflammatory effect of other agents.[56] It is believed that the action of ricin is due to inhibition of protein synthesis.[472] Van Wauwe and associates have studied the interaction of the castor bean hemagglutinin with polysaccharides and carbohydrates.[590] Both ricin and abrin have exhibited antitumor activity which has been attributed to their inhibitory effect on protein biosynthesis in the tumor cells.[607] Both toxins have shown antileukemic activity in experimental animals.[471]

Propagation, Cultivation and Harvesting:

The castor bean plant is grown from seeds sown directly in the field by hand or machine. Fresh seeds may require removal of the caruncle and piercing to speed germination. Some growers allow seed a dormancy of several months before planting.

The plant is tolerant of a wide range of soils, providing they are well-drained, and some strains stand moderate salinity. Rows must be several feet apart to permit mechanical cultivation and other operations. Irrigation is necessary in dry soils or during droughts. Fertilizers must be low in nitrogen to avoid excessive vegetative growth. Tall-growing strains have been largely abandoned in favor of recently-developed dwarf types with a high yield of non-shattering fruits which ripen uniformly. There are cultivars of short-, medium- and long-season [172] adapted to various climatic conditions. Some are ready for harvest in 4 to 5 months, others require 7 to 10 months.[415] In the tropics, the plant blooms and fruits continuously the year around. Several insect pests (red mites, borers, aphids, armyworms, leafworms, etc.) require control, and the plant is subject to fungal diseases and bacterial blight.

The capsules of dwarf types may be harvested manually with a stripping cup, or by picking machines. Some machines are capable of hulling as well.[172] With tall types, the fruiting spikes are cut before the capsules open, they are air-dried for about a week, then winnowed to separate husk and seed.[146]

Yields in the past have varied from 200 to 900 lbs. per acre (200 to 900 kg per hectare). Great advances have been made through breeding. In India, the improved variety HC-6, fully harvested in 240 days from planting, by four pickings, yielded 2,005 kg per hectare. A subsequent introduction, NPH-1, fully harvested in 150 days from seed, with only three pickings, yielded 3,083 kg per hectare.[52] World total is nearly 800,000 metric tons of castor seeds annually.[35]

The seeds must be unbroken or their oil will become rancid in storage. For the production of high-grade medicinal oil, the seeds are mechanically sorted according to size and passed through rollers to crack their coats which are then blown away, only the kernels being then immediately "cold pressed" or pressed at temperatures no higher than 90 to 100° F. Thereafter, the oil is refined, removing the albumen and traces of enzyme, to enhance keeping quality. Lower grade oil is obtained from the residue by a second pressing.

For industrial purposes, the coats need not be removed before oil extraction, and higher temperatures can be utilized to increase yield.[94] The resulting oil will be brownish and somewhat acrid in taste.[305] Solvent extraction is also practiced to obtain industrial-grade oil, as is simple crushing and boiling.[146]

Since the early 1960s the Brazilian government has prohibited the export of castor seed so that it will not compete with the oil which is extracted in Brazil and bulk-loaded into ships. Over 40 percent is exported to the United States, mainly for industrial, to a lesser extent for medicinal, consumption.[447]

Medicinal uses:

In the tropics, all parts of the plant are employed as folk remedies. The fresh leaf may be simply worn under a hat, or moistened with vinegar and bound around the head, to relieve headache and fever.[549] Fresh leaves are laid on the abdomen for internal complaints; heated and oiled leaves are placed on rheumatic joints, swellings and inflamed muscles. In Africa, a paste of the cooked leaves is applied to Guinea-worm sores, and the leaf decoction is taken as a purge, emmenagogue or galactagogue. A cold leaf infusion is used as an eye wash. Decoctions of the roots are taken as remedies for lumbago, sciatica and related ailments. The root bark is extremely purgative. One or two seeds are sometimes ingested as a drastic purge, but this is a dangerous practice, as is the African custom of mixing a few crushed seeds with peanuts or cooked cereal and eating for the purgative effect. Roasting the seeds is said to render them milder in action.[98, 117, 146, 148, 160, 378]

The freshly extracted oil is nearly odorless, has a mild flavor and is less purgative than rancid oil. It is widely employed as an application on skin diseases and on wounds of domestic animals. Mixed with turpentine, it has been taken internally to expel tapeworms.[146, 549]

In bygone years, castor oil was a universal household cathartic but has been largely replaced in the home by less repulsive products of more moderate action. Today, it is given mainly in hospitals in cases of food poisoning and to clean out the bowels before x-rays or other examinations. It is often administered in fruit juice or in the form of a sweetened, flavored emulsion. The customary dose for infants is 1 to 5 ml; for older children, 5 to 15 ml; for adults, 15 to 60 ml. It has an irritant action (after saponification by the pancreatic fluid in the intestines) and produces repeated watery movements within a few hours. Its use is not recommended during menstruation or pregnancy and is dangerous in cases of intestinal obstruction and appendicitis; also in association with fat-soluble vermifuges.[423]

The oil is made into contraceptive jellies, foams and creams,[277] is a common vehicle for ophthalmic medication and is a popular ingredient in soothing, cleansing eye drops.[423] Undecylenic acid, derived from castor oil, is an antifungal agent.[584]

Toxicity:

Reaction to castor seed ingestion may be almost immediate or delayed several hours or even two or three days.[595] The usual effects are acute gastrointestinal inflammation with a burning sensation in the throat and abdomen, thirst, hot skin, rapid pulse, blurred vision, profuse sweating, chills, vomiting and often purging as well. As few as two seeds have caused fatality in young children. Fatal doses for adults have ranged from two to twenty seeds.[605] There have been numerous poisonings of sheep, horses, pigs, dogs and poultry from consuming the leaves, seeds or press-cake. Cattle are less susceptible but have also succumbed from excessive intake. Six seeds may be fatal to a horse. In grazing animals, the usual signs of castor bean poisoning are hemorrhagic enteritis, incoordination, convulsions and elevated temperature. Coma and death may occur in three days or less.[130, 146, 272] The toxicity of ricin can be neutralized by administration of anti-ricin serum from an animal immunized by small doses of ricin. The serum may be kept refrigerated for many years.[21]

Proximity to the flowering plant may give rise to pollinosis. Gardeners often suffer dermatitis from handling castor pomace. Minute amounts of castor bean allergen in the atmosphere have caused acute respiratory and other reactions in sensitive individuals; anaphylactic shock may follow biting into a single seed[42] or exposure to broken or crushed seeds. Workers in castor oil factories are often severely affected with respiratory, cutaneous and digestive disorders,[565] as are residents of communities in the vicinity of such factories and longshoremen handling sacks of pomace.[133] In some instances, CBA-induced bronchial asthma has been fatal.[565] The seed extract is sometimes applied to the conjunctiva by malingerers to cause inflammation.[146] In parts of Africa, crushed castor seeds in food are a well-known means of infanticide.[595] Castor oil is added to leftover food which is set out to poison cockroaches.[146]

Other uses:

This weedy plant is frequently and unwisely cultivated as an ornamental. In South America and in Egypt it is planted around dwellings to repel mosquitoes.[146] Being fast-growing, it is often interplanted to furnish shade in young coffee plantations.[548] In Assam, it is grown for its leaves which are fed to a special silkworm, *Attacus ricini* Boisd.[71] Fiber from the stems can be made into paper.[549] Dry stems are weak but are used in India in constructing walls of huts, and also for fuel.[71]

Inflorescences and young, tender fruits, though bitter, are cooked as vegetables in Indonesia, and mature seeds are roasted and added to dishes as a substitute for coconut.[415] Chinese and Koreans use the seed oil for cooking.[153] In the food and beverage industry, castor oil enters into soft drinks, ice cream, candy and some baked goods.[210]

The seeds are a commercial source of *lipase* which is employed by soap-manufacturers for splitting vegetable oils into glycerine and free fatty acids, and for recovery of glycerine.[146] Castor oil as a soap ingredient makes the product transparent and improves lathering in cold water.[94]

As a hydraulic fluid and lubricant for machinery, the oil is valued in heavy equipment and locomotives and especially in aircraft, since it does not solidify in extreme cold and at high temperatures it retains its viscosity.[146] Sulfonated castor oil is the Turkey Red Oil used in dyeing and printing cotton and woolen fabrics. Castor oil serves as a fixative for dyes and is used in dressing leather. It is a source of sebacic acid for making nylon and other synthetic fabrics; it is also utilized in manufacturing linoleum, artificial leather, rubber-like foam products, typewriter inks, fly paper, candles, hair pomades and other cosmetics. The oldtime hair dressing known as "bear's oil" was basically castor oil scented with herbs.[305] Castor oil is usable as fuel for Diesel motors and has been widely employed as an illuminant.[146] The pierced seeds, slid onto slim splinters of wood, are burned for light by South American Indians.[544] The dehydrated oil has proved superior to tung oil for paints, enamels (particularly white for refrigerators) and for varnishes essential to lithography.[98, 172] In Malaya, castor oil with unslaked lime and resin is a common caulking for boats.[98]

Castor pomace, the residue after oil extraction, is much used as a soil enricher and, when detoxified by heat, has been used as cattle feed. Untreated presscake has caused numerous livestock fatalities.[197] The allergen is not entirely dissipated by storage and processing and castor meal may contain from .09 to 4.2% CBA.[467]

All around the tropics, the varied-colored castor seeds are made into necklaces and belts alone or intermixed with other seeds, despite the hazard of accidental ingestion or allergic reactions.

Rhamnaceae

Figure 56. Cascara sagrada (*Rhamnus purshiana* DC.). Photographed at the Rancho Santa Ana Botanical Garden, Anaheim, California.

Cascara Sagrada

Botanical name:

    Rhamnus purshiana DC.

Other names:

    Cascara buckthorn, sacred bark, bitter bark, chittem bark, bearwood, bearberry, coffeeberry, wild coffee, coffee-tree, wahoo.

Family:

    Rhamnaceae, the Buckthorn Family.

Description:

    Large shrub or small tree, 5 to 35 or 40 ft. (1.5 to 10 or 12 m), with trunk attaining 18 to 20 in. (45 to 50 cm) in diameter. Bark light- to dark-brown, or reddish-gray externally, yellow on the inside, superficially scaly, ⅛ to ¼ in. (3 to 6 mm) thick, bitter. Leaves deciduous, thin, broad-elliptic with short point at apex, wavy, more or less fine-toothed margins and with conspicuous veins; hairy beneath and on veins on upper surface; 2 to 7 in. (5 to 17.7 cm) long and 1½ to 2 in. (3.8 to 5 cm) wide. Flowers greenish-yellow, bell-shaped, five-lobed, 5/32 to 3/16 in. (4 to 5 mm) long, in axillary clusters of 25 or less. Fruit nearly round, ⅓ to ½ in. (8 to 12.5 mm) wide, turning from scarlet to black when ripe, with thin, juicy flesh enclosing 2 or 3 somewhat hemispherical seeds ¼ in. (6 mm) long.[401, 508, 509]

Origin and Distribution:

    Native to chaparral thickets and moist coniferous forests on mountain ranges below 5,000 ft. (1,524 m) of the Pacific Northwest, from British Columbia to Montana and south to central California and the southern slope of the Grand Canyon in Arizona. Sometimes cultivated in botanical gardens in the eastern United States and Europe and elsewhere.[508]

Constituents:

    The primary active properties are anthracene compounds, 10 to 20% being O-glycosides based on emodin, 80 to 90% being C-glycosides (aloin-like) including barbaloin (yielding the O-glycosides cascarosides A and B) and deoxybarbaloin, or chrysaloin (yielding the O-glycosides cascarosides C and D.[122, 584] The extracted cascarosides[506] are less bitter and nauseous than the aloins. Also present are the anthraquinones emodin, oxanthrone, aloe-emodin and chrysophanol in the free state and their dianthrones.[584]

These constituents have a cathartic effect through stimulating peristalsis of the large intestine.[423] Among other properties are tannin, resins, glucose and starch. Su and Ferguson have reported an improved method for extracting the anthraquinone aglycones and glycosides in pure form.[567]

Propagation, Cultivation and Harvesting:

Efforts at cultivating the tree commercially in Canada, the western United States and Kenya have been largely unsuccessful. The bark from trial plantations has proved to be of low grade. Therefore, the bark is still gathered from wild trees of Oregon, Washington and British Columbia, usually from mid-April to late August. The tree trunks are stripped and left standing or are stripped and felled leaving a stump about ½ ft. (15 cm) high, slant-cut to shed water and avoid decay. Shoots from the stump will supply a future crop of bark. The bark is then stripped from the branches of the felled trees. Production ranges from 100 to 250 lbs. (45.3 to 113.3 kg) per man daily. The bark is carried out of the woods and spread to sun-dry on platforms, with the outer surface upward. After four or five days, it is mechanically broken into pieces and shipped to dealers. It must be stored in sacks at least a year before it is considered cured for pharmaceutical use.[618] The International Pharmacopoeia specifies that artificial drying of the bark at 100° C. for one hour may be substituted for twelve months' storage.[584] The total annual harvest is about 2,000 metric tons (2 million kg).[466]

Medicinal uses:

American Indians were acquainted with the virtues of the bark of the very similar *R. californica* Esch. and valued it highly. Early settlers in the West made a cold infusion, soaking a piece of bark overnight, and took it as a tonic. They prepared a laxative potion by boiling fresh bark for several hours or by pouring boiling water over a small amount of pulverized dried bark and letting it cool.[308] *R. purshiana* was described in 1805 and the bark adopted into medical use in 1877.[584] The powdered extract of the bark (yellow-brown or yellowish-orange), the fluidextract, the aromatic fluidextract (debittered with magnesium oxide, flavored with licorice, anise oil, coriander oil and methyl salicylate and sweetened) or manufactured pills or tablets may be taken or administered as moderate laxatives in cases of chronic constipation. Doses of crude bark (rarely used) range from 10 to 30 grains; extract, 120 to 500 mg, usually 300; fluidextract, .6 to 2 ml, usually 1 ml; aromatic fluidextract, 5 to 15 ml, usually 5 for adults, 2 to 8 ml for children, 1 to 2 for infants.[423] Cascara sagrada is also valued as a

Figure 57. Buckthorn (*Rhamnus cathartica* L.). Reproduction from color painting by David Blair in *Medicinal Plants*, R. Bentley and H. Trimen (J. & A. Churchill, London; 1880).

treatment for dyspepsia and hemorrhoids.[624] The laxative effect is transmitted to nursing infants of mothers taking any of the preparations.[423]

Toxicity:

The fresh bark tends to cause griping and nausea. Chemical changes in storage render it far more acceptable to the system. However, large doses of preparations from the dried bark may cause inflammation; habitual use may induce chronic diarrhea with weakness from excessive loss of potassium. Over a long period, melanin will pigmentize the mucous membranes of the colon.[423]

Other uses:

In the food and beverage industry, bitter cascara extract has been used in liqueurs. Debittered extract serves as a flavoring in soft drinks, ice cream and some baked goods.[210]

The fruits are eaten raw or cooked but are said to give a transient reddish cast to the skin if consumed in excess.[308]

In Sonora, California, the flowers of this plant are the leading source of honey, the flowering period lasting for twenty-five days or longer. The honey is very dark, non-granulating and has a mildly laxative effect.[349]

Related species:

R. cathartica L. is a thorny shrub, native from England to North Africa, which was introduced into the eastern United States as a hedge plant and has become naturalized. The bitter fruits, known as buckthorn berries, are violently purgative, and this effect has been experienced by people eating wildfowl which have fed upon them. A sirup of the berries was long ago abandoned as a purge for humans because it caused severe griping and extreme thirst.[63] There is little use of the fruits even in veterinary practice today.

R. frangula L., of southeastern Europe and Russia is the source of frangula, or buckthorn, bark which has been official in England since 1650. It is imported into the United States mainly for veterinary use.[122] Recent studies of glycosides in the bark of this species have been made in Rumania and Paris.[150, 488, 489, 490, 510] (R)-(−)-armepavine has been found in the fresh bark but not in the dried product.[426] Aloe-emodin isolated from the seeds has been found to be active against P-388 lymphocytic leukemia in mice.[326]

Malvaceae

Figure 58. Marsh mallow (*Althaea officinalis* L.). Photographed in the Morris Arboretum of the University of Pennsylvania, Philadelphia.

Marsh Mallow

Botanical name:

    Althaea officinalis L.

Other names:

    White mallow, common althea.

Family:

    Malvaceae, the Mallow Family.

Description:

    An annual herb with perennial, spindle-shaped or cylindrical, tapering woody taproot and several erect, stout, downy stems, 1 to 5 ft. (30 to 150 cm), sparsely branched. Leaves are alternate, grayish-green, velvety, being covered with fine, star-shaped hairs. Lower leaves are nearly circular, faintly three– to five-lobed and toothed, 1¼ to 3¼ in. (3 to 8 cm) wide; upper leaves ovate or lanceolate, pointed, deeply lobed and toothed and partly folded. Flowers, borne in the axils of the upper leaves, solitary or 2 or 3 together, 1 to 2 in. (2.5 to 5 cm) wide, with 5 pink or white, obovate, notched petals and prominent staminal tube tipped with kidney-shaped anthers. The double, velvety calyx has 6 to 9 outer segments and 5 inner, triangular, long-pointed sepals. The calyx cups the downy, oblate fruit which consists of numerous carpels, each containing a single, kidney-shaped, brown seed to ⅛ in. (3 mm) long. Blooming season is from July to September.[63, 100]

Origin and Distribution:

    Native to seacoasts and borders of saline and salt marshes of southern and eastern England, and from Denmark through central and southern Europe to North Africa, Siberia and western Asia. Introduced into North America and naturalized in coastal marshes from Massachusetts to Virginia and inland along tidal rivers in New York and Pennsylvania.[618]

Constituents:

    The taproot, grayish-yellow and corky externally and white and fibrous within, is faint-scented, sweetish and astringent; when dried, it contains 25 to 35% of mucilage, about 35% starch, 10% pectin, 10% sugars and 1 to 2% asparagine.[122] There may be as much as 10% tannin.[338] The root must be thoroughly dried or it will decompose quickly and have a sour odor and taste.[63]

Propagation, Cultivation and Harvesting:

Marsh mallow is not cultivated in the United States but grown in home gardens in Great Britain and Europe and commercially in Belgium, France, Germany[466] and Yugoslavia.[122] The plant is grown from seed or offsets from old plants, the roots being harvestable the second year after planting offsets and three years from seed. Marsh mallow tolerates a variety of soils but does best in deep loam with abundant moisture.[23] The taproots are collected in the fall, freed of lateral rootlets, washed, peeled and dried[618] whole, split vertically in 6– to 8-inch (15 to 20 cm) lengths, or cut in transverse slices.[584] or into cubes 3/16 in. (5 mm) wide.[122]

Medicinal uses:

Marsh mallow root was extolled in the days of Dioscorides and Pliny, and the decoction was prescribed for asthma, bronchitis, hoarseness, pleurisy, dysentery, muscular and nerve injuries, and to counteract loss of blood and relieve inflammations. It was valued in acute ophthalmia and also used as a gargle for gum troubles and sorethroat. The powdered or crushed root was applied externally as a poultice on abrasions and eruptions.[63] Applications remain much the same in Europe today. In addition, a piece of the root is given to an infant to chew when the first teeth are appearing.[39, 513] In cystitis, retention of urine, gonorrhea, and leucorrhea, marsh mallow is taken to relax the urinary passages and relieve pain.[624] The flowers, leaves and seeds are also employed in infusions, though they are less mucilaginous than the root.[513] The dried roots are imported from Europe into the United States and infusions, fluid extracts, sirups and tinctures are prepared for pharmaceutical use as demulcents and emollients.[210]

Other uses:

The root was the original main ingredient in the marshmallow confections in which it is no longer used today. However, the tincture is employed in the food industry for flavoring non-alcoholic beverages and in formulating aromas for liqueurs.[210]

Sterculiaceae

Figure 59. Karaya (*Sterculia urens* Roxb.), a deciduous tree with large leaves, sprays of yellow-brown flowers, and red seedpods covered with stinging hairs.

Karaya

Botanical name:
    Sterculia urens Roxb.

Other names:
    Indian tragacanth, gulu, kulu.

Family:
    Sterculiaceae, the Chocolate Family.

Description:

    A tropical, soft-wooded tree, attaining 30 ft. (9 m), with erect, straight trunk and broad top. The bark is grayish-white, very smooth, glossy; the outer layer thin and easily peeled off like birch bark, the inner layer fibrous and netted. All parts of the tree exude a soft gum when injured. Leaves, mostly clustered near the end of the branches, are deciduous, alternate, palmate, shallowly five-lobed, 8 to 16 in. (20 to 40 cm) wide, velvety beneath; the petioles to 12 in. (30 cm) long, downy and armed with stinging hairs. Flowers small, yellow-brown, in large, erect terminal panicles coated with yellow, mealy, glutinous down. Many male and some hermaphrodite blooms are intermingled. There is no corolla. The calyx is bell-shaped, five-toothed, leathery; the 15 to 20 stamens are united at the base and bear large, two-lobed anthers. Fruit is star-shaped, composed of 4 to 5 united one-celled red capsules covered with yellow down and stiff, stinging hairs. The capsules split open, revealing in each 3 to 6 oblong, brown seeds attached to the margins; the inner wall of the capsule is hairy.[160, 499] Blooming time is usually February and March.

Origin and Distribution:

    Karaya is native to dry, rocky hills and tablelands[610] from the tropical Himalayas, east of the Ganges, to southern India.

Product and constituents:

    The gum, called karaya, kadira or Sterculia gum, is a polysaccharide with an acetic acid odor. It absorbs water very rapidly and remains stable for several days. It is less soluble than other similar commercial gums.[610] It is insoluble in alcohol[122] and alkali.[466] Its viscous nature decreases with storage, especially after being reduced to a powder and in hot and humid environments.[610] Partial acid hydrolysis yields *d*-galactose, 1-rhamnose (6-deoxy-1-mannose) and *d*-galacturonic acid, together with the aldobiouronic acids 2-O-(α-*d*-galactopyranosyluronic acid)-1-rhamnose and 4-O-(α-*d*-galactopyranosyluronic acid)-*d*-galactose and the acidic trisaccharide O-(β-*d*-glucopyranosyluronic acid)-1 → 3-(α-*d*-galactopyranosyluronic acid)-(1 → 2)-1-rhamnose.[610]

When hydrolyzed with 5% phosphoric acid, it has a volatile acidity of not less than 14% compared with 2 to 3% in tragacanth. Its methoxy value is 0; that of tragacanth, 30 to 40. Highest grades are nearly colorless, translucent; medium grades are pinkish; lower grades are dark and contain particles of bark.[584]

Propagation, Cultivation and Harvesting:

The tree is propagated from seed, cuttings or air-layers.[185] There are some commercial plantations in tropical Africa.[584] In India, the gum is collected from wild trees, especially in the periods of October to January and April to June, the latter producing the best quality. Gum may exude all year but is undesirably dark in color and of low viscosity in the rainy season. In order to conserve the trees, it is recommended that tapping be restricted to those exceeding 3 ft. (.91 m) at a height of 5 ft. (1.5 m) from the ground. In blazing the trees, the bark is cut away from a 12 x 12 in. (30 x 30 cm) area on each side of the trunk, with the edges of the cuts slanted to allow rain to run off and not collect in the wound. The gum flows immediately, most actively during the first twenty-four hours, and continues for several days.[269] It solidifies in the form of "tears" (some weighing several pounds)[610] or wormlike strips and is collected every few days. After the first collection, the edges of the blaze are scraped, about ⅛ in. (3 mm) of bark being removed to stimulate the flow. Trees are allowed to rest for long intervals before retapping, which should be done not more than five times in the life of the tree.[269] A single tree will yield from 2 to 10 lbs. (1 to 4.5 kg) per season.[610]

Medicinal uses:

The gum is employed primarily as a bulk laxative, second only to *Psyllium* seed in this class of product.[610] It is used as a substitute for tragacanth in preparations to relieve throat inflammation; also in lozenges. Though long employed in Indian hospitals, it entered the United States only as an adulterant of tragacanth until 1920 when it began to be imported deliberately because of its lower price. Soon karaya gum was recognized as being superior to other gums for many purposes and, since the mid-1930s, it has continued in strong demand.[269] In 1971, 3,825 tons (3,465,450 kg) of karaya gum were imported into the United States, amounting to 75 to 80 percent of the world supply.[610] Much is consumed in non-pharmaceutical industries.

Veterinary use:

The water into which the leaves and cut ends of branches are placed becomes thick like clear jelly and, in Indian veterinary practice, this mucilage is administered as a remedy for pleuropneumonia in cattle.[117]

Other uses:

In India, the gum is eaten in curries[277] and is much prized for making sweetmeats.[269] There, also, the seeds are roasted and eaten or ground as a substitute for coffee.[160] In western food manufacturing, karaya enters into French dressing, cheese spreads, sherbets, ice cream sticks, meringues, and bologna and other meat products.[610] It has important roles in toiletries and cosmetics. The powdered gum serves as an adhesive for dentures, making a firm bond and being resistant to bacterial and enzymatic breakdown.[610] It is valued in lotions which set waves in hair.[122] The mucilage is more effective as a skin softener than tragacanth. In paper-making, it serves as a binder of cellulose fibers in tissue paper.[610] A main industrial use is as a thickening agent for pigments in printing fabrics.[269] It is also valued as a binder in composition building materials.[122]

The spongy wood is made into doors, boats, packing cases, guitars and toys, and it is popular for carving.[269, 277] The bark is very astringent.[499] It provides a strong fiber for rope-making and cloth.[277]

Related product:

A very similar gum is obtained from *Cochlospermum religiosum* Alston (syn. *C. gossypium* DC.) in India. Called katira gum, karaya gum or hog gum, it is used for the same purposes as the gum from *Sterculia urens* and is of considerable economic importance.[20]

Styracaceae

Figure 60. Sumatra benzoin (*Styrax benzoin* Dry.). Reproduction from color painting by David Blair in *Medicinal Plants*, R. Bentley and H. Trimen (J. & A. Churchill, London; 1880).

Benzoin Trees

Botanical and Common names:

Styrax benzoin Dry., Sumatra benzoin, Benjamin tree, Gum Benjamin; "Benjamin" being a corruption of *"ban jabi"*, a distortion of the Arabic name, *"luban jawi"* (luban of Java). S. *paralleloneurum* Perk., (S. *sumatranus* J. J. S.), Sumatra benzoin. S. *tonkinense* Craib (*Anthostyrax tonkinense* Pierre), Siam benzoin.

Family:

Styracaceae, the Storax Family.

Description:

Graceful, fast-growing trees, ranging from 16½ to 115 ft. (5 to 35 m) and secreting, when wounded, a fragrant, acrid resin. The bark is covered with a fine, whitish down. Leaves are alternate, short-stemmed, ovate-oblong to lanceolate, broadly rounded at base, pointed apex; 2⅓ to 6⅓ in. (6 to 16 cm) long, 1 to 2½ in. (2.5 to 6.5 cm) wide. Leaves and inflorescence are more or less coated with star-shaped, white or brownish hairs. Flowers, in compound, pendent racemes, are fragrant, white, with cup-like or bell-shaped, five-toothed calyx, 5 narrow, silky petals and 10 stamens. Fruit is drupaceous, globose, 1 to 1¼ in. (2.5 to 3.25 cm) wide, dry, dehiscent, containing 1 or 2 almost round seeds, slightly concave on one side. Branchlets of S. *paralleloneurum* are often hung with slender, coiled galls.[49, 305]

Origin and Distribution:

S. *benzoin* is native to Malaya, Malacca and Java and cultivated in Sumatra and Java. It was introduced into Ceylon in 1881.[605] S. *paralleloneurum* is native to the Malay Peninsula and Sumatra and is cultivated in Java. S. *tonkinense* is native to Laos, Tonkin, Annam[423] and the province of Luang Probang in eastern Thailand.[584] It was introduced into Java for reforestation after World War II.[49]

Constituents:

Fresh resin of S. *tonkinense* contains 78% crystalline coniferyl benzoate which becomes partly amorphous with age, nearly 12% benzoic acid, 6% of a triterpenoid acid—siaresinolic acid or siaresinol—19-hydroxy-oleanolic acid ($C_{30}H_{48}O_4$), all kept liquid by 2% cinnamyl benzoate, which evaporates, allowing the exposed resin to harden. There are only traces of vanillin.

Sumatra benzoin (S. benzoin and S. paralleloneurum) contains much less benzoic acid, having in its place cinnamic acid, and instead of siaresinolic acid has sumaresinolic acid, or sumaresinol–6-hydroxyoleanolic acid. In addition to vanillin, it includes traces of styracin (cinnamyl cinnamate), styrene (phenyl ethylene) and benzaldehyde.[423]

The exposed resins are whitish, or yellowish–, grayish– or reddish-brown, hard and brittle at room temperature; they become pliable when heated or chewed.[423] They are almost completely soluble in ether and alcohol.[305]

The inferior Palembang benzoin from Sumatra is distinguished from Siam benzoin by its light weight and its irregular porous fracture when broken.[584]

Propagation, Cultivation and Harvesting:

Benzoin trees are easily grown from seeds which may be germinated in nursery beds or sown directly in the field. S. benzoin is cultivated on hillsides in Sumatra. Sometimes it is planted in advance of upland rice so that the growing grain shades the seedlings. After the rice is harvested, the young trees are shaded by weeds and receive no cultural attention. When six or seven years of age, the trees are ready for tapping. For centuries, the resin of S. tonkinense has been extracted by the primitive tribe, called Khas, in northern Laos. The resin exudes only when the trees are injured, the tree secreting resin in cavities and channels in the new wood formed to close wounds.[98] Tapping is most crudely done by slashing the trunk with a hatchet, making a series of horizontal gashes. The better method is to cut deep, triangular holes with a knife, the first holes being 15 to 16 in. (40 cm) apart vertically and the intervening bark scraped smooth.[584] After a few days, a yellowish-white latex begins to seep out, and it is allowed to continue for three months. This first flow is considered of low quality and may not be utilized. Second and subsequent tappings, made three months apart, cause the flow of resin to increase, and the product is firmer and more fragrant. Tapping proceeds up the tree for 20 ft. (6 m) or more. After the first three years, the resin flow is reduced, but it will continue for five or six years. The trees die when seventeen to nineteen years old.[98] Each tree may yield 3 lbs. (1.36 kg) per year[605] during the period of maximum production.

Six weeks after each tapping, the hardened resin is collected, generally with a sharpened stick of bamboo.[269] Bark fragments and other debris are removed and the resin graded for export. It remains in the form of tears, "almonds" or lumps, compressed into square blocks shaped by the containers in which they are packed.

The resin of S. *tonkinense* has always been exported through Thailand, which accounts for the trade name, Siam benzoin. It ranks as the highest in quality. Sumatra benzoin is used mainly as a source of natural benzoic acid.[584]

Medicinal uses:

Benzoin has a long history of medical use. Vasco de Gama was presented with a gift of benzoin in 1498. Garcia da Orta, when resident in Goa (1534-70), sent agents to Malacca to seek out the trees yielding benzoin. Cabral carried a shipment of the resin from India to Portugal in 1591, and it was a regular article of trade thereafter. It was used by the Malays to heal sores on the feet and the wound of circumcision, and to relieve ringworm, shingles and other skin afflictions. A little was sometimes added to cigarettes.[98]

Diluted with water, the alcoholic solution of benzoin becomes milky and, under the name "virgin's milk" has been used to heal cracked nipples and has also been employed in feminine hygiene.[441]

Benzoin is an important ingredient in compound benzoin tincture (friar's balsam or Turlington's balsam), each 100 ml of which contains an alcohol extract from 10 g benzoin, 8 g storax, 4 g balsam of Tolu and 2 g aloe. This tincture is commonly employed as an antiseptic and protective coating on abraded, cracked and friction-blistered skin, on facial cold sores and indolent ulcers; it is also painted on areas of the body about to receive heavy-duty adhesive tape to hold dressings in place.[423] Benzoin tincture formed the outer covering of the long-popular adhesive Court Plaster.[305]

Internally, benzoin is a carminative, expectorant and diuretic. It is sometimes given in cases of bronchitis or laryngitis, or is added to boiling water and its vapor inhaled.[423]

Toxicity:

In some individuals, benzoin produces contact dermatitis.[423] Benzoin may cross react with balsam of Peru, benzyl cinnamate, benzyl alcohol, eugenol, vanilla and alpha-pinene.[190] When heated, the resin gives off a white vapor which causes coughing. Those who pulverize benzoin resin may be afflicted with sneezing.[305]

Other uses:

In food manufacturing, benzoin is used as a gloss on chocolate eggs and is added to sirups to render them turbid or opaque. It enters into soft drinks, ice cream, candy, baked goods, gelatins and puddings, and chewing gum.[210]

Much benzoin is burned for incense in churches.[605] Benzoin tincture and resinoids serve as fixatives for fine perfumes, colognes and for the scents in lotions, soaps and other cosmetic and toilet products. It is valued as an antioxidant in fatty preparations.[41]

The wood of spent *Styrax* trees is soft and of little use except for clogs, matchsticks, matchboxes and barrels. The seeds of the trees are eaten by animals.[98]

Caricaceae

Papaya

Botanical name:

    Carica papaya L.

Other names:

    Papaw (better limited to *Asimina triloba* Dunal), tree melon and, in Spanish, *lechosa* or *fruta bomba*.

Family:

    Caricaceae, the Papaya Family.

Description:

    A perennial, herbaceous plant, with copious milky latex reaching to 20 or 30 ft. (6 to 10 m), the stem to 10 in. (25 cm) thick, simple or branched above the middle and roughened with leaf-scars. Leaves, clustered around the apex of the stem and branches, have nearly cylindrical stalks, green, purple-streaked or deep-purple, 10 to 40 in. (25 to 100 cm) long; the leaf blade has 7 to 11 main and some secondary, irregular, pointed lobes and prominent veins; leaf surface is yellow-green to dark-green above, paler beneath. Usually male and female flowers are borne on separate plants, but hermaphrodite (bisexual) flowers often occur, and a male plant may convert to a female after beheading. Flowers emerge singly or in clusters from the main stem among the lower leaves, the female short-stalked, the male with drooping peduncles 10 to 40 in. (25 to 100 cm) long. Corolla is ½ to 1 in. (1.25 to 2.5 cm) long, with 5 oblong, recurved white petals. Fruit is extremely variable in form and size; it may be nearly round, pear-shaped, oval or oblong; that of wild plants may be as small as a hen's egg, while, in cultivation, the fruit ranges from 5 in. (12.5 cm) to 2 ft. (60 cm) in length and up to 8 in. (20 cm) thick. Skin is smooth, relatively thin, deep-yellow to orange when the fruit is ripe. Flesh is succulent, very juicy, yellow to orange or salmon-red, sweet and more or less musky. The central cavity is lined with a dryish, pulpy membrane to which

---

Figure 61. Papaya latex, the source of proteolytic enzymes, oozes from shallow incisions in the unripe, green fruits. Courtesy Doctor Everette M. Burdick, Consulting Chemist, Coral Gables, Florida.

adhere numerous black, rough, peppery seeds, each with a glistening, transparent, gelatinous coating.[415, 550]

Origin and Distribution:

It is believed that the papaya originated in southern Mexico and Central America, though it was cultivated as far south as Lima, Peru, in pre-Spanish times. Today, the papaya is grown in all tropical countries and in subtropical southern Florida as a dooryard and commercial crop despite virus diseases which seriously affect the growth of the industry. Major producing regions include Queensland, the Philippines, East Africa, South Africa, Ceylon, India, Hawaii, Colombia, Venezuela and Puerto Rico. In most growing areas, the papaya is mainly or solely valued for its edible fruit, as it is in Hawaii. Only in regions where low-cost labor and land are available is it possible to exploit the papaya for medicinal purposes.

Ceylon formerly was the chief source of papain.[560] Most of the world supply now comes from East Africa.[113] Production rose and then declined in the Congo; on a lesser scale, it is carried on in India, the Union of South Africa, Mozambique, Ponape and Samoa.[245]

Products and Characteristics:

The latex which abounds in all parts of the plant and the green fruit is bitter and astringent and has a specific gravity of 1.023.[605] It contains two potent crystalline enzymes, papain and chymopapain.[355] Chymopapain is present in greatest abundance, but papain has twice the proteolytic activity. The molecular weight of crystalline papain has been reported as 20,583[324] and as 27,000.[337] Its active fragment has been determined as having a molecular weight of 8375.[324] Papain is partially soluble in water and glycerol, virtually insoluble in alcohol, chloroform, ether and most organic solvents, and slightly hydroscopic.[37] Papain in solution is inactivated by exposure to 82.5° C. for one-half hour. In dry form, papain remains stable for three hours at 100° C. Optimal activity occurs at 70° C., but it is still active at 10° C. Papain retains its potency over a wide range of pH in the reaction media.[1] However, its activity is destroyed by prolonged storaged, certain oxidizing agents, high acidity or alkalinity, and certain wavelengths of ultra-violet light, as well as by intense heat. Crystalline chymopapain is more stable in acid, and its activity persists for several weeks at pH 2 and 10° C.[337] Crystalline papain contains $15.5 \pm 0.1\%$ nitrogen. Tyrosine and tryptophane have been found in the intact molecule of papain in a ratio of 4:1.[273]

Meer's Purified Papain, a white to yellow-white powder, contains papain and chymopapain, and small amounts of amylase, lipase, pectase and other enzymes.[382] Commercial papain is capable of digesting thirty-five times its weight of lean meat.[37] Papain clots milk and has a proven clotting activity on fibrinogen, being a more complete hemostat than thrombin.[346] An anticoagulant, AC48, isolated from papaya latex, injected intravenously, prolongs clotting and prothrombin time. Large doses cause excessive dilatation and fragility of capillaries and hemorrhage. Intramuscularly, AC48 induces painful swelling and hematoma.[106] Papain cleaves normal and pathological human γ-globulins into several chromatographically and electrophoretically separable 3.5S fragments without the detectable formation of intermediates.[270]

Papaya seeds, air-dried, yield 660 to 760 mg/100 g of BITC (the aglycone of glucotropaeolin benzyl isothiocyanate) with bactericidal and bacteriostatic activity.[177]

Propagation, Cultivation and Harvesting:

Only cultivars known to be rich in papain and of long-oval or oblong form are suitable for planting for commercial exploitation of this product. Experimental work has shown that female plants of Red Panama have the highest yield, followed by Floride and next the hermaphrodite types of Red Panama.[245]

The seeds on removal from the fruit are washed to remove the gelatinous coating, then dried and dusted with a fungicide to prevent "damping-off" of seedlings. Seeds may be planted immediately but, if kept in a cool, dry place will remain viable for several months. Planting may be done at any time of year. If done in the fall, growth will stop during the winter and resume in the spring, and the flowers and fruit will be borne close to the ground.[101]

Seeds are best planted in unshaded, regularly watered nursery beds about ⅜ in. (1 cm) deep and 6 in. (15 cm) apart. Germination occurs within three to four weeks. Three months after planting, the seedlings should be 6 to 8 in. (15 to 20 cm) high. They are then beheaded to a height of 4 in. (10 cm) and transplanted to a cleared plot. Except in soils of high humus content, holes 1 x 1 x 1 ft. (30 x 30 x 30 cm) are prepared at a distance of, ideally, 8 to 9½ ft. (240 to 300 cm) each way between holes. It is customary to place 5 to 7 seedlings in each hole, a practice requiring approximately 6,000 seedlings per acre. If seeding is done directly in the field, 10 to 20 pre-soaked seeds are planted at each site. Blooming begins within five to seven months. At this time, most plants with male flowers are cut down, leaving only one male to pollinate every 10 to 20 female plants. (In some regions, the male population is lower). Thinning of female seedlings is done by first cutting down all but the first three females to bloom and then, four weeks later, destroying the two weakest. Cutting unwanted plants off at ground level is preferred to pulling out, as it avoids root disturbance.[101] The cut plants will soon die. Sometimes interplanting is begun twenty months after establishing the plantation so that the original plants can be destroyed in two to two and one-half years when they pass their productive peak or become too tall for practical purposes.[67]

The papaya plant is a heavy feeder and flourishes with a good supply of organic fertilizer and a thick mulch. In papain-producing regions, there are as yet no uncontrollable diseases or serious insect pests such as the papaya fruit fly of the Caribbean area.[531] Plants evidencing mosaic disease are quickly eliminated from the plantation. The various fungus, virus, nematode and other problems which plague many papaya-fruit growers[630] need not be outlined here.

Latex is harvested in the morning or throughout the day in overcast weather. The lowest, most mature fruits on each female plant are tapped at the age of about two and one-half months,[531] when full-grown but still entirely green. The upper fruits are tapped later as they attain full size. Some judicious trimming of the plants is done to clear the route of the tapper, who, using a knife of bone, glass or bamboo[95, 561] or a razor blade mounted in a wood-handled rubber or cork holder,[113, 530] makes up to 4 vertical, shallow incisions—no more than $\frac{1}{10}$ to $\frac{1}{8}$ in. (2.5 to 3 mm) deep—in each suitable fruit, catching the flowing latex on a tray[67] or a canvas[26] or plastic inverted "umbrella"[531] affixed to the stem of the plant. He scrapes the latex into a collecting box and carries his tray or "umbrella" to another plant. The final drops of latex which have nearly coagulated on the fruit are wiped off into a cup by a helper who follows fifteen minutes behind the tapper and cleans the incisions with a damp sponge.[67]

The same fruits will be retapped twelve to fourteen times[531] at intervals of five to eight days until the latex flow is greatly reduced or the fruits have begun to turn yellow.[67]

Tapped fruits are harvested for local human consumption and for pectin extraction,[29] any surplus being utilized for stockfeed. In India, it is believed that the seeds of tapped fruits should not be used for replanting, as they will produce weak plants.[405]

The collected latex has been dried to a paste and blended with salt as a preservative or, in the past, has been more commonly sun-dried.[561] Oven-drying is an improved method in general practice today. The latex, spread on trays and pressed to a uniform thickness, is dried down to a moisture content of 9 to 10%. It is then coarsely ground and vacuum-packed in four-gallon, paper-lined cans.[67] This material is exported as crude papain, ranging in color from ivory-white to orange.[337] India, in recent years, has adopted the practice of stirring sodium bisulphite into the latex at the rate of .6 lb. to each 100 lbs. immediately after collection. The latex is then delivered to a laboratory for alcohol-and-acetone separa-

Figure 62. Ripe fruits of the papaya (*Carica papaya* L.) in longitudinal and cross section showing thick melon-like flesh and seed-lined cavity. Photographed by Kendal and Julia Morton in Nassau, Bahamas. (*See also* color illustration 10.)

tion of the solids, which are then oven-dried for one hour at 50° C. and packed in polyethylene-lined, air-tight metal drums.[532]

Yields of fresh latex range from 80 to 175 lbs. (36.2 to 74.8 kg) per acre per year. Five lbs. (2.2 kg) of fresh latex becomes 1 lb. (.453 kg or 453 g) of crude papain.[337] It must be noted that the latex flow varies with the age of the plant. Haendler and Huet show the fluctuation over a five-year period as follows: first year, 20 to 25 kg dried papain per hectare; second year, 90 to 100 kg; third year, 60 to 90 kg; fourth year, 30 to 40 kg; fifth year, 20 kg, more or less.[245]

Techniques have been developed for extracting papain from the leaves and stems of the plant and, as such processes lend themselves to mechanization, they may be substituted in the future for the manual labor of fruit tapping. At the University of London, a new technique, covalent chromatography, has produced fully active papain from dried papaya latex and from commercial 2x crystallized partially active papain.[92]

Medicinal uses:

In tropical countries, the latex has many applications in folk medicine. It is used as a styptic and vermifuge, an anti-chigger application and a remedy for freckles, warts, corns, calluses, eczema, ringworm, infected wounds, malignant tumors, bleeding hemorrhoids, tubercles and the pain of burns.[98, 148, 275, 461] An infusion of the latex and honey is given to Asiatic children to expel roundworms. Small doses of the latex and sugar are taken as a digestive, stomachic and emmenagogue, as a treatment for enlargement of the liver and spleen, and for whooping cough. In diphtheria, the latex in a glycerine solution is applied repeatedly to the pharynx.[461] In India and Malaya, the latex is smeared on the mouth of the uterus to induce abortion.[98]

Annual imports of crude papain into the United States amount to nearly 1½ million dollars.[381] As a pharmaceutical product, papain was first employed in this country as a digestive to relieve dyspepsia and later was administered in sirup to cure chronic diarrhea in children.[273] Various trade preparations have been on the market since 1880, some of very low activity. Tablets compounded of dehydrated papaya pulp and leaves, papain and amylase are currently sold by "natural food" purveyors as an aid to digestion. A papain-containing nasal spray is sold for relief of some pollen-caused allergies.[337]

Papain has been injected into the peritoneal cavity to prevent adhesions after operations but with no positive proof of benefit. Some practitioners have claimed success in utilizing papain to overcome meat impaction in the gullet, though, in a number of cases, death has resulted from the destructive effect of papain on the esophagus. Most recent is the use of chymopapain by neurosurgeons and orthopedic surgeons for spinal injection to shrink ruptured or slipped discs and overcome the pain of pinched nerves. There are reports of deleterious effects on the cartilage and of allergic reactions. Papain has also been employed to reduce edema after operations for the removal of tumors and cysts from the head and neck areas [352] and also applied to wounds and to surgical incisions to promote healing and prevent sloughing in infected wounds.[37] It is claimed to lessen fever and dysphagia after tonsillectomies and adenoidectomies.

In veterinary practice, it has been demonstrated to be a safe and effective anthelmintic, especially in treating weak or pregnant dogs.[37]

Toxicity:

The fresh latex is acrid and will cause severe eye inflammation;[37] it can provoke irritation and blisters if allowed to remain in contact with the skin. Papaya harvesters wear gloves and aprons or coveralls to avoid dermatitis, and some wear sunglasses to protect the eyes. Under rings and bracelets, the latex will digest the tissues and cause sores. Internally, it is a severe gastric irritant and has been employed in malicious poisoning.[605]

Some people are acutely allergic to any part of the papaya plant, its pollen, fruit and latex.[523] Particularly sensitive persons react to meat tenderized by papain and to papain administered in any form or manner as medication. Pharmacists may experience rhinitis, asthma and other allergic reactions from handling papain preparations. Some people have reacted when exposed to papain-containing tooth powders.[273]

Other uses:

The ripe papaya is eaten raw, cooked or preserved; the green fruit is cooked as a vegetable or prepared as a sweetmeat. Young papaya leaves are cooked and eaten as spinach. In the tropics, tough meat is tenderized by washing it with the diluted latex or wrapping it for a few hours in the bruised leaves, or by cooking the meat with papaya leaves or with the green fruit. These practices gave rise to the development of commercial preparations of papain for tenderizing meat before cooking. In recent years, papain has been injected into the jugular vein of cattle just one-half hour before slaughter to tenderize meat more thoroughly and to make it possible to utilize a greater percentage of the animal. The liver, kidneys and tongue of a tenderized animal break down quickly on cooking and cannot be sold in their natural form but only in processed products.

Papaya latex is an ingredient in chewing gum and confections.[597] It is extensively utilized in cheese-making and also to prevent chill-haze in beer and ale, though a recoverable nylon powder is now replacing papain to some degree in the brewing industry.[113] Papain facilitates the extraction of oil from tuna liver; it is used to soften silk cocoons and to shrinkproof and improve the texture of silk and wool before dyeing. It also serves to dehair hides in the process of tanning[537] and modify *Hevea* latex in the manufacture of rubber.[419]

Papain is an ingredient not only in some dentifrices but also in cleansing creams and is incorporated into "face-lifting" preparations which peel the outer layer of skin, removing blemishes and superficial wrinkles. Papain added to detergents to improve laundering activity is a source of dermatitis, and such use is being discouraged.

Related Product:

The proteolytic enzyme ficin is obtained from the gummy latex of the tree, *Ficus glabrata* H. B. K., of the family Moraceae, which grows wild at low elevations from southern Mexico to Panama and from Colombia to Brazil and Peru. The fresh latex is locally valued as a vermifuge.[549] The commercial concentrate is ten to twenty times faster than papain in clotting milk, four to ten times as active as papain in digestion of gelatin, lean meat, etc.[254] It has many industrial uses similar to those of papain and is combined with papain in certain meat-tenderizing preparations. Because of its rapid agglutination of human blood cells, it is valued as an agent in determining the Rh factor. It is hazardous to handle, irritating the skin and eyes, and is drastically purgative if ingested in other than minute doses.[552] *See also* bromelain, under PINEAPPLE.

Apocynaceae

Arrow-Poison Tree

Botanical name:

    Acokanthera schimperi Benth. (*A. friesiorum* Markgr.)

Other names:

    Murichu, kibai, keliot, mururu, argassi.

Family:

    Apocynaceae, the Dogbane Family.

Description:

    A small, short-trunked tree, generally 15 to 25 ft. (4.5 to 7.6 m) high, sometimes to 35 or 40 ft. (10.6 to 12.1 m), with dense, spreading crown. Leaves evergreen, opposite, elliptic to obovate, 1 to 2½ in. (2.5 to 6.3 cm) long, ½ to 1¾ in. (1.25 to 4.3 cm) wide; dark-green, thick, glossy above, dull on the underside, sometimes slightly hairy, short-petioled. Flowers sweetly fragrant, white, often blushed with pink or purple tubular, five-lobed, usually somewhat hairy, ½ in. (1.25 cm) long, in small clusters in the leaf axils. Fruit ovoid, ½ to 1¼ in. (1.25 to 3.2 cm) long, red or dark-purple when ripe; flesh bitter but edible; seed kernel "rubbery," inedible.[64, 144, 284, 595]

Origin and Distribution:

    Native to East Africa and common on dry soil, in thickets, scrub and grasslands, at elevations between 4,000 and 7,000 ft. (1,219 and 2,134 m). Several efforts at introduction into Malaya have been unsuccessful.[98, 233, 595]

Constituents:

    All parts except the fruit flesh contain the cardiac glycosides ouabain (also called G-strophanthin or acocantherin) ($C_{29}H_{44}O_{12}$) and

---

Figure 63. Arrow poison tree (*Acokanthera schimperi* Benth.), photographed by James Kahurananga, East African Herbarium, Nairobi, Kenya, who estimated this tree to be about 16.5 ft. (5 m high). (*See also* color illustration 16.)

acocanthin ($C_{29}H_{44}O_{12}$).[552, 595, 605] The wood is richest in ouabain and is the part from which this drug is commercially derived.[557]

The seeds contain 1.7% acovenoside A, 0.1% ouabain, .0037% acolongifloroside G, .098% acolongifloroside H, and .024% acolongifloroside K acetate.[605]

Medicinal uses:

See use of ouabain under SERPENT-WOOD, *Strophanthus gratus*.

Toxicity:

A pitch-like extract from prolonged boiling of the chipped wood,[557] or strong decoctions of the leaves, twigs or root bark, have all been employed as arrow poison, which is lethal within a few minutes. The young leaves, branches, bark, flowers and roots are more toxic than mature growth.[595] The plant is most toxic in dry seasons and has been ascertained to be less toxic in rainy periods.[605] Ingestion of any amount of the dried leaves in excess of 15 gr may be seriously toxic to humans.[595] Inhalation of the powdered plant material may cause fatalities, and the poison is also absorbed through the skin. A small quantity of the plant extract placed in the ear of a sheep will bring about death in two hours.[605] The tree is a well-known hazard to grazing animals. Signs of poisoning in cattle include rapid, shallow respiration, diarrhea, muscular spasms, grinding of teeth and salivation; death usually occurs quickly from heart failure.[595]

Other uses:

This tree, because of its compactness and heavy shade, is cultivated as a dooryard ornamental in East Africa.[284] The fruit is made into jam.[557]

Figure 64. The periwinkle (*Catharanthus roseus* G. Don) flourishes and flowers all year in semi-tropical and tropical climates. It is a common ornamental and escape from cultivation in southern Florida. (*See also* color illustration 11.)

Periwinkle

Botanical name:

Catharanthus roseus G. Don (*Lochnera rosea* Reichb., *Vinca rosea* L., *Ammocallis rosea* Small).

Other names:

Red periwinkle, Madagascar periwinkle, old maid, church-flower, ram-goat rose, "myrtle," magdalena.

Family:

Apocynaceae, the Dogbane Family.

Description:

An erect, perennial herb to 2½ ft. (75 cm) high, becoming sub-woody at the base and profusely branched, the stems containing some milky latex; leaves are opposite, smooth, oblong-oval, blunt or rounded at the apex, 1 to 3½ in. (2.5 to 9 cm) long and ⅝ to 1½ in. (1.5 to 4 cm) wide, short-petioled. Flowers, borne all year in upper leaf axils, are tubular, ⅝ to 1½ in. (1.5 to 4 cm) long, five-lobed, flaring to a width of 2 in. (5 cm); color may be white with greenish-yellow eye (var. *albus*) white with a purple-red eye (var. *ocellatus*, or forma *oculata*) or lavender-pink with a purple-red eye (var. *roseus*, or forma *violacea*). Seedpod (*follicle*), borne in pairs, is cylindric, ribbed, downy, ⅝ to 1⅜ in. (1½ to 3.5 cm) long. Seeds, 2, black, oblong, .08 in. (2 mm) long, .04 in. (1 mm) wide, wrinkled and pitted, are scattered as the pod splits open.[3, 153, 140] There are about 350,000 seeds in 1 lb. (.453 kg).[122]

Origin and Distribution:

The periwinkle is believed to be native to the West Indies but was originally described from Madagascar. It was introduced into Europe in 1757 and reached India before 1794.[98] It is cultivated as an ornamental plant almost throughout the tropical and subtropical world, is abundantly naturalized in many regions, particularly in arid coastal locations, and occurs up to an elevation of 1,640 ft. (500 m).[3, 95, 153] In 1965, all seed for planting for pharmaceutical purposes was being imported from Japan. Commercial plantations have been established in tropical Africa, India, Australia,[584] Hungary and other areas of the Old World.

Constituents:

The plant is composed of 30% leaf, 15% root, 55% stem. All parts contain alkaloids. Seventy-three named alkaloids are listed by Willaman and Hui-lin Li.[616] Roots collected in India yielded 1.18 to 1.22% total alkaloids; 0.0195 to 0.029% reserpine ($C_{33}H_{40}N_2O_9$).[409] The root bark contains 0.01% alstonine ($C_{21}H_{20}N_2O_3$); it is said to have twice the hypotensive activity of *Rauvolfia serpentina* and to be somewhat tranquilizing.[445] The alkaloid called ajmalicine, vinceine or vincaine ($C_{21}H_{24}N_2O_3$), found mainly in the root, is believed to be identical to δ-yohimbine.[552]

Two alkaloids, vincaleukoblastine or vinblastine ($C_{46}H_{58}N_4O_9$), the sulfate of which is marketed as Velban®, and leurocristine or vincristine ($C_{46}H_{56}N_4O_{10}$), the sulfate of which is marketed as Vincovin® or Oncovin® —indole and dihydroindole in character—have been employed for the past sixteen years in cancer chemotherapy.

The leaves contain, in addition to alkaloids, a volatile oil containing aldehyde, sequiterpenes, furfural, sulfurous compounds, an alcohol called lochnerol, $C_{24}H_{35}O_2(OH)$ ½H_2O,[461] several monoterpene glycosides including adenosine and roseoside ($C_{19}H_{30}O_8$), as well as deoxyloganin and loganin.[78, 79] A thorough review of "The Phytochemistry and Pharmacology of *Catharanthus roseus* (L.) G. Don" is presented by G. H. Svoboda and D. A. Blake, pp. 45-83, in the book, *The Catharanthus Alkaloids*.[570]

Propagation, Cultivation and Harvesting:

The periwinkle grows readily from seeds and cuttings. It becomes straggly after two years and is improved by cutting back occasionally to promote new growth and flowering.[275] The first large-scale plantings in the United States for research and pharmaceutical purposes were at Coudersport, Pennsylvania, and at the University of Florida in Gainesville. In these areas, the plant dies back in the winter and must be replanted annually. For a two-acre plot in Gainesville in 1964, Doctor Carl H. Johnson, Professor of Pharmacy, planted some seeds in flats in March in order to provide frost protection and, later in the spring, sowed more seeds directly in the ground. When the seedlings were 2 to 3 in. (5 to 7.6 cm) high, they were transplanted to the plot and spaced 1 ft. (30 cm) apart in rows 36 in. (.91 m) apart. They were watered when set in the field but thereafter received no water nor fertilizer. Heavy rains in July, August and September caused much yellowing and shedding of leaves, but the plants continued to bloom and set much seed. At the end of September, the plants were taken up and the roots and upper parts separated and air-dried for pharmacognostic study at the University of Illinois. The seeds were planted immediately, and by the end of November, the new crop of plants had attained sufficient size to harvest. Both plantings were unmolested by insects.[285]

In South Florida, the periwinkle, planted and allowed to self-multiply solely as an ornamental, is attacked by the ambrosia beetle which bores the stems. The larvae create tunnels that become inhabited by fungi, causing the plant to die.[131]

Colchicine-induced tetraploids are vigorous growers with larger than normal flowers.[279]

The vincaleukoblastine content is highest in young plants. From 220 lbs. (100 kg) dried leaves, Javanovics *et al.* extracted 700 g of combined "vinblastine, vinleurosine, and vincristine."[281] The yield of leurocristine is about .0002% of the dried plant.[584] Approximately 1,100 lbs. (500 kg) of plant material are required to gain 1 g of this alkaloid.[122] Various techniques have been reported for the separation of the alkaloids.[48, 120, 282]

Medicinal uses:

The plant has many uses in tropical folk medicine. In Cuba, the flowers are sold by herb vendors, and a decoction of a few flowers with a few drops of alcohol is much used as an eyewash for infants.[485] Puerto Ricans use the decoction of the white, red-centered flowers to refresh tired, inflamed eyes. In several of the West Indian islands, an infusion of the flowers of all three varieties is regarded by laymen as a very effective treatment for diabetes.[554] The bitter and astringent leaves are emetic and diaphoretic.[441, 545] In Central America and Colombia, the plant infusion is gargled to relieve sorethroat, laryngitis and chest complaints.[440, 546] Juice squeezed from the leaves is applied to wasp stings in India.[461] Hawaiians take the plant decoction internally to arrest hemorrhages.[42] In southeastern Asia it is given as a depurative after childbirth, also to halt lactation, and taken as a cough remedy and employed for washing the head.[441]

The leaf decoction has long been esteemed in South Africa and the Philippines as antidiabetic. A patent medicine, "Vinculin," was sold for many years in Great Britain as a cure for diabetes. A South African product called "Covinca" was investigated in 1926 and found to have no effect on the blood-sugar level of fasting rabbits. It did show a slight digitalis-like action. Similar findings resulted from a study of a leaf tincture in 1928. A leaf extract tested in fifteen diabetic patients in 1930 was judged of no hypoglycemic value but declared to be an ideal purgative in chronic constipation.[461] Injections of a concentrated aqueous extract lowered blood-sugar level in cats and tended to abate symptoms in a group of human diabetics.[605]

In Mauritius, the leaf infusion is taken for indigestion and dyspepsia.[461] Residents of the Eastern Transvaal believe that it alleviates rheumatism.[605]

Herb vendors in China and North Vietnam recommend a decoction of the stems and leaves for regulating menstruation. It is considered beneficial also to the kidneys. If the urine is scanty and reddish, the decoction clarifies it and promotes an abundant flow. In the region of Cap St.-Jacques, South Vietnam, the decoction is taken as a remedy for malaria.[441]

The roots are regarded as purgative, vermifugal, depurative and hemostatic, and are used to relieve toothache.[545] In the Philippines, the root is valued as an emmenagogue and is reputedly abortifacient.[95] Mauritians employ the root to control dysentery.[461] Vietnamese choose the roots of the white-flowered variety to treat malaria.[441]

It was the local reputation of "periwinkle tea" as a diabetes remedy in Jamaica that inspired simultaneous investigations by scientists at the Lilly Research Laboratories and a team at the University of Western Ontario. Neither group observed antidiabetic action, but both noted leukopenia in their experimental animals. In 1957, extracts were tested against leukemia in mice and found effective. The American and Canadian research groups joined together in the development of vincaleukoblastine sulfate, marketed as Velban by Eli Lilly, and this was the beginning of nearly universal study of and experimentation with *Catharanthus* alkaloids for cancer chemotherapy.

Vincaleukoblastine is administered mainly in the treatment of Hodgkin's disease and choriocarcinoma. Leurocristine is employed mainly in childhood leukemia and breast cancer. Vincaleukoblastine and leurocristine, combined and injected weekly, produced beneficial results in 43 percent of a group of patients with malignant lymphoma.[563]

A combination of leurocristine and the synthetic drugs prednisone and methotrexate has been more successful in achieving regression of adult leukemia than leurocristine alone. Dramatic remission of leukemia in children has been attained with leurocristine, prenisolone, 6-mercaptopurine and methotrexate administered together over thirteen months. More recently, leurocristine, prednisone and the antibiotic daunomycin have given encouraging results.[598] Many other combinations with periwinkle alkaloids have been tried with varying degrees of success in multiple-drug cancer therapy.[344, 568, 569]

Vincaleukoblastine has been prepared as an ointment for treatment of psoriasis.[287] There is extensive literature on the current applications of these drugs, and much of it is cited by DeConti and Creasey in their chapter, "Clinical Aspects of the Dimeric Catharanthus Alkaloids," in *The Catharanthus Alkaloids.*"[152, 577]

Toxicity:

The periwinkle plant is reported to have caused poisoning in grazing animals in India and New South Wales.[272]

Frequent or prolonged administration of small doses of Catharanthus alkaloids in cancer therapy has caused platelet damage.[633] Patients may exhibit transient side-effects such as various degrees of leukopenia[562] (more likely to occur with vincaleukoblastine than with leurocristine), hair loss, nausea, vomiting, abdominal cramps, phlebitis, dermatitis, loss of deep-tendon reflexes, temporary mental depression, headache, nosebleed, loss of appetite, stomatitis, diarrhea and, rarely, numbness of fingers and toes and bleeding of old peptic ulcers. In experimental animals[105] and in humans, excessive dosage over long periods has resulted in convulsions, damage to the central nervous system[387] and sometimes fatality. Neurological effects may continue for several months following cessation of chemotherapy.[339, 423] Leurocristine sulfate, given intravenously, has been observed to inflame and retard healing of wounds or lesions in leukemia patients.[198]

Other uses:

The periwinkle is widely appreciated as a garden plant and cut flower. There is no record of its having been employed elsewhere as a fumatory, but in 1967, health authorities in Bradenton, Florida, found that juveniles were drying and smoking periwinkle leaves as a fancied substitute for marihuana.[72]

Figure 65. Serpent-wood (*Rauvolfia serpentina* Benth. ex Kurz) in bloom. Photographed in the Morris Arboretum of the University of Pennsylvania, Philadelphia.

Serpent-Wood

Botanical name:

    Rauvolfia (or *Rauwolfia*) *serpentina* Benth. ex Kurz.

Other names:

    Serpentine root, serpentina root, rauvolfia root.

Family:

    Apocynaceae, the Dogbane Family.

Description:

    An erect subshrub, generally ½ to 1½ ft. (15 to 45 cm) high, seldom as much as 3 ft. (90 cm), having a nearly vertical, tapering taproot becoming as much as 6 in. (15 cm) thick at the crown and up to 20 in. (50 cm) long, occasionally branched or tortuous, and developing several smaller, fibrous side roots; grayish-yellow externally, pale-yellow within; acrid in odor when fresh, odorless when dried; very bitter. The plant stem when broken exudes a pale, sticky sap. Leaves, borne in whorls of 3 or 4, are deciduous, elliptic-lanceolate or obovate, pointed, green on the upper surface, pale-green on the underside, 3 to 8 in. (7.5 to 20 cm) long. Flowers, numerous, in terminal or axillary, long-stalked clusters, are tubular, five-lobed, ⅜ to 1⅛ in. (1 to 3 cm) long, the lobes white, the tube pink; calyx is yellowish at first; at the onset of fruit-setting, the calyces, pedicels and flowering stalk become bright-red. Fruit, usually in pairs united for half their length; obliquely ovoid, 5/16 in. (7.5 mm) long, purple-black and glossy when ripe, scantily fleshy, with a hard stone containing 1 or 2 seeds.[20, 165, 241, 425]

Origin and Distribution:

    The plant is native to moist deciduous forests (often associated with bamboo) from sea level to 4,000 ft. (1,200 m) in northern India, East Pakistan, Burma, Thailand, Ceylon, Malaya, the Andaman Islands and Indonesia. It is of recent cultivation in East Pakistan, India and the Philippines.[425] Trial plantings in Puerto Rico were abandoned after one year; a Squibb plantation was initiated in Mexico in 1955 and discontinued in 1960.[135]

Constituents:

The stem and leaves have a low alkaloid content. Alkaloids abound primarily in the root bark which constitutes 40 to 56% of the whole root and contains 90% of the total alkaloids. The fibrous roots are also more active than the interior of the main taproot.[165, 241] The following alkaloids have been reported: ajmalicine, or δ-yohimbine ($C_{21}H_{24}N_2O_3$); ajmaline, also called rauwolfine ($C_{20}H_{26}N_2O_2$); ajmalinine ($C_{20}H_{26}N_2O_3$); alkaloids A, C and F; alloyohimbine ($C_{21}H_{26}N_2O_3$); chandrine ($C_{25}H_{30}N_2O_8$); deserpidine ($C_{32}H_{38}N_2O_8$; isoajmaline ($C_{20}H_{26}N_2O_2$); yohimbine ($C_{21}H_{26}N_2O_3$); γ-yohimbine; isoyohimbine; 11-methoxy-δ-yohimbine ($C_{22}H_{26}N_2O_4$); methyl reserpate; neoajmaline ($C_{20}H_{26}N_2O_2$); papaverine ($C_{20}H_{21}NO_4$); corynanthine, also called rauhimbine, isorauhimbine and 3-epi-α-yohimbine ($C_{21}H_{26}N_2O_3$); raunatine; rauwolfinine ($C_{19}H_{26}N_2O_2$); rauwolscine, also called α-yohimbine ($C_{21}H_{26}O_2N_3$); reserpiline ($C_{23}H_{28}N_2O_5$); reserpine ($C_{33}H_{40}N_2O_9$); rescinnamine ($C_{35}H_{42}N_2O_9$); reserpinine ($C_{22}H_{26}N_2O_4$); reserpoxidine ($C_{33}H_{40}N_2O_{10}$); sarpagine, also called raupine ($C_{19}H_{22}N_2O_2$), the only phenolic alkaloid in this species; serpine ($C_{21}H_{26}N_2O_3$); serpinine, identical to tetraphyllicine ($C_{20}H_{24}N_2O$); serpentine ($C_{21}H_{20}N_2O_3$); serpentinine ($C_{42}H_{44}N_4O_6$, H_2O); thebaine ($C_{19}H_{12}NO_3$); unnamed alkaloids I and II.[108, 187, 427, 616, 617]

Total alkaloid content, ranging from 1.49 to 2.38%, is apt to vary with location,[135] season and other factors. Pharmacologically, the alkaloids are placed in two main divisions, the reserpine group (tertiary indoles), which release the sympathomimetic amines, and the ajmaline group (tertiary indoline alkaloids).[108]

Reserpine (3,4,5-trimethoxybenzoyl methyl reserpate), isolated and identified at CIBA Pharmaceuticals in Switzerland in 1952, is the main active alkaloid and may represent .05 to .20% of the dried root. According to the British Pharmaceutical Codex, the reserpine content must be at least .15%. Reserpine reduces the amounts of serotinin (5-hydroxy tryptamine) and catechol amines in the brain and affects the concentrations of glycogen, acetylcholine, γ-aminobutyric acid, nucleic acids, antidiuretic hormone and substance P. It inhibits respiration, stimulates peristalsis and myosis, relaxes nictating membranes and affects temperature regulation; it also promotes and acidifies gastric secretions and sometimes stimulates prothrombin activity. In experimental animals, it interrupts vaginal cycles, inhibits ovulation and induces false pregnancy, inhibits androgenic secretions of gonads in males and decreases compensatory hypertrophy of testes. It is bacteriostatic and lessens skin homograft rejection in laboratory rats.

Rescinnamine, 3,4,5-trimethoxycinnamic acid methyl reserpate, isolated in 1954, is second in therapeutic importance as a hypotensive drug from this source.[584]

Deserpidine, 11-desmethoxyreserpine, is nearly as potent as reserpine as a hypotensive and sedative. Reserpinine has only slight hypotensive action. Serpentine induces hypotension and inhibits intestinal activity. Reserpine and serpentine together are more hypotensive than either alkaloid alone. Serpentinine is mildly hypotensive and purgative. Ajmaline, present in larger quantities than the other alkaloids, is not sedative, stimulates respiration and intestinal movements, and depresses cardiac tissues. Isoajmaline is hypotensive and causes drowsiness. Ajmalicine is a central nervous system depressant. Ajmalinine is hypotensive. Rauwolfinine is hypertensive. Raujemidine is half as tranquilizing as reserpine. Chandrine is antiarrhythmic. Rauwolscine is hypotensive, a cardiovascular depressant and hypnotic, and is comparatively toxic.[20]

Propagation, Cultivation and Harvesting:

Most of the roots in the pharmaceutical trade and derived from wild plants, and many natural stands, especially in Java,[253] have been depleted. As a means of conservation, restrictions have been placed on root collection. In some regions of India, it is done only by forestry departments or by licensed private collectors. In Orissa, the collector is required to replant a portion of the root to assure a future supply. Commercial cultivation is being encouraged, and horticultural methods have received considerable attention in India.

In propagating experiments, stem cuttings have sprouted in three to four days but usually do not strike root until seventy-five days after planting unless treated with β-indolyl acetic acid which shortens the rooting period of hardwood cuttings to fifteen days.[20] Experimenters in Mexico found that tip cuttings rooted most easily, and two-node cuttings were only slightly less satisfactory.[135] At the University of Kerala, 78 to 92 percent success was obtained in rooting single-leaf cuttings treated with 3-indole acetic acid,[219] but workers in East Pakistan declare that leaf cuttings "have no propagative value."[253]

Root cuttings ½ in. (1.25 cm) in width put out roots and shoots in ten to fifteen days, and 220 lbs. (100 kg) of such cuttings are sufficient to plant a hectare. Unfortunately, the alkaloid content of roots raised by this method is lower than that of roots produced from seeds.[20] Also, planters are reluctant to use salable roots for propagation purposes.[241]

Generally, seeds have a poor rate of germination, varying with the time of year,[253] and many lack normal embryos, probably because of inadequate pollination, though the plant is apparently self-fertile.[312] Seeds for planting can be tested by flotation. Those that sink in a 10% brine solution will show the highest percentage of germination.[541] Viability is not of very long duration. Seeds planted immediately after removal from the fruits have a germination rate of 58 to 74 percent. If held two to three days, the rate may decline to 25 or even 10 percent, and few if any seeds will germinate after six months unless stored in airtight containers.[20] However, properly stored, some seed collections have displayed no loss of viability after eighteen months.[312] In the soil, seeds may be longer-lived, for growers maintain that volunteers from fallen fruits may spring up in plantations a year later.

In the middle of May (in India), seeds are soaked overnight and placed $3/16$ in. (5 mm) deep in well-tilled soil enriched with rooted manure and leaf mould. If germination is good, 12 lbs. (5.5 kg) of seeds will produce enough seedlings to plant a hectare. Growth is slow. In August, at the onset of the rainy season, the plants, which should be 3 to 4¾ in. (7.5 to 12 cm) high with 6 to 8 leaves,[253] are set out in holes 6 to 8 in. (15 to 20 cm) deep, in rows 2 ft. (60 cm) apart. R. serpentina in its natural habitat flourishes in the shade, but in cultivation, sunlight has been shown to promote strong plant growth and heavy roots with maximum alkaloid content.[592] Regular irrigation, weeding and manuring are required for optimum growth.[20] Spraying of young plants with gibberellic acid greatly stimulates plant development but lowers the alkaloid content of the roots.[299]

In brown clay soils, there is likely to be serious nematode infestation. Very dark soils are nematode-free.[592] Among diseases that attack the plant are *Cercospora* leaf spot and blotch, leaf browning and blight caused by *Alternaria*, virus mosaic, *Fusarium* wilt, powdery mildrew and target spot caused by *Corynespora*.[20] Sap-sucking insects attack the flowers and fruits. In the Philippines, aphids, mealy bugs and red ants are troublesome.[425] In Mexico, leaf-cutting ants defoliated some of the plants.[135] Caterpillars feed on the leaves, and further seedling damage is caused by grubs and rats, though cattle and other grazing stock leave the plants alone.[20]

Flowering begins within three to six months and may last eight to nine weeks, occur twice a year or continue all year in warm climates. In northern India, the plants are dormant during the winter. Fruits mature thirty to forty days after flowering. Since ripening of fruits is not uniform, collections must be made twice a week to obtain maximum seed harvest—approximately 220 lbs. (100 kg) for a three-year-old planting. Inasmuch as the fruit pulp inhibits germination, it is scrubbed off with old burlap sacks or other coarse material.

Root harvesting may take place when the plants are fifteen months old but may be delayed until the end of the second or third year in order to obtain greater root volume. The Indian Pharmacopoeia, in 1946, required that roots be gathered from three– to four-year-old plants, but studies have shown that there is little change in alkaloid content after fifteen months. In Taiwan, total alkaloid content was found to be 1.94 to 2.12% in roots harvested at sixteen months,[330] 2.12% in those taken up at eighteen months, and 1.94% at twenty and twenty-two months.[629] Reserpine-rescinnamine content ranged from 0.0603[629] to 0.0808%.[330]

The roots are taken up carefully in December, when the plants have shed their leaves and the alkaloid content is highest. The cleaned roots are air-dried, reducing moisture to 12 to 20%, then artificially dried to lower the moisture to 8% to improve keeping quality. Packing in airtight containers prevents mould and helps to conserve alkaloid content.

Seeded plantations may yield 2,600 lbs. (1,175 kg) of air-dried roots

66.   67.

Figure 66. *Rauvolfia serpentina* in fruit. Courtesy Doctor Rajendra Gupta, Project Co-ordinator (Medicinal and Aromatic Plants), National Bureau of Plant Introduction, I. A. R. I. Campus, New Delhi.

Figure 67. *Rauvolfia serpentina* roots. Courtesy Doctor Rajendra Gupta, Project Co-ordinator (Medicinal and Aromatic Plants), National Bureau of Plant Introduction, I. A. R. I. Campus, New Delhi.

per hectare. Root cuttings, in contrast, will produce about 760 lbs. (345 kg) per hectare and stem cuttings only 385 lbs. (175 kg).[20] Intercropping, even with relatively non-competing vegetables, tends to reduce yields.[241]

Most of the Indian production is exported, primarily to Japan, France and the United Kingdom. Former high exports to the United States have declined.[20]

Medicinal uses:

In India, the root has a 4,000-year history of use[277] in the treatment of snakebite, insect stings, nervous disorders, mania and epilepsy, and has been much used as a vermifuge and remedy for diarrhea, dysentery, cholera and fever, and to promote uterine contractions in childbirth.[117] In 1949, R. J. Vakil reported in the *British Heart Journal* that, as a result of ten years of testing various hypotensive drugs and canvassing fifty physicians throughout India, he was convinced that *R. serpentina* was the best hypotensive agent available.[391] Only since 1953 has it been adopted universally as a tranquilizer and treatment for mild hypertension, anxiety and schizophrenia. In severe hypertension, it is given with more potent drugs.[423] It has been considered helpful in dealing with menstrual tension and menopausal disturbances.[386]

The powdered root is administered as a tranquilizer in doses of 100 mg twice daily. The full effect may be delayed one to three weeks and may persist for a month after medication has been discontinued. In chronic cases, doses of 50 to 300 mg may be given daily.

Reserpine is prescribed in doses of .05 to .75 mg daily, orally or by injection, the higher dosage being safe only for the neuropsychotic patient under hospital care.[423] Synthetic production of reserpine has been achieved in France but is not economically practical so long as there is sufficient supply of the natural product.[122, 423]

Rescinnamine is given at the rate of 500 μg once or twice a day for two weeks, and thereafter 250 μg daily, and is believed to be somewhat safer than reserpine.[122] Ajmaline has been injected (in a single 50 mg dose) during diagnostic cardiac catheterization with no untoward effects but cannot be used in all patients.[503]

Toxicity:

The whole crude drug is hypnotic and purgative, and may cause impotence. Pure reserpine was formerly believed to be of low toxicity,[24, 386, 391] but experience has revealed serious side-effects and it is being administered with greater caution. It should not be given to people with bronchitis, asthma or gastric ulcers.[277] Even in small doses and in only two weeks, it may cause gastritis, aggravation and perforation of ulcers and hemorrhage.[386, 608] Other reactions include dizziness, nasal congestion, respiratory difficulty, skin irritation, joint and muscular pain, frequent bowel movements and increased weight.[386] In higher doses, reserpine may induce edema, cardiac depression, pseudoparkinsonism, insomnia, nightmares, despondency and suicidal moods. In toxic amounts, reserpine produces unconsciousness and death from respiratory paralysis.[117] Adverse reactions to rescinnamine are usually less frequent and milder.[423]

Other uses:

Reserpine increases the appetite and is added to feed to enhance the development of chickens and turkeys.[20]

Adulterants:

In the past, the exported roots have been adulterated with stems (having only .3% alkaloids), roots of *R. micrantha*, *R. densiflora*, *Ophiorrhiza mungos* and *Clerodendron* spp.[463]

Figure 68. American serpent-wood (*Rauvolfia tetraphylla* L.) in fruit. Photographed in the Fairchild Tropical Garden, Coral Gables, Florida.

American Serpent-Wood

Botanical name:

 Rauvolfia tetraphylla L. (which has been misapplied to R. nitida Jacq. in the West Indies); syns. R. canescens L., R. hirsuta Jacq., R. heterophylla Roem. & Schult.

Other names:

 Devil pepper, boboró, borrachera, veneno, guataco colorado, comida de culebra, viborilla, amatillo, yerba de San José, señorita, matacoyote, chalchupa, sarna de perro, cocotombo.

Family:

 Apocynaceae, the Dogbane Family.

Description:

 A subshrub or shrub, 1½ to 4½ ft. (.5 to 1.5 m) tall, much-branched, with milky latex and cylindrical, curved or straight roots to 3.3 ft. (1 m) in length; young growth is downy; twigs have yellowish bark. Leaves arranged in whorls of 3 to 5, usually 4, vary in form from narrow-oblong to broadly ovate or ovate-elliptic, usually pointed at both ends; the smallest are ⅜ to 1½ in. (1 to 4 cm) long, the largest ¾ to 6 in. (2 to 15 cm long), thin and softly woolly beneath or, in var. *glabra*, having no more than a few hairs on the lower surface. Flowers, borne in compact clusters of 3 to 8, are downy, tubular, ⅛ to 3⁄16 in. (3 to 5 mm) long, greenish-white, sometimes dappled with pink. Fruit, borne in abundance, consists of 2 fused carpels, 3⁄16 to ½ in. (5 to 12.5 mm) in diameter; turns from red to purple and finally black when ripe; contains 1 or 2 ovoid, rough stones. The bush in fruit resembles a cherry pepper bush (*Capsicum* sp.).[3, 359, 619]

Origin and Distribution:

 Native from southern Mexico to Colombia and Venezuela, Trinidad, Tobago, Barbados, St. Thomas, Jamaica and Cuba, occurring in sandy thickets and along roadsides in damp soil. This species was introduced to the coast of eastern India and spread inland, becoming naturalized in moist, hot lowlands from Bihar to Mysore. It is common in the Howrah district near Calcutta.[20, 117] It also occurs semi-wild in Queensland, Australia.[606]

Commercial cultivation has been undertaken in Orissa. In 1957, plantings were made at the Estación Experimental Agronómica de Santiago de las Vegas, Cuba.[486]

Constituents:

The major alkaloid, present in the entire plant, is rauwolscine ($C_{21}H_{26}N_2O_3$)—.5% in the leaves, .2% in the stem bark and .1% in the root.[117] This alkaloid is strongly hypotensive but not sedative.[109]

Deserpidine, also called canescine and recanescine ($C_{32}H_{38}N_2O_8$), is extracted commercially from the roots.[122] Reserpine content ranges from .04 to .17%; sarpagine (raupine), .01%.[80]

The alkaloid heterophyllin, or aricine ($C_{22}H_{26}N_2O_4$), was isolated from the roots in 1955.[227] Air-dried roots also contain the following: ajmalicine ($C_{21}H_{24}N_2O_3$); ajmaline ($C_{20}H_{26}N_2O_2$); alstonine ($C_{21}H_{20}N_2O_3$); corynanthine ($C_{21}H_{26}N_2O_3$); isoraunescine ($C_{31}H_{36}N_2O_8$); raunescine ($C_{31}H_{36}N_2O_8$); raujemidine ($C_{33}H_{40}N_2O_9$); as well as reserpiline, reserpinine, reserpoxydine; serpine, serpentine; yohimbine and β-yohimbine or amsonine ($C_{21}H_{26}N_2O_3$). Other properties include an anti-rheumatic principle, serposterol.[20, 616, 617]

In addition to rauwolscine, the leaves contain ajmalicine, ajmaline, chalchupine A ($C_{14}H_{21}N_3O_{12}$); chalchupine B ($C_{15}H_{24}N_6O_{11}$); reterophyllin; isoreserpiline, also called elliptine ($C_{23}H_{28}N_2O_5$); isoreserpinine ($C_{22}H_{26}N_2O_4$); reserpiline and serpentine.

Alkaloids other than rauwolscine reported in the stems are ajmalicine, ajmaline, heterophyllin, sepentine and yohimbine.[617]

Indian analyses show total alkaloid content of dry leaves, .4 to 1.8%; stems, .10 to 5%; roots, .06 to .2%; and the alkaloid level seems to be highest from October to January.[20]

Propagation, Cultivation and Harvesting:

The plant grows readily from seeds. In Cuba, it was observed that a majority of the seeds which fell from the parent plants onto clear ground germinated in seasons of frequent rainfall.[486]

Vegetative propagation has proved feasible in India, where 86 percent success has been obtained in rooting hardwood cuttings after treatment with β-indolyl acetic acid for twelve hours. Root formation in root– and shoot-cuttings has been stimulated satisfactorily by β-indole butyric acid.[20]

In Cuba, 91.2 percent success was achieved in rooting stem cuttings in river sand. Best overall results, however, were realized by rooting and replanting the main stem of plants after root removal. These "trunks" rooted readily, showed a higher yield of roots at harvest, and the total alkaloid content of the roots was 2.1% compared with 1.4% in roots grown from stem cuttings and 1.23% in roots grown from seed. Total root yield per hectare was calculated as 8,035 to 10,000 lbs. (3,635 to 4,530 kg).[486]

Few diseases or pests were encountered by Cuban horticulturists, mainly a leaf rust and a leaf-eating pest, *Pachyarches aureocostalis* Fab. Songbirds presented a problem, seriously depleting the crop of ripe fruits.[486]

Medicinal uses:

In Panama, decoctions of the bark and leaves of this species are much used as gargles and taken internally for intestinal disorders. In cases of syphilis, the bark decoction is mixed with castor oil, and the combination is considered dangerous.[378] The bark extract is also used to relieve chronic skin diseases and to destroy parasites. In Mexico, peasants apply the crushed root to erysipelas; the leaf decoction is gargled and is rubbed on sore gums. The latex is applied to granulated eyelids and is employed as a diuretic, expectorant, emetic and purgative. Small pieces of the bark are placed on aching teeth. The crushed fruit is a remedy for mange of dogs.[378, 485]

When supplies of *R. serpentina* from India were cut off in 1955, Chas. Pfizer, Merck, Squibb and other pharmaceutical concerns sent agents to locate and collect *R. tetraphylla* in Central America, and chemical studies showed that it could be used as a substitute. It has become a commercial source of reserpine[88] and deserpidine. The latter, which may be given orally in daily does of .25 mg as a tranquilizer and hypotensive, produces little of the side effects associated with other *Rauvolfia* preparations.[122]

Toxicity:

The latex blisters the skin of sensitive individuals.[12] The fruits are very poisonous to humans; if ingested, they cause pain in the mouth, constriction of the pharynx, intense thirst, violent inflammation of the entire alimentary canal, causing acute burning sensations; nausea, vomiting, bloody diarrhea, convulsions, coldness in the extremities and death in extreme cases.[20, 378, 545]

The plant extract, injected into guinea pigs, caused paralysis and death by asphyxiation. Autopsy showed severe lung congestion.[88]

Other uses:

The purple juice of the fruit has been used in Central America as a substitute for ink[548] and for dyeing cloth.[544]

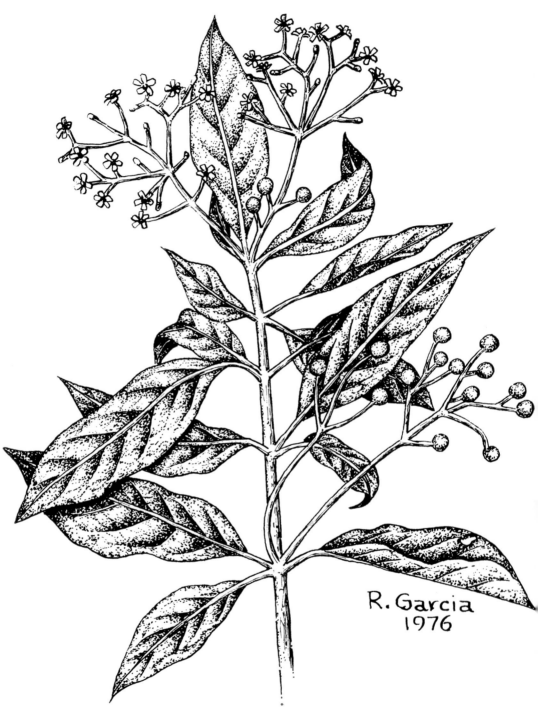

Figure 69. African serpent-wood (*Rauvolfia vomitoria* Afzel) has superseded *R. serpentina* as a source of reserpine.

African Serpent-Wood

Botanical name:
> *Rauvolfia vomitoria* Afzel.

Other names:
> African rauwolfia, penpen, swizzle stick, papaw table.

Family:
> Apocynaceae, the Dogbane Family.

Description:
> A shrub or small tree to 15 ft. (4.5 m), seldom to 20 ft. (6 m), with bitter, white latex. Young branches are quadrangular. The wood is white, turning rose-red on exposure. Roots may attain a diameter of 2 in. (5 cm) or more. Leaves, in whorls of 3 or 4, may be oblong-obovate, lanceolate or elliptic, pointed, 3 to 7 in. (7 to 17.7 cm) long and 1 to 3 in. (2.5 to 7 cm) wide; glossy. Flowers are fragrant, tubular, five-lobed, somewhat hairy at the mouth, to ⅓ in. (8.5 mm) long, white, numerous, in long-stalked, open, terminal clusters. Fruit, paired or single, is ovoid, smooth, ⅓ in. (8.5 mm) long, red when ripe.[146, 169, 275, 419, 584, 603]

Origin and Distribution:
> Native and common on edges of forests or in secondary forests from Senegal and the Congo to Mozambique.

Constituents:
> Attention was called to this species as an alkaloid source in 1943, but it received little further notice until 1954. Thereafter, it became the object of much study. The following alkaloids have been reported:
> ROOT: ajmaline, alstonine, isoreserpiline, mitoridine ($C_{20}H_{22}N_2O_2$), purpeline ($C_{21}H_{24}N_2O_2$), rauvanine ($C_{23}H_{28}N_2O_5$ or $C_{23}H_{28}N_2O_9$); rauvomitine ($C_{30}H_{34}O_2N_5$), rescidine ($C_{35}H_{38}N_2O_9$ or $C_{34}H_{40}N_2O_9$), reserpilene, reserpine (.01 to .02%), rescinnamine, sarpagine, seredamine ($C_{21}H_{26}N_2O_2$ or $C_{22}H_{26}N_2O_2$), serepentenine and yohimbine.
> LEAVES: alkaloid A ($C_{20}H_{22}N_2O_2$), heterophyllin, carapanaubine ($C_{23}H_{28}N_2O_6$), isoreserpiline, picrinine ($C_{20}H_{22}N_2O_3$), rauvoxine ($C_{23}H_{28}N_2O_6$), rauvoxinine ($C_{23}H_{28}N_2O_6$), tetrahydroalstonine ($C_{21}H_{24}N_2$

O_3) and vomifoline.[605, 617] Among flavonoids found in the leaves are two heterosides of kaempferol.[428]

Total alkaloid content of root: .79%; stem tips, .11%; stem bark, 0.35%; leaves, .16%.[275]

Propagation, Cultivation and Harvesting:

There are no commercial plantations of *R. vomitoria* in Africa. Roots and bark for pharmaceutical use are gathered from the wild plants. Large-scale collection began in 1954, when 700 tons of roots were exported from the Congo.[395]

Medicinal use:

Various parts of the plant have multiple uses in indigenous medicine in Africa. The powdered root is made into a paste and placed on snakebites. Macerated bark and roots, mixed with palm oil, form an ointment valued for killing lice and vermin. The root bark is strongly purgative and emetic in large doses. It is taken as an abortifacient. Decoctions are given as a remedy for leprosy, gonorrhea, jaundice and convulsions, and are employed as eye drops; also, mixed with spices, they are used as enemas.[146] Gouros of the Ivory Coast claim that a warm infusion, injected as an enema, acts as an aphrodisiac after evacuation.[275] A poultice of root bark or leaves, mixed with banana plant sap, lime juice and spices, is applied to sprains and swellings.[146] The mashed leaves, simmered in butter, are applied on contusions, dislocations and rheumatic joints. The stem bark is employed to relieve indigestion and as an emetic and cathartic, either as a decoction or mixed with overripe banana pulp.[603] Infusions of the crushed bark and leaves are believed to relieve fever in children. The powdered bark is spread on sores, skin rash, herpes and the eruptions of measles.[275] A leaf infusion is rubbed on the face to overcome dizziness; taken internally, it causes vomiting and purging. Vomiting is also induced by swallowing two or three raw leaves. The latex derived by mashing young leaves is taken as an antidote for poisoning. Either the latex or a leaf decoction may be employed on parasitic skin diseases, yaws and head lice. To relieve lung congestion, the crushed bark is rubbed on the chest and the leaf decoction is added to steam baths.[275] The root decoction is given to control spasms of tetanus and to subdue maniacal behavior. It causes deep sleep lasting several hours.[146, 187]

In the pharmaceutical trade of Europe and America, *R. vomitoria* has superseded *R. serpentina* as a source of reserpine,[605] the drug being extracted from either the roots or the bark. Exported roots are in cylindrical or flat pieces to ⅝ in. (1.5 cm) wide and up to 1 ft. (30 cm) long. They differ from roots of *R. serpentina* by groups of sclereids in the bark in as many as five interrupted bands and also by large vessels in the wood. R. vomitoria roots also have alternating lignified and non-lignified cork cells which serve to differentiate them from roots of R. caffra, another African species.[584]

Toxicity:

Large doses of the root bark or stem bark may cause paralysis and death. Crushed leaves, thrown into shallow pools or streams, will poison animals that drink the water.[187]

Other uses:

The tree is popular as an ornamental in West Africa and is planted also as a living fence.[146] It is grown, too, as a support for vanilla vines and to provide shade for young cacao plantations,[603] though the wild plants are alleged to be hosts of Collar Crack disease of cacao.[275] In Sierra Leone, pieces of papaya are stuck on the branches as bait in trapping birds which come in numbers to feed on the berries.[146] Metallic blackbirds are particularly fond of these fruits.[603] Small branches with four clipped twigs left attached are twirled as swizzle sticks in mixing drinks. Larger branches are similarly employed by native dyers for stirring indigo.[146]

Figure 70. *Strophanthus gratus*, principal source of ouabain, is valued also as an ornamental because of its showy white-and-purple flowers. The twin seedpods are typical of the genus and of the dogbane family.

Smooth Strophanthus

Botanical name:
>Strophanthus gratus* Franch.

Other names:
Sawai, isha, onaye.

Family:
Apocynaceae, the Dogbane Family.

Description:
A stout, climbing vine, attaining 30 ft. (9.1 m); bark of young branches red-brown dotted with white lenticels, that of older growth corky and furrowed; leaves evergreen, thick, leathery, oblong to elliptic, pointed at the apex, rounded at the base, to 6 in. (15 cm) long and to 3 in. (7.5 cm) wide; smooth, glossy; midrib often reddish, petiole purple. Flowers in terminal clusters are fragrant at night, showy, bell-shaped, five-lobed, purple in bud; when expanded are rose-purple externally, white to pink internally with 5 pairs of erect purple appendages arranged around the mouth of the corolla; 2 to 2⅜ in. (5 to 6 cm) or more in length, with a spread of 2⅜ in. (6 cm). Seedpods, in divergent pairs, 6 in. (15 cm) long, 2 to 3 in. (5 to 7.5 cm) wide, develop over a twelve-month period; when mature, they split open and release numerous sharp-edged, curved or twisted, brown, hairless seeds, minutely pitted, .11 to .19 in. (3 to 5 mm) wide, attached to silky pappus. The dried seeds are odorless but very bitter.[275, 603]

Origin and Distribution:
Native and common in deciduous forests of west tropical Africa from Sierra Leone to the Congo. It was introduced into cultivation in 1845 and distributed to gardens in East Africa, throughout the Old World tropics and to some islands of the West Indies.

Constituents:
The main active principle is ouabain (G-strophanthin), containing not less than 95% $C_{29}H_{44}O_{12} \cdot 8H_2O$.[423] It constitutes 4 to 8% of the seeds. Other cardenolide glycosides are present in small amounts. Among those

isolated are sarmentoside E, sarmentoside A, sarmentoside D, acolongifloroside K, bipindoside, tholloside and sarhamnoloside.[276] The seeds also contain the alkaloid trigonelline ($C_7H_7NO_2$).[617]

Propagation, Cultivation and Harvesting:

Though easily grown from seeds or cuttings, the plant is not widely cultivated for pharmaceutical purposes. The seeds are generally harvested from wild vines that are protected from destruction in and around the native compounds.[146] There is a commercial plantation at Bipinde in the Cameroons that was established by the botanist Zenker. The vine is trained on the trunks of ambarella trees (*Spondias dulcis* Forst., syn. *S.*

Figure 71. Smooth strophanthus (*Strophanthus gratus* Franch) growing vigorously in the vine collection of the Fairchild Tropical Garden, Coral Gables, Florida.

cytherea Sonn.). At the Ebolowa Experimental Station, *S. gratus* seedlings are grown in nursery beds and transplanted to the field with avocado trees for support. The vine begins to bloom when three years old but requires six to ten years for maximum fruit production. Many of the flowers do not set fruit. Average yield of seeds per vine is approximately 1 kg.[146]

Medicinal uses:

In Nigeria, fever patients are rubbed with the crushed leaves or a leaf infusion. The leaf decoction is taken as a remedy for gonorrhea. An extract of the crushed stems is given for extreme debility. Though it is known to be hazardous, some patients are said to make good recovery.[146] The mashed leaves or their juice are applied to wounds, ulcers and Guinea worm.[275]

The seeds of *S. gratus* are the principal source of ouabain, which is official in the International Pharmacopoeia and the Pharmacopoeias of Germany, England, France and the United States.[419] In pharmaceutical practice, ouabain is utilized as a cardiac stimulant, administered intramuscularly or intravenously.[446] It is not given orally in the United States, as the dose is considered too large and its effects uncertain. Ouabain acts more quickly than other cardiac glycosides (beginning in 5 to 10 minutes and taking full effect in 20 to 60 minutes) and wears off most rapidly (usually in 12 to 24 hours, though the improved circulation may persist 4 or 5 days). It is particularly valued in emergency treatment of acute heart failure and pulmonary edema.[423] Dosage is .25 to .5 mg, followed by .1 mg every half hour.[423] Toxicity develops as the heart rate and calcium exchange increase.[220] If the patient has recently received digitalis, the initial dose must be reduced to .1 or .2 mg.[423] Ouabain does not cause the peripheric vasoconstriction that is induced by digitalis,[419] but it cannot be given to patients having had acute or recent myocardial infarction. It may be administered intravenously to relieve hypotension during anesthesia and surgery.[423]

Toxicity:

The seeds and sometimes the wood, mixed with the sap of various plants, especially *Colocasia*, *Aframomum* or *Palistota*, are used to poison arrows for hunting game (including elephants) or for warfare. Internal antidotes are the sap of *Alstonia congensis* or seeds of *Garcinia kola*; externally, the powdered bark of *Erythrophleum guineense* is used.[603] The root bark of *S. gratus* is maliciously employed on the Ivory Coast to poison food of enemies.[275]

Ouabain is twenty to thirty times as toxic intramuscularly as orally, forty-three to eighty-six times as toxic when injected intravenously.[146] Intraventricular administration of small doses of ouabain in rodents produces catalepsy and hypothermia.[158]

Other uses:

This species is the handsomest of the genus and admired for its profusion of blooms.[603] Superstitious gamblers or accused persons rub a leaf on the palm of the hand for good luck.[146]

Figure 72. Green strophanthus (*Strophanthus kombé* Oliv.). The five lobes of the yellowish, red-streaked flowers have elongated, dangling tips. Seeds of all *Strophanthus* species bear feather-like, hairy awns.

Green Strophanthus

Botanical name:

   Strophanthus kombé Oliv.

Other names:

   Poison vine, kombé, kombé arrow poison plant, mbolo.

Family:

   Apocynaceae, the Dogbane Family.

Description:

   A woody, perennial vine, climbing to 10 or even 30 ft. (2.75 to 9.1 m); stems gray or black, rough, dotted with white lenticels, becoming 4 to 6 in. (10 to 15 cm) thick at the base; young branches hairy; leaves opposite, elliptic or ovate-elliptic, 4 to 5½ in. (10 to 14 cm) long, 2½ to 3½ in. (6.5 to 9 cm) wide, rough-hairy, the upper surface crinkled. Flowers in small terminal clusters on short branches; calyx long-pointed, hairy, the corolla yellowish-white with red markings, bell-shaped, five-lobed, each lobe having an extended slender, dangling tip 3 to 5½ in. (7.5 to 14 cm) long. Seedpods, in divergent pairs, smooth, somewhat woody, slender, to 12 or 15 in. (30 or 37.5 cm) long, containing numerous flat, lanceolate seeds, ½ to 1⅛ in. (12.5 to 28 mm) long, ⅛ to 3/16 in. (3 to 5 mm) wide, coated with fine silky, greenish-brown hairs and bearing long-stalked, feather-like tufts of pappus.[124, 275, 620, 624] The seeds have a faint odor and are very bitter.[624] They give a green reaction to sulphuric acid.[584]

Origin and Distribution:

   Native to dry, hilly woodlands of central and eastern Africa, mainly Mozambique, Northern Rhodesia, Nyasaland and Tanzania.[605] It is cultivated commercially in the Cameroons.[466] Nyasaland is the main source of supply of seeds for the pharmaceutical trade.[620]

Constituents:

   The major active principle, amounting to 5 to 10% of the seeds, is the glycoside K-strophanthoside (or strophoside), formerly called strophanthin or K-strophanthin. It is a natural compound of strophanthidin and a trisaccharide composed of cymarose, β-glucose and α-glucose.[122] Makarevich has documented eleven other cardenolides including cymarin, periplocymarin, periplocin, K-strophanthin-β, and erysimoside.[364] Fat represents 33⅓% of the seed and must be extracted before the glycosides can be isolated.[466] Herrmann has perfected a fermentation process for isolating and purifying cymarin for pharmaceutical use.[256, 257] Other properties include the alkaloid trigonelline, choline, saponin, resin and mucilage.[284, 605]

Propagation, Cultivation and Harvesting:

Collection and preparation of the seeds is a long-established native industry.[595] Formerly, the pods were gathered and merely scraped and dried for shipment. The modern practice is to scrape and lightly pound the pods to promote opening, dry them till they split and then remove the seeds and eliminate the pappus so that only cleaned seeds are marketed.

Medicinal uses:

In Africa, the root decoction is a remedy for bronchitis.[605] *K-strophanthoside* was formerly official in the United States and given as an emergency measure in acute cardiac asthma, acute congestive heart failure and pulmonary edema.[423] In small amounts, injected intramuscularly or intravenously, it serves as a cardiac stimulant, strengthening and slowing the beat, and it also acts as a diuretic.[595] The usual dose is 60 mg.[122] Because it is unsuitable for oral administration, being slowly absorbed from the gastrointestinal tract, the drug also varying in potency (losing strength in storage),[605] it is no longer used in the United States, though it is still listed in foreign Pharmacopoeias.[423] Cymarin is employed in Germany in the treatment of heart insufficiency.[257]

Toxicity:

In Africa, a gum obtained by boiling the seeds is used as arrow and spear poison. The venom acts directly on the heart, paralyzing it.[595]

Related species:

Brown strophanthus (*S. hispidus* D.C.) is a similar vine, reaching 50 ft. (15.24 m), that is native to West Tropical Africa. The young branches are covered with stiff, rough hairs; the flowers are yellow with purple spots in the throat, and the "tails" of the corolla lobes are much longer than those in *S. kombé*. The seedpods range up to 18 in. (45 cm) in length, are red-brown, streaked and spotted with white lenticels. The seeds are almost spindle-shaped and are coated with brownish hairs which are readily rubbed off.[419] They give a green reaction to sulphuric acid.[584]

This vine was formerly widely planted in West Africa for its seeds, an extract of which was much valued as the main ingredient in arrow poison, until local governments prohibited its cultivation.[146] Today the seeds are accepted as substitutes for those of S. *kombé* in pharmaceutical use abroad and formerly appeared in the U. S. Pharmacopoeia as an acceptable source of the tincture. They are exported from Senegambia and Guinea.[122]

Chemically the seeds are similar to seeds of S. *kombé* but they are considered only 40 percent as toxic. They yield 4 to 8 percent of active glycosides.[146]

The roots are used in native medicine in the treatment of rheumatism, venereal disease, malaria and dysentery.[146, 605]

Labiatae

Figure 73. Japanese mint (*Mentha arvensis* subsp. *haplocalyx* Briquet var. *piperascens* Holmes). Photographed at Longwood Gardens, Kennett Square, Pennsylvania.

Japanese Mint

Botanical name:

    Mentha arvensis subsp. *haplocalyx* Briquet var. *piperascens* Holmes.

Other name:

    Corn mint.

Family:

    Labiatae, syn. Lamiaceae, the Mint Family.

Description:

    A perennial, subtemperate to subtropical herb, with slender, creeping rootstocks sending up erect, stiff, quadrangular, branched, purple stems 1 to 3 ft. (.3 to .91 m) high, slightly hairy at the lower nodes. Leaves are aromatic, opposite, broad-lanceolate, gradually tapering toward the base, 1¼ to 2½ in. (3.1 to 6.7 cm) long, ¾ to 1¼ in. (1.2 to 3.1 cm) wide, blunt-toothed, minutely hairy, especially near the midrib on the face. Flowers, clustered in the leaf axils, are white or ivory, sometimes tinged with purple, tubular, with spreading lobes and hairy, bell-shaped, toothed calyx. Seeds are minute and smooth. They do not develop in plantings in the United States.[20, 110, 164, 240]

Origin and Distribution:

    Native to China, this mint was introduced into Japan in the third century A.D., was soon cultivated in temple gardens and parks, escaped and ran wild in wetlands, and subsequently was adopted as a major Japanese crop.[240] It was introduced into India about 1954 by the Regional Research Laboratory, Jammu, with the aid of UNESCO. It has been planted commercially in Uttar Pradesh and other favorable locations of India up to an altitude of 4,000 ft. (1,260 m). In the mid-1930s, experimental culture was begun by Japanese immigrants in Brazil, particularly in the states of São Paulo and Paraná. The results being highly favorable, plantations were expanded with government support, and large-scale cultivation extended into the adjacent state of Minas Gerais.[22] It is now grown commercially also in Argentina, California, South Africa, Angola, western Australia, Formosa and Korea.[41, 164]

Constituents:

The leaves yield 1.28 to 2% of volatile oil consisting mainly (70 to 95%) of menthol.[422] Other properties include menthyl acetate, menthone and small amounts of piperitone, α-pinene, furfural, l-limonene, camphene, caryophyllene, d-3-octanol, dl-sesquiterpene alcohol, β, y-hexenyl phenyl acetate, α, β-hexenic acid and other fatty acids.[20]

Menthol stimulates nerve perception of cold but subdues sense of pain. It is more germicidal than phenol.[423] Internally, it has a depressing effect on the heart.[164]

Propagation, Cultivation and Harvesting:

Seeds are used only for breeding and selection purposes, inasmuch as seedlings are weak and show considerable variability.[164] "Cultivar 701," developed in Brazil, has been the only type grown in that country since about 1956 because of its resistance to rust and its high oil content.[395]

Stem cuttings can be used for quick multiplication,[164] but propagation is usually effected with rooted suckers taken from selected mature plants which have been especially mulched with compost or manure to promote development of stolons. After separation, the suckers may be kept in well-moistened nursery beds until they put out new sprouts.[444] For optimum growth, the crop requires well-drained sand or loam high in organic matter, sunny summers, winters free of frost and annual rainfall of at least 32 in. (81.2 cm). The field is disked and enriched with fertilizer in advance. Suckers are set out, usually 6 to 8 in. (15 to 20 cm) apart in rows spaced at 2 to 4 ft. (.6 to 1.2 m). In northern India, suckers are planted in late winter when they are dormant. In South India, planting is done at the beginning of spring rains. The fields are heavily irrigated in the summer, and weeds must be thoroughly eliminated. Supplemental dressings of fertilizer are applied soon after planting and after harvesting—entirely organic, a mixture of organic and superphosphate, or a blend of equal parts of ammonium sulphate, superphosphate and potassium sulphate.[30]

Principal diseases of Japanese mint are rust (caused by *Puccinia menthae* Pers.) and anthracnose, which are controlled by copper-lime sprays.[240] Other problems which may appear are powdery mildrew (caused by *Erysiphe cichoracearum* DC.), leaf blight (caused by *Alternaria* sp.) and wilt (caused by *Verticillium alboatrum* Reinke & Berth. and *Sclerotium rolfsii* Sacc.). The fungus *Macrophomina phaseoli* (Maubl) Ashby sometimes rots the stolons in new plantations. Leaves are attacked by the larvae of the moth *Syngamia abruptalis* Wlk.; also by the caterpillar *Autographia nigrisigna* Walk. and the red pumpkin beetle *Radiopalpa foevicollis* Lucas. In northern India, a termite, *Odontotermis obesus* Ramb., may completely destroy plantations in the summer months unless the soil has been flooded or pretreated with effective insecticides.[164] Other insect pests include mint loopers, leaf hoppers, cutworms, spider-mites and aphids.[240]

Harvesting may take place ninety to one hundred days after planting when the plants are beginning to bloom. Oil content is highest at the stage between the development of the inflorescence and the opening of the flowers. The tops are cut off close to the ground, and cutting may be repeated once or twice during the year when regrowth has reached the proper stage. In Kerala, three cuttings yield a total of 5 to 6 metric tons (5,000 to 6,000 kg) of tops, 44 to 61 lbs. (20 to 28 kg) of oil per acre annually.[444]

Plantations in northern India cannot be maintained more than one year unless thinned out and the soil reworked and refertilized, and the maximum life is two years due to soil exhaustion.[164] In most growing areas, a field can be kept in full production for three to four years, and it is then best to plow it up and use the ground for another crop (beans, rice, late wheat, fodder corn, senna or mustard) for two years[164] or merely sow a leguminous cover crop to restore fertility. Such rotation has been found to enhance oil yields and menthol content.[30]

Harvesting is performed in the morning on clear, sunny days after the dew is gone. The fresh plant material may go direct to the distillery (after field wilting for 2 to 4 hours to reduce volume), but it is still excessively bulky, requires double the time for distillation and the oil is clouded with chlorophyll.[164] Sun-drying has been practiced but causes much oil loss through resinification and evaporation.[293] For best results, the tops are spread out in a thin layer on the floor of a shed or tied in small bunches and hung up to dry in the shade. Water content should be reduced from 73% in the fresh material to approximately 20%. Excessive drying causes the leaves to fall off and break up and lose volatile constituents during handling and transportation. If necessary, properly dried mint may be stored for fifteen months with no more than 10 percent loss of oil. At the distillery, dried leaves are threshed from the stems (which constitute 60 to 70% of the herbage and contain little oil) and subjected to steam processing, which usually takes about one and one-half hours. Sometimes the plant material is chopped to enhance yield.[444]

The liberated oil is filtered to remove impurities and put into clean, galvanized drums of 5¼ to 13 gal. (20 to 50 l) capacity for shipment. Menthol is separated in commercial laboratories first by gradual chilling in three stages—from 14° C. to 10° C. and finally 5° C. The crystals obtained represent 40 to 50 percent of the total menthol. Further recovery can be achieved by saponification of the decanted oil with aqueous alkali and conversion of liberated menthol to solid borate ester. Thus, from 220 lbs. (100 kg) of oil, 88 lbs. (40 kg) of menthol crystals may be derived by chilling, and 28½ lbs. (13 kg) of menthol crystals plus 16½ lbs. (7.5 kg) of liquid menthol by saponification.[164]

Prior to 1940, Japan supplied 70 percent of the world demand for Japanese Mint oil (also called Arvensis oil)—between 600 and 800 tons. Brazil, by 1963, assumed first place with an annual production of more than 2,400 tons.[164]

Medicinal uses:

Originally, in Japan, this mint was valued as a home remedy for coughs and colds.[240] In modern pharmaceutical products, menthol serves mainly as an ingredient in cough drops and in nasal inhalants meant to provide a sense of relief from congestion (though not of real benefit). It is used to a lesser extent, but widely, in lotions, ointments, creams, etc., for the treatment of skin diseases. In dermatology, menthol serves as an antipruritic, antiseptic, counterirritant, stimulant and anaesthetic. Mentholated after-shave lotions soothe sensitive skin. Menthol is much used in cases of dermatitis accompanied by itching, which it replaces with a comfortably cool sensation. An ether solution is applied to boils and carbuncles. Menthol dabbed on the temples sometimes lessens the pain of ordinary headache and neuralgia.

Internally, menthol has been prescribed for gastric distress.[423]

Toxicity:

When combined with vasoconstrictors in nasal inhalants, menthol may nullify the latter and increase congestion. If added to hot water for vapor inhalation by people with acute bronchitis, it can induce a feeling of comfort but cause irritation. It has brought on asphyxiation when applied to infants. Taken internally in quantity, menthol acts on the central nervous system. In experimental animals it has produced ataxia, increased respiration, convulsions, paralysis and death from asphyxia. The fatal dose is judged to be 1 g/kg bodily weight. Allergic reactions to externally used menthol may include flushing, urticaria and headache.[423]

Other uses:

Most of the menthol imported into the United States goes into cigarettes. Outside the United States, Japanese Mint oil is frequently used to adulterate peppermint oil.[41, 422] Japanese Mint oil, under the name of Japanese Peppermint Oil, is used as a substitute for true peppermint oil in low-priced food and beverage products only, as it is inferior in flavor and quality. Chewing gum flavored with Japanese Mint oil may have a "mask" of true peppermint oil in the outer coating.[41] No food use of Japanese Mint oil is allowed in the United States.[210]

Dementholized oil, still containing about 10% menthol,[422] is used in inexpensive perfumes and also serves as flavoring in mouthwashes, toothpaste and other toiletries.[20]

The spent leaf residue from the distillery can be returned to the field as fertilizer or dried and ground up for cattle feed.[164]

Substitutes:

Menthol can be obtained also from *Mentha piperita* (which contains 35 to 50%), but this oil has more important uses in flavoring foods, candies, liqueurs, chewing gum and toothpaste and is very little used in pharmaceutical preparations.[164]

Synthetic menthol is created from starting materials produced by other plants—including α-pinene from *Pinus* species [see SLASH PINE (*P. elliottii*) and LONGLEAF PINE (*P. palustris*)]; also citronellol from *Eucalyptus citriodora* and *Cymbopogon nardus*, thymol from *Thymus vulgaris* [see THYME] and *Carum copticum*, piperitone from *Cymbopogon jwarancusa* and pulegone from *Mentha pulegium*. In India, it is considered uneconomic to utilize these plants for synthetic menthol production when natural menthol of better flavor is obtainable more simply from Japanese Mint.[164]

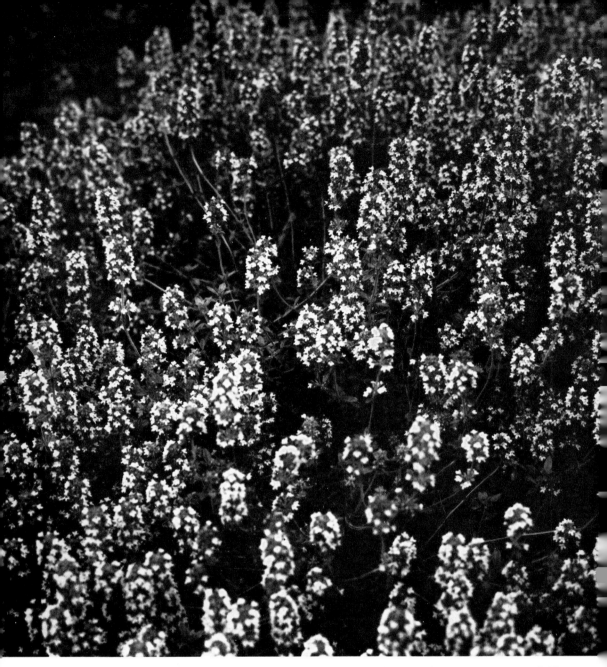

Figure 74. Thyme (*Thymus vulgaris* L.) in bloom. Photographed in the Beal-Garfield Botanic Garden, Michigan State University, East Lansing, Michigan.

Thyme

Botanical names:

    Thymus vulgaris L. and *T. zygis* L.

Other names:

    Common thyme, garden thyme (*T. vulgaris*), wood marjoram, Tomillo salsero (*T. zygis*).

Family:

    Labiatae, the Mint Family.

Description:

    Perennial subshrubs; *T. vulgaris* ranges from 6 to 12 or sometimes 18 in. (15 to 30 or 45 cm) in height, has slender, wiry, spreading, white-hairy branches and stiff, vertical shoots. Leaves are aromatic, nearly evergreen, opposite, nearly sessile, oblong-lanceolate to linear, pointed, 1/5 to 3/5 in. (5 mm to 15 mm) long, 1/10 in. (2.5 mm) wide, with recurved margins, minutely downy, gray-green and gland-dotted. Flowers are lavender, tubular, two-lipped, resinous, about 1/5 in. (5 mm) long with hairy, glandular calyx, borne with leaflike bracts in loose whorls in axillary clusters on the branchlets or in terminal oval or rounded heads. Seeds are round and tiny—170,000 per ounce (28.3 g), 1 qt. (1.1012 l) of seeds weighing 24 oz. (680.3 g). They remain viable for three years.[115, 233, 305] *T. zygis* is similar but smaller, hairier; the narrower, shorter, sessile leaves are clustered at the nodes; the flowers are white or purple and in clusters spaced at intervals in an elongated inflorescence.

Origin and Distribution:

    T. vulgaris is native to southern Europe, from Spain to Italy and was introduced into England about 1548. It is commonly cultivated there and in most mild-temperate and subtropical climates which include southern and central Europe, including Bulgaria,[551] the southern U.S.S.R., Morocco, Asiatic Turkey, Syria, Israel, northern India—up to altitudes of 3,000 ft. (914 m)—and in southern Africa, New England, California, Colombia, Brazil and at upper elevations in some of the West Indies.[41, 485] *T. zygis* is indigenous to Portugal, Spain and the Balearic Islands and is little known elsewhere. Commercial cultivation is most extensive in

France, Germany and Greece (*T. vulgaris*) and Spain (*T. zygis*). Of the latter species, var. *gracilis*, thriving at low elevations, is the more important, var. *floribunda* being alpine.[233]

Constituents:

Thyme contains saponins, thiamine, ursolic acid, caffeic acid, resin, tannin, gums and 1 to 2.6% of a volatile oil. Thyme oil (formerly called "Oil of Origanum") from *T. vulgaris* grown at different locations may vary considerably in color, odor, viscosity and chemical composition.[5] Generally it may yield 25 to 42% of phenols, primarily 20 to 25% thymol ($C_{10}H_{14}O$), and lesser amounts of carvacrol, *p*-cymene, camphene, l-borneol, linalool, α- and β-pinene, tricyclene, myrcene, α-terpinene, limonene, 1,8-cineol, caryophyllene and other volatile constituents.[502] Thyme oil from *T. zygis* grown in Spain yields 50 to 70% phenols, mainly carvacrol.[122] Thyme oil may be red, yellow or colorless through redistillation.

The physical-chemical constants of red thyme oil from *T. zygis* in Spain are as follows: specific gravity at 20° C., 0.911 to 0.954; refractive index at 20° C., 1.4940 to 1.5100; optical rotation at 20° C., $-5°$ to $+1°$; phenol content (as thymol) 30 to 75%; solubility 2:5 in 70% alcohol, 1:2 in 80% alcohol.

White thyme oil, pale-yellow in color, results from rectification of the distilled red thyme oil and is milder in odor and flavor.[210]

Two antibiotic substances were isolated from thyme in Italy in 1953, the activity of one being gram-positive, the other, gram-negative.[116]

Propagation, Cultivation and Harvesting:

Thyme may be grown from seed or propagated by division, layering of side shoots or by cuttings taken in early summer. Seedlings are raised in beds and set out when 2 or 3 in. (50 to 75 mm) high, or seeds may be drilled directly in the field no more than ¼ in. (6 mm) deep and at a spacing of 2 or 3 ft. (60 to 90 cm), 6 lbs. (2.7 kg) being required to plant an acre. Rooted cuttings are set out 1 to 1½ ft. (30 to 45 cm) apart in rows 2 ft. (60 cm) apart. Weeding is performed by hand, and fertilizer is applied in fall and spring. In cool climates, the plants are usually covered with earth or mulch during the winter. Thyme is long-lived in light, dry, calcareous soil, short-lived and less aromatic in heavy, wet soil. It thrives in full sun. Old plants become woody and straggly and are usually replaced in two to three years.[23, 233, 493]

Thyme blooms from June through August, and the flowering tops are harvested in one or two cuttings per season and delivered fresh or partially dried for distillation. In Egypt, highest thymol and carvacrol concentrations were detected in the early stage of blooming.[295] An acre will yield

1 to 2 tons of fresh thyme, providing 20 to 40 lbs. (9.07 to 18.14 kg) of thyme oil.[560, 618]

The oil, obtained by water-and-steam distillation, was formerly produced in the south of France and on Cyprus[233] but is now mainly exported from Spain.[210] Thymol for medical and other purposes is derived from thyme oil or obtained synthetically by dehydrogenation of *l*-piperitone.[466] In Europe, other natural sources are *Monarda didyma* L. and *M. punctata* L., the horsemints. In India, thymol is obtained from oil of ajowan (*Carum copticum* Benth.), the seeds of ajowan being formerly exported to Germany for distillation.[584]

Figure 75. Thyme (*Thymus vulgaris* L.), a well-known culinary herb is also a commercial source of pharmaceutical thymol. Photographed in the Botanischer Garten im Belvedere, Vienna, Austria.

Medicinal uses:

Thyme was valued as an antiseptic and fumigating herb in the classical era. Modern investigations have shown the essential oil to be active against *Salmonella typhi*, *Corynebacterium diphtheriae*, *Escherichia coli*, *Diplococcus pneumoniae*, *Staphylococcus aureus* and *Streptococcus pyogenes*.[11, 73] Formerly, a cold infusion of the whole plant was prescribed in dyspepsia and in convalescence from "exhausting diseases," a warm infusion to relieve hysteria, dysmenorrhea, flatulence, colic, headache and as a sudorific.[305] In Colombia and Cuba, a decoction of the fresh or dried herb is taken as a stomachic, sudorific and remedy for ordinary coughs and whooping cough.[440] The steam from the boiling herb may be inhaled, and the decoction has also been injected as an enema.[605]

Red thyme oil, official since the 16th century, is germicidal and has been much used in mouthwashes, gargles, dentifrices, cough drops and various disinfectant preparations. In whooping cough and bronchitis, it has been given in doses of .6 to .3 ml. Externally, it is used as a rubefacient and counterirritant in liniments for relief of neuralgia and rheumatism.[423, 605]

Thymol has been employed as an antiseptic on wounds and sores, but it is strongly irritant and its effectiveness is lessened in contact with an abundance of proteins. Its main role in modern practice is in treating fungus diseases of the skin, either in the form of a 1% alcohol solution or 2% strength powder. Thymol iodide was for some time used by dermatologists as a bacteriostatic and fungicidal agent in powders, lotions and ointments but has been abandoned.

Thymol is sometimes prescribed for blastomycosis, moniliasis and coccidioidal infections. It is no longer given as an antiseptic for the stomach and intestines, as it causes diarrhea and is absorbed from the intestines, 50 percent being excreted in the urine as the glucuronate or sulfate.[423]

Toxicity:

In addition to diarrhea, thymol taken internally may cause dizziness, headache, nausea, vomiting and muscular weakness, and may have a depressing effect on the heart, respiration and body temperature. Its use as a vermifuge, particularly in cases of hookworm, has been discontinued because doses strong enough to be effective can cause fatality.[423]

Other uses:

Thyme, fresh or dried, is extensively utilized as a culinary herb. Dried thyme and thyme extract have been long employed for flavoring liqueurs. Thyme oil is important in the food industry for seasoning meats, soups, sauces, condiments, pickles, bakery products, soft drinks and ice cream.[210] It also enters into colognes and aftershave lotions, soaps and detergents.[41] Thymol serves as a meat preservative, insect repellant and as a fungicide in mildew preventives.[493] The herb was burned by the Greeks as an incense in temples. The flowers are favored by honeybees, and wild thyme honey was extolled by ancient writers, that produced on Mount Hymettus near Athens being of finest flavor and sweetness.[233]

Solanaceae

Figure 76. Belladonna (*Atropa belladonna* L.) in flower and fruit. Photographed in the Beal-Garfield Botanic Garden, Michigan State University, East Lansing, Michigan. (*See also* color illustration 12.)

Belladonna

Botanical name:

    Atropa belladonna L.

Other names:

    Deadly nightshade, sleeping nightshade, English belladonna, dwale, poison black cherry.

Family:

    Solanaceae, the Nightshade Family.

Description:

    An erect, herbaceous plant to 6 ft. (1.8 m) high, with a long, cylindrical, branched, fleshy, creeping, perennial root, varying from 1 in. (25 mm) to several inches in width;[305] yellow-brown externally, white within.[624] The plant is shrub-like, with spreading, slender, cylindrical, sometimes slightly downy, often purplish, branches. Its leaves are alternate, but on the upper branches, each leaf is usually accompanied by a smaller, stipule-like leaf.[130] The blade is ovate, pointed, entire, soft, with conspicuous veins and frequently with a purplish midrib which may be hairy beneath; 3 to 9 in. (7.6 to 22.8 cm) long, 2 to 4 in. (50 to 100 mm) wide, on ¼ to 1½ in. (.5 to 4 cm) petioles. The flowers, borne singly in the leaf axils or in the forks of the branches, are drooping, to 1¼ in. (3 cm) long, the bell-like corolla five-lobed, dull-purple within, yellowish on the outside, cupped at the base by a five-pointed, green calyx. After the corolla falls, the calyx persists on the oblate, glossy fruit which turns from green to dark-purple and finally black as it ripens. The fruit reaches ½ in. (12.5 mm) in diameter, is two-celled and contains many kidney-shaped seeds embedded in pulp, which is somewhat sweet[286] and yields a purple juice.[400] In variety *lutea*, the leaves and stems are yellow-green and the flowers and fruits are yellow.[584]

Origin and Distribution:

    Belladonna is indigenous to southern and central Europe and naturalized in southern England. It is cultivated for drug use in central Europe, England, the United States and northern India. It is grown also, in temperate climates, as an ornamental in flower gardens and as a specimen in botanical collections.[116, 130]

Constituents:

All parts of the plant contain alkaloids, chiefly *l*-hyoscyamine ($C_{17}H_{23}O_3N$) plus a little atropine and hyoscine ($C_{17}H_{21}O_4N$); also certain volatile bases, pyridine and N-methyl-pyrroline. Succinic acid, asparagin and β-methylaesculetin (scopoletin) have been found in the leaves.[584] Hellaridine, resembling atropine in action, was reported in 1956 in the roots of belladonna grown in Greece, the proportion being .002%.[116]

Alkaloid content is highest in the fresh ripe fruit, fresh seeds and fresh leaves. When dried, the unripe fruit is the most potent, secondly the leaves and thirdly the ripe fruit.[116] The alkaloid content of the root is highest just before the flowers appear. The entire plant is deficient in alkaloids during prolonged periods of cool and cloudy weather. Maximum total alkaloid content is 0.8% in seeds, 0.7% in roots, 0.6% in leaves.[272] The old, large stems have a low alkaloid content.[624] For the drug trade, the leaves must contain at least 0.35% hyoscyamine USP, 0.30% BP.[584] Roots must contain at least 0.45% USP, 0.4% BP.[23, 618]

Propagation, Culture and Harvesting:

Belladonna may be propagated by cuttings taken from new growth and by rootstock division in spring but is grown commercially from seed. Germination rates may not exceed 60 percent because of the hard seed coat and the frequency of seeds with undeveloped embryos.[118] Nevertheless, one to two ounces (28 to 56 g) of the tiny seeds will produce more than the 10,000 plants required for an acre of ground at a spacing of 2 × 2½ ft. (.6 × .75 m).[23]

Highest alkaloid content is attained on fully open, newly cleared land or burned-over forest areas.[586] In India, high-yielding mutants have been developed by treating seed with methyl nitrosoguanidine (NG) and ethylenimin (EI).[300] The plant requires rich, moist soil, ample fertilization and thorough weeding. It is subject to a wilt disease, attacks by the potato beetle and flea beetle, and also occasionally root rot caused by *Phytophthora erythroseptica* Pethybr.[23]

Harvesting is done when the plants are in full bloom. A single crop is obtained the first year, two or three crops during each of the following two, three or four years,[23] after which new plantings are established. Before 1942, only the leaves were plucked for the highest grade of product. At that time, restrictions were removed on the amount of stems in the harvested material, and thereafter the plants could be cut or mowed, leaving only 2 in. (50 mm) of stem above ground, from which new shoots soon emerge. One acre yields a crop of approximately 5 to 6 tons. The cut tops are immediately dried on wire racks in shade or with artificial heat, with or without fans for air circulation. One hundred lbs. (45.3 kg) of fresh material is reduced to roughly 18 lbs. (8.1 kg) of dried. Prolonged drying reduces alkaloid content through enzyme action.[116] Leaves stored at a relative humidity of 100% for one year have showed a moisture content of 61.7%, only 3.3% degradation of hyoscine, 11.8% decomposition of atropine.[354]

When the plants are at least two or three years old, they are mowed down in the fall; the roots are plowed up, washed, cut into 4 in. (10 cm) lengths, split lengthwise if thick, and shade- or sun-dried.[560] The yield varies from 150 to 300 lbs. (68 to 136 kg) per acre. The alkaloid content is higher (up to 0.72%) in year-old roots, but they are too small to harvest.[20]

Medicinal uses:

Belladonna leaves were introduced into the London Pharmacopoeia in 1809.[584] They have been official in the United States since 1820. The root became official in the United States and in Great Britain in 1860[584, 618] but is no longer official in the United States. Belladonna leaves are used only for the preparation of tincture, extract (in powdered form or pilular for making ointments) and fluidextract. The tincture contains 30 mg alk./100 ml; the extract, 1.25 g alk./100 g; the fluidextract, 300 mg alk./100 ml.[423] These belladonna preparations are administered in carefully measured, subtoxic doses, the drug serving as an anodyne, relaxant, sedative, antigalactagogue, antidiuretic or means of limiting other glandular secretions,[513] or as a mydriatic or antiasthmatic, as required. It is employed in ophthalmology, bradycardia, Parkinson's disease, psychiatry; it is used in treating convulsions, epilepsy,[305] whooping cough and spermatorrhea, night sweats, gastric ulcers, complaints of stones in the kidneys or gallbladder,[624] and also as an antidote for depressant poisons such as opium, muscarine and chloral hydrate.[20, 513]

Where root preparations are official, they are used primarily in the external treatment of gout and rheumatism and other afflictions.[624]

Toxicity:

Belladonna poisoning in grazing animals is rare, though occasionally calves[197] or horses are victims. Sheep, pigs, goats and rabbits are less susceptible.[130] Human poisoning has resulted from eating the flesh of rabbits and birds that have fed on the plant. Children have been fatally poisoned by eating as many as three fruits.[272] If belladonna poisoning is suspected, a few drops of urine from the subject may be applied to a cat's eye. Exposure to light after one-half hour will show a fully dilated pupil if belladonna has been ingested.[197] Toxic effects may be experienced by those who have handled the plant, the poison being absorbed through the skin.[631] The juice of the fruits irritates the face; the sap of the plant causes dermatitis of face and hands of harvesters. Pharmacists may suffer skin irritation from contact with the dried plant material.[68]

Symptoms of internal poisoning by the plant or fruit include dilated pupils (near-blindness from large doses), flushed skin, dryness of mouth and throat, difficulty in swallowing,[305] nausea, vomiting, muscular weakness, trembling, incoordination, prostration, excitement, delirium, hallucination, loud and rapid heart beat,[197] difficult urination, constipation, CNS depression,[631] painful respiration, drowsiness, sometimes paralysis and coma; death may follow from respiratory and cardiac failure.[130] Poisoning by accidental overdoses of atropine sulfate (600 mg orally) produced mainly sustained CNS excitation in three children treated in England. The atropine was absorbed rapidly, most of it excreted within forty-eight hours, and recovery was complete.[360]

Figure 77. Indian belladonna (*Atropa acuminata* Royle ex Lindl.) has bell-shaped flowers, brownish-yellow with greenish veins.

Indian Belladonna

Botanical name:

    Atropa acuminata Royle ex. Lindl. (*A. belladonna* var. *acuminata, A. belladonna* C. B. Clarke NOT Linn.)

Other names:

    Indian Atropa, sagangur.

Family:

    Solanaceae, the Nightshade Family.

Description:

    The plant closely resembles *A. belladonna*. It has a large taproot with lateral rootlets 6 in. (15 cm) or more in length and ⅜ to ¾ in. (1 to 2 cm) thick. The leaves are 3 to 8½ in. (7.5 to 21.5 cm) long, 1½ to 3½ in. (4 to 9 cm) wide, olive-green, elliptic-lanceolate, pointed at both ends. The flowers are brownish-yellow with greenish veins and are borne singly or in 2s or 4s, or as many as 5 together in the axils of the upper leaves.[117] As with *A. belladonna*, the fruits are purple-black when ripe.[20]

Origin and Distribution:

    Indian belladonna is indigenous from Simla to Kashmir and adjacent Himalayan regions of northern India, growing in abundance at altitudes between 6,000 and 11,000 ft. (2,000 to 3,500 m). During World War I and the early years of World War II, roots were gathered from the wild and exported in large quantities, mainly to England. Since 1942, all of the crop has been utilized in India. Depletion of the wild sources stimulated interest in cultivation, and there are now large plantations in the Kashmir Valley.[116]

Constituents:

    The hyoscyamine content of *A. acuminata* leaves is comparable to that of *A. belladonna* leaves, ranging from 0.13 to 0.94%.[116] In comparative assays of *Atropa* leaves from plantings in northern India, two samples of *A. belladonna* showed 0.48 and 0.51% total alkaloids; *A. acuminata*, 0.42 to 0.55%.[399] In *A. acuminata*, the percentage is highest when the blooms are in the primordial bud stage and decreases as flower development progresses. However, at the time of highest alkaloid content, the plant is not in full leaf, and harvesting must be delayed until the "advanced flower-bud stage," when the plant has attained 75 to 90 percent of its maximum growth.[520] The alkaloid content increases markedly after the first year.

Hyoscyamine content of the roots varies from 0.29 to 0.8%, but they show also a larger proportion of non-hyoscyamine volatile bases than do those of *A. belladonna*. The reputed low quality and low alkaloid content of Indian belladonna leaves and roots is blamed largely on the percentage of stem bases or immature roots, or adulteration with leaves and roots of a somewhat similar but hollow-stemmed plant, *Phytolacca acinosa* Roxb.[117] *Ailanthus* leaves have also been discovered as adulterants.[584] Another cause of low rating of Indian belladonna leaves has been the indiscriminate collection at all stages of growth rather than at the ideal time.[117]

Propagation, Culture and Harvesting:

B. acuminata seeds germinate in two to three weeks at a daytime temperature of 70° F. (21° C.), even though nights may be cool. Commercial plantations have proved successful at elevations as low as 1,000 ft. (305 m). Root rot has not appeared in this crop in India, but much destruction of transplanted seedlings is caused by cutworms if transplanting is done in the dry season. Downy mildew and various caterpillars and beetles which attack the foliage, flowers and fruits require strict control measures.

Two crops are harvested per season, the plants being cut by hand 2 in. (5 cm) above ground level. Prompt irrigation of the field thereafter stimulates growth for the second harvest. The plant tops are chopped and spread in the sun to dry, with repeated raking to ventilate and prevent heating.[117]

Medicinal uses:

Same as for *A. belladonna*. Indian belladonna is accepted as official in India, the United States and Great Britain.

Toxicity:

All parts of the plant are poisonous. Chopra *et al.* relate that travelers in the Kagan Valley sometimes cook and eat the leaves of *A. acuminata*, mistaking it for *Phytolacca acinosa* Roxb., and are poisoned as a result.[116]

Figure 78. Corkwood (*Duboisia myoporoides* R. Br.). Typical leaves and flowers, as illustrated in Fig. 1, Bulletin 51, *Hybrids between Duboisia myoporoides and D. leichhardtii,* by H. M. Groszman, G. P. Kelenyi and C. N. Rodwell; Department of Primary Industries, Queensland, Australia, 1949.

Corkwood

Botanical name:

    Duboisia myoporoides R. Br.

Other names:

    "Elm," "eye-plant," orungurabie, ngmoo.

Family:

    Solanaceae, the Nightshade Family.

Description:

    A subtropical shrub or small tree, occasionally attaining 40 ft. (12 m) in height; when cut or burned down, producing numerous suckers from the roots and forming dense clumps.[62] The bark is gray, corky, fissured, and the wood is light in weight. Leaves are alternate, 2½ to 4 in. (6 to 10 cm) in length, ovate-oblong to broad lanceolate, blunt-tipped, tapering at base; petiole 3/16 to ½ in. (5 to 12.5 mm) long. Flowers borne in terminal, much-branched panicles, white or pale-lavender or lavender-streaked, bell-shaped, with 5 rounded lobes; 5/32 in. (4 mm) long. Fruit abundant, nearly round, 3/16 in. (5 mm) wide, black and juicy when ripe, containing 6 to 12 seeds.[44, 50, 104, 272]

Origin and Distribution:

    Corkwood grows wild on the borders of thickets and in open woodlands along the east coast of Queensland and northern and central New South Wales. It is indigenous also to New Caledonia.[62]

Constituents:

    The leaves contain a complex of tropane alkaloids including hyoscine (scopolamine), hyoscyamine, norhyoscyamine, valeroidine, tigloidine, poroidine, isoporoidine, isopelletierine and anabasine. Nicotine and nornicotine have been found in seedlings as well.[617] The constituents vary not only with the stage of growth but also with the geographical location. Hyoscine is the major alkaloid in the leaves of plants growing in the northern part of the natural range, while hyoscyamine dominates in the leaves from southern areas.[606] Analyses of sixteen samples of leaves from a single commercial bale showed hyoscine variation from 0.209 to

0.587% and hyoscyamine from 0.067 to 0.244%.[235] During World War II, investigators in Australia discovered some specimens with concentrations of hyoscine four to five times greater than had previously been found in any plant, and more than 7,000 oz. (128.45 kg) of the drug were produced from *Duboisia* during the acute shortage.[260]

In the wood of the young root, tropine is the principal alkaloid, with lesser amounts of valtropine, hyoscyamine and valeroidine; tthe root bark possesses hyoscine, atropine, tropine and apohyoscine. Wood of mature roots contains hyoscyamine, hyoscine, apohyoscine, tropine and tetramethylputrescine. These alkaloids and valtropine occur in the mature root bark.[136]

Propagation, Culture and Harvesting:

Interest in cultivation of corkwood was greatly stimulated by the demand for hyoscine during World War II. There are seedling plantations in Australia, and seeds have been distributed to other countries where the plant has been found to tolerate a diversity of soils and climatic conditions, even surviving severe frosts.[62] *D. myoporoides* is relatively tolerant of "wet feet," needs good air flow but must be protected from strong winds. Some plantations consist of rows of "volunteer" regrowth of *Duboisia* after land has been cleared and disked in alternate strips. More often, seeds treated with gibberellic acid are planted in sterilized soil in nursery beds under 60% shade. The seedlings are set out when twelve weeks old or when the stem is well pigmented. The young plants will usually be 6 to 8 in. (15 to 20 cm) high at that stage. Spacing may be 12 × 10 ft. (3.6 × 3 m) or, in drier conditions, 15 × 12 ft. (4.5 × 3.6 m).[104] Field planting usually takes place in the fall, and care is taken to set the seedlings at the same depth as in the nursery bed to avoid crown rot. Clean cultivation is requisite. Weeds are controlled mechanically or by herbicides. Grass cover crops are sown after leaf harvest.

Major pests of *Duboisia* trees (especially of *D. Leichhardtii*, q.v.) are flea beetles, leaf-eating ladybirds, brown olive scale and a trunk borer. Pesticides are employed sparingly to avoid depredation of biological controls.[537]

In favorable sites, seedlings grow rapidly, as much as 1 in. (25 mm) a day for a period of three to four weeks,[62] attaining 4 ft. (1.2 m) in eight months.[260]

Leaves are first harvested when the plants are seven to ten months old[104] and thereafter by repeated trimming of the plants at seven- to eight-month intervals, removing no more than 70 to 80 percent of the foliage each time to encourage quick regrowth. Thus, the plants are maintained in the form of bushes rather than trees.[62] The lopped branches from each tree are tied in a bundle and hung on the tree until the end of the working day, when all bundles are gathered and taken to an open-sided shed and stacked for further drying, which takes two weeks to a month in warm weather. Removal of the leaves cannot be done by manual threshing, as this releases alkaloid-laden dust injurious to the laborers. It is customary to crush the bundles by running a light tractor over them, then the dry material is sieved through rectangular screens to separate the branches and twigs from the leaves and the leaves are baled or, less often, bagged for transport. Handlers are advised to wear protective clothing, face masks and goggles to avoid intake of alkaloids by respiration and through the skin.[104]

Plantings of 300 to 400 trees per acre (740 to 980 per hectare) yield about 400 lbs. dried leaves per acre (498 kg per hectare).[104] Annual yields of 1 ton of dry leaves per acre have been reported.[62]

Medicinal uses:

From the time of the discovery of the mydriatic activity of this plant by Doctor Joseph Bancroft in 1877, an extract either of the leaves or of the fruit has been employed in opththalmia in Australia and Europe. It acts more swiftly than atropine, producing wide dilation of the pupil in twenty minutes. The extract has been administered to relieve night sweats of consumptives and painful tenesmus caused by inflammation of the urethra and neck of the bladder. It has also been employed to treat goiter, to dispel maniacal delirium and, in the form of an ointment, has been applied as a sedative in cases of inflammation of the cornea.[363]

The leaves of this plant and of *D. Leichhardtii*, q.v., and hybrids between these two species, are now second only to *Hyoscyamus muticus* as commercial sources of hyoscine and of atropine derived from hyoscyamine. The unusual demand for hyoscine during World War II was because of its effectiveness in controlling motion sickness and seasickness as well as shock cases in military personnel.

Toxicity:

The Australian aborigines have long been aware of the toxic properties of this plant. They put branches in shallow water to stun the eels and bring them to the surface for capture. They also made holes in the main stem, poured in a liquid, let it remain overnight and then imbibed it as a narcotic the next morning.

Emanations from branches hung in a closed room have caused dizziness and nausea. The extract employed in eye drops (4 grains to the ounce) has produced dizziness, delirium and dryness of the mouth.[50, 363] Ingestion of the plant material causes, in addition, impairment of vision (even blindness), loss of taste, loss of muscular control and gastroenteritis. Children have displayed nervous and muscular symptoms with delirium after chewing the dried leaves, and in 1933, a two-year-old child died from eating the leaves. The plant is rarely grazed by livestock, but ingested leaves have fatally poisoned cattle and horses in New South Wales.[272]

Other uses:

The soft wood of *D. myoporoides* is popular for carving and for engraving woodcuts.[363]

Figure 79. Leichhardt corkwood (*Duboisia Leichhardtii* F. v. M.). Typical flowers as illustrated in Fig. 2, Bulletin 51, *Hybrids between Duboisia myoporoides and D. leichhardtii*, by H. M. Groszman, G. P. Kelenyi and C. N. Rodwell; Department of Primary Industries, Queensland, Australia, 1949.

Leichhardt Corkwood

Botanical name:
>Duboisia Leichhardtii F. v. M.

Other name:
>Corkwood.

Family:
>Solanaceae, the Nightshade Family.

Description:
>A shrub or small tree to 17 ft. (5 m) high. Leaves are generally shorter and more slender than those of *D. myoporoides*; 2 to 4 in. (5 to 10 cm) long, oblong-lanceolate to obovate, blunt-tipped, tapering at base, entire. Flowers, borne in broad, terminal panicles, fairly leafy at the base, bell-shaped with flaring, narrow, long-pointed lobes, much more conspicuous than the lobes in *D. myoporoides*. Calyx small, with short, broad teeth. Fruit oval, ¼ in. (6 mm) long, with 6 to 12 kidney-shaped, pebbly-surfaced seeds.[50, 272]

Origin and Distribution:
>This species is found wild only in two limited areas of southeastern Queensland, a short distance inland from *D. myoporoides*. Hybrids between the two species are sometimes observed. *D. Leichhardtii* is most luxuriant in the Nanango-Yarraman-Kingaroy district on red soils, and here its numbers have been reduced by land-clearing for agricultural and ranching activities. The existing wild trees are mainly on roadsides or in neglected pastures. The species was first described in 1867, and it was correctly named in 1877.

Constituents:
>Doctor F. M. Bailey recommended investigation of its properties in 1880, but it did not receive attention until 1917, when the total alkaloids of the dry leaves were reported as amounting to 1.4%, mainly nor-hyoscyamine, *l*-hyoscyamine, and *l*-hyoscine, with lesser quantities of atropine and nor-atropine. There was no further interest in this species until 1940. Leaf analyses by Barnard and Finnemore in 1941-42 showed

total alkaloids as 2.7 to 3.0%, with hyoscyamine predominating and readily extractable with less difficulty than from *D. myoporoides*. Similar results were reported by later investigators who discovered, in addition, that a clear, crystalline product could best be obtained in autumn and winter, free of sirupy bases which are present in spring and summer.[62]

Propagation, Cultivation and Harvesting:

This species is being propagated by seed and by tip cuttings of types that give a high and consistent yield. It is the leading species in the South Burnett area of Queensland, where commercial cultivation is concentrated. There the rainfall averages 30 in. (760 mm) annually. *D. Leichhardtii* is sensitive to waterlogging. Though it occurs naturally in a range of soil types, it is grown commercially in red-brown sandy loam having a pH from 4.5 to 5.5[104] For cultivation and harvesting operations, see *D. myoporoides*.

Medicinal uses:

D. Leichhardtii is being more and more utilized as a source of hyoscine and atropine which, in selected types, occur in economically

Figure 80. Young, well-foliaged tree of *Duboisia Leichhardtii* (or possibly a hybrid). Photo courtesy Mr. A. R. Carr, District Adviser in Horticulture, Department of Primary Industries, Queensland, Australia.

favorable proportions.[62] Hyoscine on isolation is converted to its *N-butyl* derivative.[235]

Toxicity:

Feeding tests in Queensland have demonstrated that the foliage of this plant is highly toxic to grazing animals.[272]

Hybrids:

D. myoporoides × *D. Leichhardtii* hybrids are now being grown in the more advanced plantations in preference to either of the parents. They are easily grown from cuttings and lack the nicotine which is often present in *D. myoporoides*, hampering the isolation of hyoscine. Some selected hybrids yield 1 to 3% hyoscine.[234]

Production:

In 1971-72, the total crop of dried *Duboisia* leaves from about 250 farms (2,500 acres) planted to the species and hybrids was 900 tons; in 1972-73, because of labor shortage and competing, more profitable, crops, it was 830 to 850 tons; in 1974, it was expected to be no more than 750 tons. The dried leaves are valued at $1,000 (Australian) per ton. A small amount of the crop is processed in Australia, the bulk being exported to West Germany and Japan for alkaloid extraction.[235]

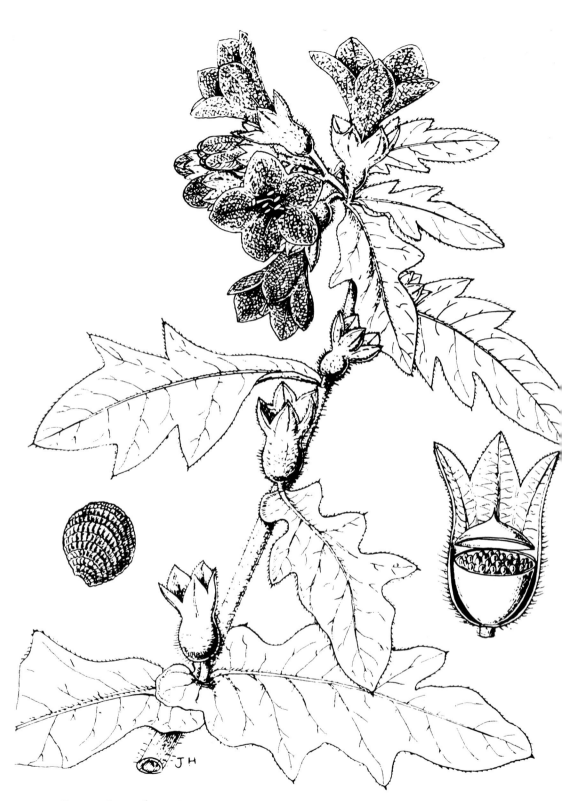

Figure 81. Henbane (*Hyoscyamus niger* L.). Reproduction of Fig. 2 in *Culinary and Medicinal Herbs*, Bulletin 76, 2nd ed, 1951, by permission of the Controller of Her Britannic Majesty's Stationery Office, London. (*See also* color illustration 13.)

Henbane

Botanical name:
>Hyoscyamus niger* Linn.

Other names:
Indian henbane, black henbane, fetid nightshade, sickly-smelling nightshade, stinking nightshade, hog bean, poison tobacco.

Family:
Solanaceae, the Nightshade Family.

Description:
An annual or biennial, much-branched, herbaceous plant, attaining a height of 3 to 6 ft. (1 to 2 m) and having a viscid-hairy, stout stem, woody at the base, and a spindle-shaped, somewhat branched taproot, brown outside, whitish inside, and slightly sweetish.[63] Leaves are gray-green, alternate, 3 to 8 in. (7.5 to 20 cm) long, or to 1 ft. (30 cm) long, in the basal rosette; oblong to oblong-ovate with conspicuous veins, nearly entire or coarsely and irregularly toothed or sometimes deeply lobed, slightly downy, with long, glandular, black-tipped hairs beneath the midrib and veins. The lower leaves are petioled, the bases of upper leaves half-clasping the plant stem. Biennial plants do not normally flower the first year.[117] Flowers are 1 to 2 in. (25 to 50 mm) wide, short-stalked on lower parts, sessile on upper branches,[197] borne in the leaf axils or in short nodding spikes at the tips of the branches. The corolla is funnel-shaped, with 5 rounded lobes, greenish-yellow, pale-yellow or pinkish, with purple throat and minutely reticulated with dark-purple in the biennial type, light-purple in the annual. The five-pointed, green, tubular, ribbed, hairy calyx 1 in. (25 mm) long, extends and persists, completely enclosing the fruit, and becomes dry and stiff. The fruit is an ovate, two-celled capsule ½ in. (12.5 mm) long, with a groove on each side and a "lid" which opens on maturity, releasing numerous brown or gray, hard, pitted seeds, slightly kidney-shaped and $\frac{1}{18}$ in. (1½ mm) long. The fresh plant has a heavy, unpleasant odor said to repel rats. When burned, the leaves smell like tobacco and sparkle.[63]

Origin and Distribution:
Henbane is native from Scandinavia and southern England to the Mediterranean and Siberia, also from North Africa and western Asia

to northern India, where it flourishes up to altitudes of 14,000 ft. (4,270 m). In England, it is usually found wild on sandy seacoasts and as a weed of disturbed ground.[100] It is naturalized in the northern United States, particularly the northwestern states, and in Canada.[211] It has been cultivated to a limited extent in the United States, particularly during World Wars I and II, but requires too much hand labor to be economically produced in this country.[618]

Constituents:

All parts of the plant contain alkaloids. In the young stage of growth, the plant contains more hyoscine ($C_{17}H_{21}NO_4$) than hyoscyamine ($C_{17}H_{23}NO_3$). In the mature plant, hyoscyamine is the main active principle of the flowering tops and leaves[423] (especially the petiole[117]), also in the roots;[423] and there is some hyoscine (l-scopolamine) and perhaps atropine ($C_{17}H_{23}NO_3$). The plant material is more potent fresh than dried.[116] In India, the alkaloid content was found to vary from 0.04% in leaves produced at 1,000 ft (305 m) to 0.08% in leaves produced at 7,500 ft. (2,287.5 m). Contrary to former beliefs, the alkaloid of annual and biennial plants is virtually the same.[117] Minimum alkaloid content for commercial henbane in the United States is 0.04%; in England, 0.05%. Seeds contain 0.06 to 0.10% alkaloids, mainly hyoscyamine.[117]

Propagation, Culture and Harvesting:

While much of the henbane in the drug trade of the past was derived from wild plants, the supply today comes largely from plantations in England, Turkey and India.[117] Cultivation was begun in India as far back as 1839, expanded, and then declined. Extensive commercial plantings were not established until the mid-1950s. Plants are raised from seed either in nurseries or directly in the field. Great care must be taken in transplanting. Germination is improved by pre-treatment with sulphuric acid and thorough rinsing, and takes place in twelve to fifteen days. The plants require protection from the potato beetle and the henbane flea beetle (*Psylloides hyoscyami*), and they are highly susceptible to virus diseases.[23] With annual types, harvesting of leaves and flowering tops is done by hand at the time of full bloom. If biennial types are grown, the non-flowering tops may be cut the first season and again early in the second season, the third harvest taking place at the flowering period.[23] Plant growth, development and seasonal alkaloid content have been discussed at length by Chu.[119] The yield of plant material is about 600 lbs. (272 kg) per acre.[560] The crop must be immediately spread out to dry in the shade. If seeds are desired, only the leaves are plucked. Then, when the fruit matures but before the capsule opens, the plants are cut down and carried to sheds for seed removal, each plant yielding about 10,000 seeds. According to the British Pharmacopoeia, plant material for the drug trade may contain no more than 3% of stems exceeding ⅕ in. (5 mm) in diameter.

Medicinal uses:

Henbane has been employed medicinally and superstitiously since ancient times. It was the principal drug employed in sorcery. Pieces of the root were strung and worn around infants' necks to prevent fits and to ease the pain of teething. Seeds, capsules, and leaves have been smoked to relieve toothache. Henbane was omitted from the London Pharmacopoeias of 1746 and 1788 and restored in 1809.[63] In the United States, henbane tincture and extract were formerly administered mainly to control urinary spasms, but the drug is no longer official in this country. Though similar to belladonna in action, it contains less alkaloid and must be given in larger doses. Inasmuch as its action is modified by the central depressant effect of hyoscine,[423] it is less of a cerebral excitant than belladonna.[584] It has been employed to calm nervous irritation in hysteria[423] and may be combined with purgatives to avoid griping.[408] Henbane is used in Europe as a mydriatic, sedative and anodyne, and it is administered in cases of earache and rheumatism.[513]

Toxicity:

The poisonous properties throughout the plant, but particularly in the root,[197] are not destroyed by drying or by boiling. Severe intoxication has resulted from cooking and eating the root believing it to be the root of chicory or parsnip[286] or from eating the young leaves as salad greens.[63] Children have suffered from ingesting the seedpods or seeds.[412] Convulsions and mania have been occasioned by smoking the leaves or seeds as mentioned above. The odor of the fresh leaves may provoke dizziness and stupor.[286]

Among the symptoms of henbane poisoning are dilated pupils, disturbed vision, salivation,[306] loss of speech, convulsions, paralysis, abdominal pains, diarrhea, delirium and stupor. Autopsy shows inflammation of stomach and bowels. Years ago, nine people who were poisoned by the root reported that "all objects appeared scarlet for 2 or 3 days" after their recovery[457]

Grazing animals usually avoid the plant because of its odor and harsh texture, the flavor being rather mild[63] and only slightly acrid.[305] There have been losses of poultry from eating the seeds.[114] Henbane is said to be toxic to deer, fish and most birds,[63] but pigs are apparently immune.[286]

Egyptian Henbane

Botanical name:

    Hyoscyamus muticus L.

Other names:

    Indian henbane, mountain hemp.

Family:

    Solanaceae, the Nightshade Family.

Description:

    A perennial, erect, herbaceous, unpleasantly odorous plant, 1 to 3 ft. (30 cm to 1 m) high, glabrous or viscid-hairy, with a thick, simple, often warty stem, unbranched or having few branches. Roots creep and spread extensively. Leaves range from 2 to 5 in. (5 to 12.5 cm) in length and from 4 to 8 in. (10 to 20 cm) in width, the lower, larger leaves having winged petioles 1½ to 3 in. (4 to 7.5 cm) long, the upper, nearly or quite sessile. The blades are thick, fleshy, glabrous or pubescent, ovate-oblong to broad lanceolate, pointed, entire or variously lobed or toothed. Flowers (10 to 30) are funnelform, 1 to 1½ in. (25 to 40 mm) long, brownish-yellow or greenish-white, purple-veined or spotted and with dark-purple throat; lower flowers are borne in leaf axils, the upper in a terminal raceme. The five-toothed, hairy calyx, ⅔ to 1 in. (16 to 25 mm) long, enlarges and embraces the subglobose seed capsule, ¼ in. (6 mm) wide and ⅜ in. (10 mm) long.[16, 116, 454] The numerous seeds are kidney-shaped, 1/24 in. (1 mm) wide, yellowish- or grayish-brown and finely reticulated. They are oily and of bitter and acrid flavor.[266] In the Sahara and Himalayas, the plant blooms in March and fruits in June.[605]

Origin and Distribution:

    This species grows wild in tufts on rocky soil in desert regions of Egypt and the Sudan and from Syria to Afghanistan, western Pakistan and northern India. It is cultivated on a small scale in Kashmir and neigh-

---

Figure 82. Egyptian henbane (*Hyoscyamus muticus* L.), now the leading plant source of hyoscyamine from which is derived pharmaceutical atropine sulfate.

boring areas; also in Algiers.[116] It is a crop of some importance in Nigeria.[419] At one time, it was grown in southern California.[584]

Constituents:

Hyoscyamus muticus is more narcotic than *H. niger*.[277] All parts, including the roots, contain alkaloids, Egyptian plants more than those grown in India. Dried leaves and flowering tops contain 0.35 to 1.39% total alkaloids, mainly (90%) hyoscyamine. There may be .02% hyoscine (scopolamine) and considerable potassium chloride.[246] Some analyses have shown 0.5 to 0.6% hyoscyamine in the stems, 1.4 to 1.7% in the leaves, 2.0% in the flowers[116] and 0.9 to 1.3% in the seeds.[20] The International Pharmacopoeia specifies a minimum of 0.5% hyoscyamine in the drug product.

Commercial supply:

Egypt is the principal source of the crude drug material collected from wild plants by shepherds. The exported product consists of dried leaves, stems, flowering tops and some fruits, all matted together.[20] The seeds, harvested separately, are shipped from Iran to India, where they are sold in bazaars.[266] Egypt prohibits the export of viable seeds.[116] Seed germination is strongly influenced by temperature and light.[176]

Medicinal uses:

In Nigeria, *Hyocyamus muticus* is employed as an antispasmodic, is administered to prevent griping, as a remedy for sea sickness and to relieve asthma.[419] In Iran, the smoke of burning seeds is inhaled to relieve toothache.[266] It was reported by W. R. Dunstan and H. H. Brown at the Imperial Institute, London, in 1899, that hyoscyamine in a crystalline form could be isolated from *H. muticus* more easily than from *H. niger*, and they predicted that this species could become an important commercial source of that drug. They were correct, and this species is still the major plant source of the alkaloid, followed by *Duboisia myoporoides*, q.v. In the extraction of *l*-hyoscyamine from plant material, it is partly racemized to atropine, which is subsequently purified, then neutralized with sulfuric acid and administered as atropine sulfate.[426]

Toxicity:

In Syria and Iran, goats and sheep,[266] and in Africa, gazelles graze on this plant with impunity, but it is toxic to camels and horses.[605] In India, the dried leaves and flowers are smoked to induce intoxication.[277] The effect is so powerful that the users appear insane, discarding their clothes and dancing wildly.[604] Parts of the plant are occasionally used for malicious poisoning, causing dryness and constriction of the throat, delirium and violent actions.[116] For criminal purposes, Arabs powder the dried plant and mix it with dates or milk. If an antidote made of pulverized dates, water, butter and pepper is taken in time, the victim will perspire profusely and will not hallucinate.[605]

Scrophulariaceae

Figure 83. Foxglove (*Digitalis purpurea* L.). Photographed in the Maynard W. Quimby Medicinal Plant Garden, School of Pharmacy, University of Mississippi. (*See also* color illustration 14.)

Foxglove

Botanical name:

   Digitalis purpurea L.

Other names:

   Purple foxglove, throatwort, fairy finger, fairy cap, lady's thimble, Scotch mercury, dead men's bells and many others.

Family:

   Scrophulariaceae, the Figwort Family.

Description:

   A biennial or rarely perennial herbaceous plant to 6 ft. (1.8 m), with erect, thick, cylindrical, downy stem, seldom branched. Leaves in a dense rosette the first year, thereafter alternate; those on the lower part of the stem are long-petioled; in the upper leaves, the base tapers into a winged petiole; blades are 6 to 12 in. (15 to 30 cm) long, 3 to 4 in. (7.5 to 10 cm) wide, ovate-lanceolate to elliptic, toothed, wrinkled, conspicuously veined, green and woolly on the upper side, covered with whitish or grayish hairs beneath. The leaves are very bitter in taste. Flowers, borne the second year, in erect, showy, usually one-sided spikes, are pendent, tubular-bell-shaped (like the finger of a glove), 1½ to 3 in. (4 to 7.5 cm) long, faintly five-lobed; they are rarely white in the wild, usually lavender or cerise with a hairy, purple-blotched white area on the inner surface of the lower side; 4 stamens, tipped with large yellow, two-lobed anthers. The calyx is divided into 5 pointed, ribbed segments; it persists on the fruit—a dry, ovoid, downy capsule to ⅝ in. (1.5 cm) long, with 2 cells containing many small, red-brown, pitted seeds.[63, 197, 305]

Origin and Distribution:

   Native throughout the British Isles, temperate western Europe and Morocco. Widely grown as an ornamental in temperate climates, it was introduced into North America as a garden flower and has become naturalized in Newfoundland and British Columbia; also the northwestern United States and parts of the East along roadsides and edges of woods. It has run wild also in Mexico and in Central America at elevations of 7,217 to 9,842 ft. (2,200 to 3,000 m); also at sea level in southern South

America, in highlands in a few of the islands of the West Indies, in northern India and in New Zealand.[130]

Foxglove is cultivated as drug crop in England, Germany, Austria, Hungary, Japan and northern India, especially Kashmir at altitudes of 5,000 to 6,000 ft. (1,515 to 1,830 m). There has been limited cultivation in the United States from New England to Minnesota,[618] and, at one time, 90 percent of the American crop was produced by the S. B. Penick Co. at Oley, Pennsylvania.[122] In Argentina, commercial cultivation was placed under government control in the mid-1940s, and exportation was prohibited. Plantations in warmer areas of India (as in Darjeeling and the Nilgiri Hills) produced leaves of inferior quality and have been abandoned, though the plant has become thoroughly naturalized in these areas.[117]

Constituents:

Ornamental strains of D. purpurea usually have a low level of active principles. The leaves of wild plants and strains selected for pharmaceutical use contain at least thirty glycosides (in total quantities of 0.10 to 0.63%),[194] primarily the two tetraglycosides known as purpurea glycoside A (yielding digitoxin by one molecule loss during extraction) and purpurea glycoside B, the precursor of gitoxin. A third major glycoside is glucogitaloxin, discovered in 1956.[122] On hydrolysis, digitoxin and gitoxin each lose three molecules of the alpha-deoxymonosaccharide D-digitoxose ($C_6H_{12}O_4$), thus producing their respective aglycones— digitoxigenin and gitoxigenin. Jakovljevic has presented a review of the physical properties, other aspects and biological activity of digitoxin.[278] Glucogitaloxin is the precursor of gitaloxin, yielding gitaloxigenin, and of verodoxin, which is reportedly three times as toxic as gitaloxin but present in small quantity.[584] A new cardiac glycoside, digicorin, was isolated from the leaves in Japan in 1949.[116]

The seeds contain digitalin ($C_{36}H_{56}O_{14}$), which on hydrolysis is converted to dianhydrogitoxigenin. In both the leaves and seeds, there are several saponins including digitonin, gitonin and tigonin.[423] The seeds also yield 31.4% of a yellow, bland oil.[117]

Propagation, Culture and Harvesting:

The plant is grown from seeds which are so fine they are often mixed with sand to achieve even distribution in the nursery beds. Sowing may be done as early as February. The seedlings are transplanted to the field when 2 to 3 in. (5 to 7.5 cm) high, in late spring. The soil may be acid or calcareous;[77] it must be loose, well-drained and with an adequate level of manganese.[584] Weed control is essential, and a winter mulch may be necessary to avoid cold damage. In India, 10,800 plants are set per acre, self-sown seedlings being taken up and transferred to a new field. The plant flourishes best in light shade.[117] It was formerly believed that first-year leaves were deficient in glycosides, but tests have shown that they may be equal to or superior to second-year leaves. Therefore, leaves can be harvested in late fall of the first year, even though the plants are not old enough to bloom.[77] In Pennsylvania, this was the only leaf harvest, and new plants were set out each spring.[122] In milder climates, the next harvest of leaves takes place the second year when the flowers are in bud or throughout the period of full bloom.[117] Some investigators have found the glycoside content highest from May to July, prior to blooming.[77] After fruiting, the plant dies down, but new growth may arise from the root and survive another year or two. Immediately after collection, the leaves are spread on racks to wither for thirty-six hours and then are shade-dried for seven to ten days[117] or oven-dried without delay, recommended temperatures varying from 104° to 176° F. (40° to 80° C.). Moisture content of dried leaves may range from 8 to 12%.[97] The U.S. Pharmacopoeia specifies 140° F. (60° C) and reduction of moisture to no more than 8%. Finally, the leaves are packed in airtight drums, as excess humidity will cause a marked decline in potency.[77]

Yields of 450 to 600 lbs. (204 to 272 kg) dry leaves per acre have been obtained.[560] Glycoside yield varies with the cultivar grown as well as with the stage of growth, cultural conditions[195] and methods employed for extraction.[298, 340, 350, 434]

Medicinal Uses:

The therapeutic use of foxglove has been traced back to the 10th century. It was long employed in England, Wales and Europe as an expectorant and remedy for epilepsy and was applied externally on scrofulous swellings. A leaf infusion was taken to relieve sorethroat.[457] The plant became official in the London Pharmacopoeia in 1650.[584] Its diuretic, dropsy-relieving activity was proclaimed in 1775 by William Withering who learned of it from a Shropshire woman.[423] In 1877, the plant's reputation was that of a cardiac sedative and diuretic, and it was commonly so employed. Subsequently, it was almost abandoned because of occasional fatalities from overdoses or accidental ingestion,[63] but it soon became an indispensable remedial agent.

In South America today, infusions or pills made of the powdered leaves are given to relieve asthma or as diuretics, sedatives or cardiotonics. In Colombia, epilepsy in infants is treated by daily doses of a sweetened "tea" made of three or four corollas of the white-flowered type, over a period of nine days.[440]

The United States Pharmacopoeia specifies that only powdered digitalis—finely ground leaves with a potency of one U.S.P. unit per 100 mg—or preparations containing pure digitoxin or gitalin (a mixture of glycosides—mainly gitaloxin, gitoxin and digitoxin) may be employed. The most important use of *Digitalis purpurea* is in the treatment of congestive heart failure because it increases the force of systolic contractions. It also provides more rest between contractions. A secondary effect is the lowering of venous pressure in hypertensive heart disease. It elevates blood pressure in cases of low pressure due to impaired heart function and reduces the size of dilated hearts. Diuresis and reduction of edema result from improved circulation.[423]

Digitoxin is 1,000 times more active than the powdered leaves,[117] is completely and rapidly absorbed from the gastrointestinal tract and causes little or no gastric irritation.[423]

In India, an ointment containing digitalis glycosides is valued for healing wounds and burns.[117] Physiological effects and clinical uses of digitalis are amply reviewed by Smith and Haber.[538]

Toxicity:

All parts of the plant are toxic. Horses, cattle, swine and turkeys have been fatally poisoned by feeding on the fresh plant when forage was scarce; horses and cattle have succumbed to eating the dried leaves in hay.[197] Children have been made ill by sucking the flowers or ingesting the seeds or parts of the leaves. Toxic doses of the fresh leaves are reported as 6 to 7 oz. (170 to 198.4 g) for an ox, 4 to 5 oz. (113.3 to 141.7 g) for a horse, 1 oz. (28.3 g) for a sheep and ½ to ¾ oz. (14.17 to 21.25 g) for a pig. Lethal doses of digitalin are 2½ g for a horse, ⅓ g for a dog and ⅙ g for a cat.[116]

Digitalis glycosides are cumulative in the system and excreted slowly; therefore, administration must sometimes be discontinued if adverse effects appear. Digitalis intoxication is common.[504] The incidence in hospitalized patients has been reported as ranging from 8 to 35 percent.[539] A survey of patients receiving digitalis was conducted in 1967-68 and showed that 21.4 percent experienced intoxication. After dosage guidelines were set forth in an educational program on digitalis therapy, instances of intoxication and fatality markedly decreased.[416]

Signs of poisoning by the plant or overdoses of the drug (often delayed several hours) include contracted pupils, blurred vision, strong but slow pulse, gastroenteritis, nausea, vomiting, dizziness, dullness, drowsiness, abdominal pain, bloody diarrhea, excessive urination, acute headache, tremors; in severe cases, stupor, mental confusion, delirium, sometimes convulsions, and death.[117]

Grecian Foxglove

Botanical name:
    Digitalis lanata Ehrh.

Other name:
    Woolly foxglove.

Family:
    Scrophulariaceae, the Figwort Family.

Description:
    An annual, biennial or perennial herb, 2 to 3 ft. (.60 to .91 m) high; stem unbranched, upper part downy-white but less hairy than that of *D. purpurea*. Leaves linear– to oblong-lanceolate, to 1 ft. (30 cm) long and 1½ in. (4 cm) wide, pointed, entire or only slightly toothed, gray-green with a conspicuous network of veins, somewhat hairy or glabrous (less woolly than those of *D. purpurea*), sessile, the base continuing briefly down the stem. Flowers in erect, dense, many-flowered, hairy raceme; buds very woolly; corolla to 1 in. (25 mm) long, partly urn-shaped, downy, ivory, yellow or purple with brown reticulations, the lower lobe whitish, pointed and projecting. Calyx has narrow, downy lobes, shorter than the floral bracts. Blooming occurs the first year in regions of high light intensity, the second year elsewhere.[51, 54, 533, 584]

Origin and Distribution:
    D. lanata is native to central and southern Europe, especially along the Danube. It is cultivated as a drug crop in Switzerland,[466] Greece,[51] Russia, Egypt and Kashmir at an elevation of 7,000 ft. (2,133 m) or more; also in Holland,[584] England and to a limited extent in Canada[117] and southern California.[122]

Constituents:
    The principal glycosides of *D. lanata* are the tetraglycosides lanatoside A ($C_{49}H_{76}O_{19}$), lanatoside B ($C_{49}H_{76}O_{20}$), lanatoside C

Figure 84. Grecian foxglove (*Digitalis lanata* Ehrh.) has a downy-white stem and gray-green leaves; ivory, yellow or purple flowers with brown reticulations.

($C_{49}H_{76}O_{20}$), lanatosides D and E, from which are derived, by loss of an acetyl group, the purpurea glycosides A and B, desacetyllanatosides C and D and glucogitaloxin. Removal of one glucose molecule from the tetraglycosides yields the triglycosides acetyldigitoxin, acetylgitoxin, acetyldigoxin, acetyldiginatin and acetylgitaloxin. Removal of both an acetyl group and a glucose molecule leaves the triglycosides digitoxin, gitoxin, digoxin ($C_{41}H_{64}O_{14}$), diginatin and gitaloxin.[584]

Minor constituents are the diglycosides digitalinum verum and gitorin, the monoglycoside strospeside,[584] saponins and the enzyme digilanidase.[513] Lanatosides are easier to extract in pure form than are their counterparts from *D. purpurea*.[436, 466] The seeds contain glycosides and 30% of a yellow-green, viscous fatty oil.[20]

Propagation, Culture and Harvesting:

D. lanata is grown from seed and thrives in calcareous soil. Experiments in Egypt have shown that irrigation and manganese have no effect on glycoside content. Plants grown in full sun exposure bloomed at the age of one year (6 months after transplanting) and produced more flowering shoots than plants grown in semi-shade, but there was no significant difference in glycoside level. Spring plantings flourished better than fall plantings and yielded more leaves per plant.[54] Russian investigators found the highest levels of lanatosides in upper and middle leaves of biennial plants collected at the "beginning of stemming." Leaves of annual plants were found to be richest in glycosides in September in the southern Ukraine.[536] Lenkey *et al.*, in Hungary found a rapid increase in lanatoside C in June-July followed by a decrease, while an opposite change occurred in the content of lanatosides A and B.[334] Because of the smaller leaves, yield of dried leaves per acre is approximately half that of *D. purpurea*. In India, the yield has amounted to only 240 lbs. (108.8 kg) per acre annually.[20]

Artificial and natural drying methods have been compared in Egypt. No appreciable difference was found in the glycoside levels in leaves dried in sun for two to three days, in shade for three to four days, or in a circulating hot-air oven at 55° to 60° C. for three to four and one-half hours, but leaf color was best in the shade-dried material. Oven drying at 90° to 125° C. for three-quarters to one hour caused a decrease in total glycosides. Shade drying was preferred as least expensive.[53] An improved method of lanatoside extraction has been reported by Russian chemists.[215]

Medicinal uses:

In the United States, *D. lanata* is not utilized in the form of digitalis powder but mainly as a source of lanatoside C and digoxin. The latter was adopted in the British Pharmacopoeia in 1948. Digoxin is 300 times stronger than digitalis powder prepared from *D. purpurea*, more stable and is especially valued for its prompt action and quick elimination. Orally, its maximum effect is achieved in six hours; intravenously, in one to two hours.[20] In Europe, tinctures, infusions, pills and suppositories made of the leaves have many applications, and the leaves are used as poultices on wounds.[513]

Toxicity:

While the fresh or dried plant is far more toxic than *D. purpurea*, digoxin is considered a relatively safe drug. The first sign of overdosage is nausea, and the toxic effects usually disappear in twenty-four to forty-eight hours.[423] In two attempted suicides by ingestion of large amounts of digoxin, atrial arrhythmias and atrioventricular block were the primary manifestations during the early stages of intoxication. No ventricular arrhythmias appeared.[76]

Related species:

STRAW FOXGLOVE (*D. lutea* L.) is native to limestone hills of middle and southern Europe. It is a biennial or perennial plant, 2 to 3 ft. (.6 to .91 m) high, non-hairy, with oblong or lanceolate, toothed leaves; flowers in a one-sided, dense raceme, yellow to white, ¾ in. (2 cm) long; smooth. It is equal to *D. purpurea* in potency, possibly less irritating and has served as a substitute.[122]

YELLOW FOXGLOVE (*D. grandiflora* Mill., syn. *D. ambigua* Murr.) grows wild on rocky hills of southern Europe and western Asia and is cultivated for drug use in the U.S.S.R. It is a biennial or perennial herb, 2 to 3 ft. (.6 to .91 m) high, hairy, with ovate-lanceolate leaves 2¾ to 10 in. (7 to 25 cm) long, ¾ to 2 in. (2 to 5 cm) wide, green above and below, soft, downy on the underside, thin, sessile or bases clasping stem, finely toothed. The middle and upper parts of stem are slightly hairy.[533] Flowers are 2 in. (5 cm) long, yellow with brown markings.[51] The leaves are half as potent as those of *D. purpurea*.[116]

RUSTY FOXGLOVE (*D. ferruginea* L.) is native to rocky hills of middle Europe and western Asia and cultivated as a drug crop in the U.S.S.R. It is a biennial or perennial plant, 4 to 6 ft. (1.2 to 1.82 m) in height and very leafy. Leaves are spatulate, 1½ to 16 in. (4 to 40 cm) long, ⅜ to 1³⁄₁₆ in. (1 to 3 cm) broad, sessile, thick, leathery, deeply veined, entire or slightly toothed and hairy only on the margin. The stem bears few or no hairs.[533] Flowers are broadly bell-shaped, ¾ to 1½ in. (2 to 4 cm) long, externally downy, yellowish with brick-red reticulations on the inside.[115] The leaves are twice as potent as the leaves of *D. purpurea*.[122]

Plantaginaceae

Figure 85. Psyllium (*Plantago ovata* Forsk.), with three of its unhusked seeds enlarged.

Psyllium

Botanical name:

    Plantago ovata Forsk. (*P. isphagula* Roxb.)

Other names:

    Blonde psyllium, ispaghula, spogel, Indian plantago.

Family:

    Plantaginaceae, the Plantain Family.

Description:

    A stemless or short-stemmed annual herb; leaves in a rosette or alternate, clasping the stem, strap-like, recurved, 3 to 10 in. (7.5 to 25 cm) long, narrow, varying from less than ¼ in. (6 mm) to ½ in. (12.5 mm) in width, tapering to a point, three-nerved, entire or toothed, coated with fine hairs. Flowers are white, minute, four-parted, in erect, ovoid or cylindrical spike ½ to 1½ in. (12.5 to 37.5 mm) in length. Capsule ovate, 5⁄16 in. (8 mm) long, two-celled, the top half lifting off when ripe, releasing the smooth, dull, ovate seeds, 1⁄16 to ⅛ in. (1.8 to 3.3 mm) long, pinkish-gray-brown or pinkish-white with a brown streak on the convex surface. Each seed is encased in a thin, white, translucent membrane ("husk") which is odorless and tasteless. When soaked in water, the whole seeds appear much swollen because of the expansion of the mucilage in the husk. The husked seeds are dark-red and hard.[20, 117, 148, 160, 624]

Origin and Distribution:

    Plantago ovata is native from the Canary Islands and Mediterranean regions of southern Europe and North Africa to West Pakistan. It is cultivated in West Pakistan, in northern and western India and has been experimentally planted in the United States, especially Arizona.

Constituents:

    The husk yields a colloidal mucilage consisting mainly of xylose, arabinose and galacturonic acid. Also present are rhamnose and galactose. Two polysaccharide fractions have been derived from the mucilage.[59, 60] One, soluble in cold water, contains 20% uronic acid; on hydrolysis, it yields 46% *d*-xylose, 40% aldobiouronic acid, 7% *l*-arabinose and 2% insoluble residue. The second fraction, having 3% uronic acid, dissolves in hot water and, when cool, forms a viscous gel yielding, on hydrolysis, 80% d-xylose, 14% l-arabinose, .3% aldobiouronic acid and a trace of d-galactose. In the Indian Pharmacopoeia it is specified that the husk should contain no more than 2% foreign organic matter, 2.9% ash, 0.45% acid insoluble ash; also that 1 g of husk, agitated intermittently over a four-hour period in 20 ml water and allowed to set for one hour, should form 20 ml gel.[20] The swelling factor of the husk is 40 to 90 compared with 10.25 to 13.50 for the seeds.[584]

The mucilage constitutes over 30% of the whole seed. The husked seeds possess 5% of a yellow, semi-drying oil, a small amount of the glycoside aucubin, $C_{13}H_{19}O_8 \cdot H_2O$, considerable tannin[117] and an active principle resembling acetylcholine. Fatty acid content of the oil is as follows: linolenic, 0.2%; linoleic, 47.9%; oleic, 36.7%; palmitic, 3.7%; stearic, 6.9% and lignoceric, 0.8%. From the seed embryo there has been obtained 14.7% oil rich in linoleic acid. Alcoholic extracts of the seed show cholinergic activity.[20]

Propagation, Culture and Harvesting:

Seeds are sown at the rate of 13.2 to 28.6 lbs. (6 to 13 kg) per hectare in rows 9 in. (22.5 cm) apart, in a well-tilled field, and blended into the soil by means of weed-brooms. Planting is done from mid-October to mid-December in India and is immediately followed by irrigation. Germination should begin in four days. If delayed, it is stimulated by a second watering, and thereafter the crop is irrigated every week to ten days. It is usually not fertilized, but, if the soil is of low fertility, well-rotted manure may be applied during the field preparation.[20] In trial plantings in Arizona, the plants suffered damage by frost and from attack by grasshoppers.[618]

Blooming begins two months after sowing, and the crop should be ready to harvest in March and April. The plants are cut 6 in. (15 cm) from the ground, using hand sickles. This work is performed in early morning so that the moisture of dew will keep the spikes from shattering. The plants are threshed and winnowed and the seeds repeatedly sifted until clean. Yield ranges from 1,100 to 2,420 lbs. (500 to 1,100 kg) per hectare. The seeds may be marketed whole, or the husks may be sold separately.

To remove the husk, the cleaned seeds are passed six or seven times through stone or emery grinders, sieved and screened through several grades of mesh to sort according to fineness of grind. Most of the exported husk is of the grade designated 70 mesh. The highest quality husk is white, with no particles of the red kernels.

The United States is the principal consumer of psyllium, annual imports from India exceeding 800 metric tons (800,000 kg) of seed, 3,000 tons (3,000,000 kg) of husk.[584] The crude material is subject to purification and processing after arrival in this country.[122] Iran exports considerable amounts of psyllium to Bombay.[266]

Seeds of the lesser-known Indian species *P. amplexicaule* Cav., larger and darker in color, are sold as a substitute in the Punjab. Psyllium seeds are sometimes adulterated with the mucilaginous seeds of *Salvia aegyptiaca* L., and pulverized parched starch has been found as an adulterant.[20]

Medicinal uses:

Psyllium was adopted into the Indian Pharmacopoeia in 1868. Psyllium seeds are demulcent and mildly astringent. Astringency is increased by roasting.[148] Infusions or decoctions of whole seeds or the husks are administered to relieve constipation because the swelling of the mucilage stimulates intestinal peristalsis by providing a smooth, bulky mass which is unaltered by enzymes or bacterial flow. The mucilage acts as a soothing lubricant and absorbs toxins in the digestive tract.

It overcomes chronic diarrhea and dysentery by relieving intestinal irritations. Psyllium may be taken also as a diuretic and to alleviate kidney and bladder complaints, gonorrhea, urethritis and hemorrhoids.[148]

Recommended doses (accompanied by plentiful water) are 5 to 15 g seeds or .5 to 2 g husk. In India, powdered seeds are mixed with aniseed and sugar, or combined with fruit juice or stewed fruit. Roasted seeds have been given to children to halt diarrhea.[233] The husks are preferred to whole seeds in acute cases or for treating children and are commonly combined with sodium bicarbonate, monobasic potassium phosphate, anhydrous dextrose or other substances. They are common ingredients in laxative products of various manufacturers.[122]

A solution of the sodium salts of the liquid fatty acids after hydrolysis of the fixed oil of the seeds is known as Sodium Psylliate Injection and is administered as a sclerosing agent.[122] In Argentina, psyllium seeds are boiled in water for fifteen minutes, the liquid strained off, chilled and used as eyedrops to dispel inflammation.[368] The decoction is taken with honey as a remedy for sorethroat and bronchitis.[148, 368]

In India, pulverized psyllium seeds with vinegar and oil are applied to rheumatic and gouty swellings. A paste or lotion of the crushed seeds and vinegar is smeared on the head to reduce fever.[148]

Toxicity:

The ingestion of unsoaked psyllium seeds may cause gastrointestinal irritation, inflammation, mechanical obstruction and constipation.[277, 423] Powdered or chewed seeds release pigment which is deposited in and harmful to the kidneys.[423] Psyllium seed cookies, consumed by the unwary, have produced profuse diarrhea.

Other uses:

Psyllium mucilage is sometimes employed as a substitute for agar-agar. It may serve as a stabilizer in ice cream, a filler for wheat starch, an ingredient in chocolates, a sizing for textiles and in the formation of pharmaceutical tablets and cosmetics. The seed kernels are mixed with guar (*Cyamopsis tetragonoloba* Taub.) for feeding cattle.[20, 618]

Figure 86. Black psyllium (*Plantago psyllium* L.). Photographed in the Beal-Garfield Botanic Garden, Michigan State University, East Lansing, Michigan.

Black Psyllium

Botanical name:

    Plantago psyllium L.

Other names:

    French psyllium, Spanish psyllium, clammy psyllium, fleawort, flea seed, zaragatona.

Family:

    Plantaginaceae, the Plantain Family.

Description:

    A stout, erect or spreading, much-branched, glabrous or glandular-hairy annual herb, ½ to 1½ ft. (15 to 45 cm) high. Leaves opposite or whorled, linear, ¾ to 4 in. (2 to 10 cm) long, entire or faintly toothed, lower leaves having short, leafy shoots in their axils. Flowers 3/16 in. (5 mm) wide, rose-pink or brownish in numerous ovoid or globular spikes 3/16 to 3/8 in. (5 to 10 mm) long, on axillary peduncles exceeding the leaves. Capsule smooth, two-celled, seeds ovate to oblong, light- to dark-brown, smooth, very glossy, grooved on the convex surface, 1/20 to 1/10 in. (1.3 to 2.7 mm) long, odorless and nearly tasteless.[20, 83, 454, 574]

Origin and Distribution:

    Native to grassy fields of central and southern Europe and from North Africa to southwest Asia; naturalized in disturbed ground in southern England[100] and South Australia;[83] has been found growing as an escape in Ecuador and Bolivia;[356] cultivated in southern France and Spain, mainly for the European market;[20] since 1943, has been grown to some extent in Cuba.[485]

Constituents:

    The seeds are not as rich in mucilage as those of *P. ovata*, possessing only 10 to 12%. Agitation of 1 g of seeds in 20 ml of water during a twenty-four hour period followed by one hour of rest produces a maximum volume of 14 ml. Other properties include protein, oxalic acid, mucic acid, invertase, emulsin; 7% of fatty oil containing 5.04% free acids (as oleic), 0.82% soluble acids (as butyric) and 94% insoluble acids;[20] also

β-sitosterol, stigmasterol, several alkaloids[55] and 0.14% of the glycoside aucubin.[294]

Analyses of the seed husk reveal 8.3% moisture, 1.4% protein, 3.4% ether extr., 84.55% carbohydrates, .35% fiber, 1.98% ash, 57.8 mg phosphorus, 305.9 mg calcium, 18.3 mg iron, 4,000 I.U. vitamin A activity per 100 g.[20]

Propagation, Culture and Harvesting:

In France, seeds are sown in March, and harvesting takes place in August before the capsules are fully ripe. The cut plants are spread in the sun to partly dry, then are threshed and the separated and cleaned seeds are thoroughly dried. Yields average 1,000 lbs. (454 kg) per acre.

Medicinal uses:

P. psyllium has the same pharmaceutical uses as *P. ovata*, q.v. The swelling factor of the seeds is 12.75.[584]

Other uses:

Husks of *P. psyllium* are eaten raw or cooked in West Pakistan. Their mucilage is employed in skin lotions and hair applications; also in sizing silk, cotton, and paper and as waterproofing for certain explosives. The seed oil is combined with linseed oil in making varnish, and the presscake may be used as fodder.[20]

Related species:

Sand plantain, or branched plantain (*P. arenaria* Wald. & Kit.; syn. *P. ramosa* Asch.), is a low, tufted, stiff-branched, roughly hairy annual with linear to filiform leaves and greenish-white flowers in dense, long-stalked spikes. The ovate seeds are black, smaller and less glossy than those of *P. psyllium*.[233] Swelling factor is 14.50.[584] The plant is native to Mediterranean Europe and is common in sandy, coastal and inland deserts of Egypt and Palestine. The seeds may be accepted in the pharmaceutical trade as "Spanish" or "French" psyllium without distinction from those of *P. psyllium*, though Wren says they possess little mucilage and are used mostly for feeding birds.[624]

Rubiaceae

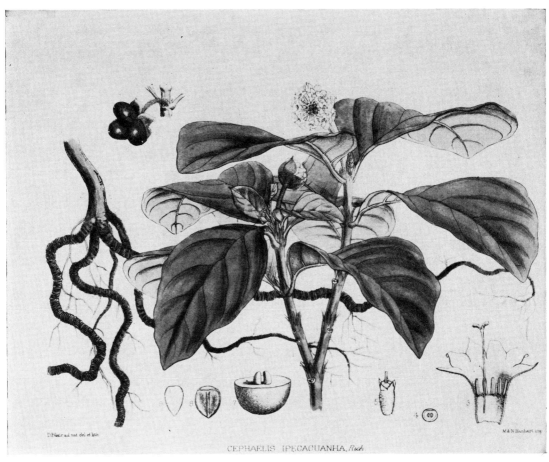

Figure 87. Ipecac (*Cephaelis ipecacuanha* A. Rich.). Reproduction from a color painting by David Blair in *Medicinal Plants*, R. Bentley and H. Trimen (J. & A. Churchill, London; 1880).

Ipecac

Botanical name:

    Cephaelis ipecacuanha A. Rich. (*Psychotria ipecacuanha* Stokes).

Other names:

    Rio or Brazilian ipecac, golden root.

Family:

    Rubiaceae, the Madder Family.

Description:

    A tropical, perennial, herbaceous plant to 1.6 ft. (.5 m) high, with slender, cylindrical, corky, sinuate underground stem (or false rhizome) from which extend several horizontal roots. When mature, the roots have brick-red to dark-brown "bark," smooth or more or less prominently transversely ridged or ribbed. They possess a musty odor and a bitter taste. The above-ground stem is quadrangular, somewhat downy, with few or no branches. Leaves are opposite near the top of the plant, alternate below,[305] 2 to 4 in. (5 to 10 cm) long, 1 3/16 to 2 in. (3 to 5 cm) wide, dark-green on the upper surface, light-green below, rough-hairy with short, downy petioles. Flowers are white, funnelform, sessile, less than 3/8 in. (1 cm) wide and are borne in dense, globular clusters of 8 to 20 on a purple-green stalk. The fruit is oval, purple, becoming nearly black when ripe, and contains two hemispherical stones each enclosing a single white seed.[161, 239, 305]

    In Brazil, the plant blooms in January and February and the fruit ripens in May.[305]

Origin and Distribution:

    This plant is native to the humid forests of Bolivia and Brazil, especially around Mato Grosso and Minas Gerais. Most of the roots for the drug trade come from wild plants.[90] In the mid 1960s, experimental culture of *C. ipecacuanha* was initiated in Bolivia[102] and in Nicaragua, where it is preferred over the native *C. acuminata* Karst. for commercial exploitation.[237] Within the past few years, a plantation of a half million plants has been established in Mato Grosso, Brazil, and Indians are being taught to develop small farms. Exports from Brazil are approximately 80 tons of roots annually.[395] C. ipecacuanha is cultivated to a small extent in Malaya, Burma,[192] the Darjeeling district of West Bengal and parts of Assam and Sikkim in India.[239] Efforts are being made to expand ipecac culture into tea and coffee plantations to meet the high domestic requirements of that country.[20, 277] The total annual Indian crop at present amounts to no more than 25 or 30 tons.[239]

Constituents:

The "rhizome" and roots contain several alkaloids (isoquinoline derivatives), ranging up to 2.5%—primarily emetine ($C_{29}H_{40}N_2O_4$), secondarily cephaëline ($C_{28}H_{38}N_2O_4$). Psychotrine ($C_{28}H_{36}N_2O_4$) is mainly in the bark. There is also psychotrine methyl ether, emetamine, a glucosidal tannin named ipecacuanhin or ipecacuanic acid, ipecoside (a neutral monoterpenoid isoquinoline glycoside),[584] a crystalline coloring matter called erythrocephaleine, saponin, resin, fat,[20] some calcium oxalate[584] and 40% of starch.[122]

The International Pharmacopoeia specifies a content of not less than 2% alkaloids, at least 60% of which must be non-phenolic[584] (55% in the British Pharmacopoeia,[20] 50% in the Indian[239]). Ipecac experimentally produced in Rangoon showed 1.88% nonphenolic alkaloids and 0.22% phenolic.[417]

The leaves contain 0.45% emetine[98] and the above-ground stem less; these parts are not worth processing.[239] A simplified method for assaying emetine and cephaëline in crude ipecac and in the tincture and liquid extract by thin-layer chromatography was described by Habib and Harkiss in 1970.[244]

Propagation, Cultivation and Harvesting:

Cultivation of ipecac is handicapped by the plant's need for shade, wind protection, high humidity [a minimum of 80 in. (200 cm) of rainfall well-distributed over the year[90]] and deep soil, rich in humus and potash and well-drained.[239] It has been found most practical to intercrop it with a major crop such as rubber (*Hevea*) or *Cinchona*. Otherwise, it is necessary to supply artificial shade. In Nicaragua, it is grown in beds under a canopy of interlaced branches.[237]

Propagation is by seed, cuttings of old stems, petioles, tip-cuttings with 4 to 6 leaves or sections of roots bearing adventitious buds. Seed germination rate is often poor and may take three to five months. Vegetative propagation is preferred for reproducing high-yielding clones and advancing the harvest by eight to ten months and is expedited by the use of intermittent mist. The young plants are raised in nursery beds and transplanted two or three times. To eliminate weeding, setting the plants in small holes in black polyethylene sheets has proved effective in India.[239] It is important to fumigate the soil to destroy root-attacking nematodes and other deleterious organisms before planting.[237]

The roots may be harvested when the plants are two years old, but four-year-old plants give maximum yield and highest alkaloid content.[239]

In Brazil, harvesting of ipecac in the wild is done by workers who go into the forests with a month's supply of food and build huts in which they stay for the collecting season. Agents bring them food each month and take away what roots have been gathered, production ranging from 8 to 30 lbs. (3.6 to 13.6 kg) daily per man. Harvesting takes place from November through March. The plants are ripped from the soil by hand but with the aid of a trowel-like tool thrust beneath to free the mass of "rhizomes" and roots. Then the plant is reset in the earth, usually with enough broken-off roots left to resume growing.[161, 584] After shaking off the soil, the extracted roots and "rhizomes" are first dried over a slow fire and then sun-dried for two to three days before packing in sacks for pick-up. At the marketing centers, they are broken into small pieces[239]—1¼ to 6 in. (3 to 15 cm) in length and to $^{3}/_{16}$ in. (4 mm) in diameter[122]—then sieved to eliminate particles of soil and packed for export or transport to factories for alkaloid extraction.[239]

Each plant develops 8 to 12 usable roots, averaging 3 to 4 oz. (85 to 113 g) in weight when dried. Yields of 50 to 60 lbs. (22.6 to 27.2 kg) have been realized in Malaya.[161]

Medicinal uses:

The Indians of Brazil knew and valued ipecac as a dysentery remedy, and knowledge of its virtues was gained by a Portuguese priest, Manoel Tristaon late in the 15th century.[98] The roots were introduced into Europe in 1649 by a Dutch physician by the name of Piso.[596] They were being imported and sold in Paris in 1672 by a Doctor Legras and their efficacy tested on condemned prisoners. In 1682, Doctor Adrien Helvetius of Paris utilized ipecac root in treating Louis XIV for dysentery and was honored and rewarded for his success. The plant was properly identified and named in 1803, and it remained in demand until the development of antibiotics. Its status declined and then rose again when it was found that antibiotics could not control all forms of dysentery.[192]

One of the most widespread uses of ipecac has been, in the form of a sirup, as an emetic to stimulate the vagus nerve, cause vomiting and empty the stomach of poisons. It is contraindicated when lye, petroleum products or strychnine have been ingested;[423] also if the patient is unconscious or in convulsions.[596] It has been successfully administered to relieve tachycardia. The emetic action is caused mainly by the cephaëline, emetine serving largely as an expectorant.[584]

A mixture of ipecac powder and opium was prescribed as a sudorific at the onset of influenza but is no longer official. Sirup of ipecac is often given in the early stages of acute bronchitis and in croup but is ineffective in chronic bronchitis.[423] The hydrated hydrochloride of emetine, by injection, is of great value in the treatment of amoebic dysentery[122] but of no benefit in cases of bacillary dysentery.[423] In India, emetine has been successfully employed against bilharziasis, Guinea worms and oriental sores.[20] In very small doses, ipecac is given to stimulate appetite and improve digestion.[423] Emetine hydrochloride was reported in 1918 and 1929 as an effective agent in cancer treatment, but, in experimental use since 1960, it has proven to be of no value at tolerable dose levels.[390]

Toxicity:

If given over a prolonged period or if the total dosage exceeds 1 g, the cumulative effect of emetine may give rise to myositis at the injection site, diarrhea and vomiting, neuromuscular symptoms, hypotension, dyspnea and heart palpitations; hematuria and circulatory failure may follow. It is considered a safe procedure to give adults 60 mg per day for no more than ten days and to refrain from repeating the treatment for at least one month.[596] Children may be given 1 mg per kg of body weight per day.[361] Fluid-extract of ipecac, formerly in use, was abandoned after the occurrence of fatal overdoses when mistakenly given instead of the sirup.[361, 423] The phenolic cephaëline is too toxic for employment as such and is usually converted into emetine by methylation.[20] The nonphenolic emetine is less irritant, and psychotrine is virtually nontoxic.[423]

If contacted while preparing formulations or deliberately applied externally, emetine causes skin inflammation, itching and rash or more severe eruptions.[20, 68]

Powdered ipecac is a respiratory irritant causing coughing and sneezing. Pharmacists and assistants sensitized by repeated exposure may experience rhinitis and asthma.[74] The offending allergen is reported to be a mixture of glycoproteins.[75]

Synthetic, 2,3-dihydroemetine is an effective substitute for natural emetine in amebiasis but presents the hazard of cardiotoxicity and can be administered only when the patient is under hospital care.[192]

Related species:

Cephaelis acuminata Karst., called Cartagena, Nicaragua, Savanilla or Panama ipecac, is native to Central America and Colombia. The plant is similar to the preceding but has white fruits.[440] The roots are grayish-brown or reddish-brown externally and less distinctly ridged. They are generally thicker than those of *C. ipecacuanha*, varying from $3/16$ to $5/16$ in. (4 to 7 mm) in diameter. They are higher in cephaëline, the ratio of the two major alkaloids being about $1/3$ to $2/3$ cephaëline.[122] The roots from wild plants are exported mainly from Colombia, Nicaragua and Costa Rica.[584] *C. acuminata* is recognized in the British Pharmacopoeia as well as *C. ipecacuanha*.[244]

Figure 88. *Cinchona calisaya* Wedd. Photographed in the Morris Arboretum of the University of Pennsylvania, Philadelphia.

Cinchona

Botanical name:

 Cinchona spp. Of forty species in this genus, about a dozen have served as commercial sources of the drug. The most important are commonly cited as *C. calisaya* Wedd., *C. ledgeriana* Moens & Trimen (possibly a variety of *C. calisaya* or of *C. officinalis* L. or a hybrid between one of these and other species), *C. officinalis* L. and *C. succirubra* Pav. ex Klotzsch (possibly a variety of *C. pubescens* Vahl.) Also, various hybrids are cultivated.

Other names:

 Peruvian bark, Jesuits' bark, Cardinal's bark, Jesuit's powder, fever tree.

Family:

 Rubiaceae, the Madder Family.

Description:

 Shrubs or trees to 100 ft. (30 m), with bitter, astringent, slightly aromatic bark, corky and vertically or horizontally fissured when old. Leaves are evergreen, opposite, elliptic to ovate-lanceolate, simple, entire, 4 to 20 in. (10 to 50 cm) long, thin to thick, with prominent midrib and conspicuous, oblique lateral veins. Leaves of *C. succirubra* are hairy on the veins beneath. Flowers, in terminal panicles, are pleasantly fragrant, pink or whitish, .39 to .78 in. (1 to 2 cm) long, tubular, with 5 flaring lobes, hairy on the margins. Seed capsule is oblong to ovoid, .31 to 1.1 in. (.8 to 3 cm) long; when ripe, it splits open beginning at the base.[95] It contains 40 to 50 flat, slender, winged seeds up to .39 in. (1 cm) long, so light that there may be 75,000 or more in one ounce (28.3 g).[450]

Origin and Distribution:

 All species are native to the higher altitudes in tropical America. *C. calisaya* grows wild in southern Peru and Bolivia at elevations between 3,500 and 5,000 ft. (1,067 and 1,524 m); *C. officinalis* from northern Peru to Colombia, between 7,000 and 9,000 ft. (2,133 and 2,743 m); *C. succirubra* from Costa Rica to Bolivia, between 3,000 and 11,000 ft. (914 and 3,350 m).

In 1840, a specimen of *C. calisaya* was raised in Paris from Bolivian seed. This plant, in 1851, was transferred to the Buitenzorg Botanic Gardens, Java, but could not tolerate the low altitude. A cutting taken from it was planted in the mountains, survived and was multiplied. The Dutch, in 1865, acquired young plants direct from South America and later purchased seeds collected by Charles Ledger in Bolivia.[458] Thus began the Dutch dominance in production of cinchona and their crop-control practices which kept the price high.[199, 302, 576]

In 1859, the British government sent Clements R. Markham on an expedition to South America to obtain seeds and small trees. The first plantings were made in 1860 in Jamaica, but this effort at cultivation failed, as did the first British plantings in India. In 1861, seeds which Markham sent to Kew were dispatched to India and Ceylon, and these became the basis of the successful British plantations. German interests developed plantings in Tanganyika in 1900, and cinchona was introduced into Uganda in 1918. In 1868, seeds from Ceylon were planted in Guatemala where cinchona culture languished until stimulated by realization of the need for an American source of quinine. The Guatemalan government, in 1883, engaged the Englishman W. J. Forsyth to collect seeds in Ceylon and convey them to Guatemala for planting, mainly on land owned by President Justo Barrios, but the seeds were apparently from the low-yielding *C. succirubra*.[451]

The Philippine Bureau of Forestry started growing cinchona in 1927, with seed brought from Java.[95] In the early 1930s, Japanese teams had created cinchona nurseries and plantations in Peru and these were taken over by the Peruvian government in 1943.[151] In 1934, Merck & Co., Inc., secured *C. ledgeriana* seeds from Java and began to establish nurseries in Guatemala and make selections of high-alkaloid strains. Some of the Java seed was germinated by the U.S. Department of Agriculture in greenhouses in Maryland, and the seedlings were shipped to Guatemala for planting. By 1942, 90 percent of the world supply of ground cinchona bark was exported from Indonesia.[450] World War II cut off this trade and greatly intensified efforts to produce more and better cinchona in this hemisphere. Seeds desperately salvaged by Colonel Arthur Fischer just before the loss of the Philippines to the Japanese were germinated by the U.S. Department of Agriculture, and the seedlings were planted mainly in Costa Rica.[151] Teams of botanists surveyed wild stands of cinchona trees and took bark samples for laboratory analysis and microscopic identification.[262] In 1948, there were estimated to be about 1,000 acres (calculated at 1¾ million trees four to five years old) in Guatemalan plantations.[451] Experimental work was also being carried on in Brazil, Bolivia,[151] Venezuela, Mexico[450] and Puerto Rico.[249] Years before, plantings of *C. succirubra* had been made in Colombia and Ecuador, but by this time, these low-yielding trees had been more or less abandoned.

Constituents:

Cinchona species contain over thirty quinoline alkaloids in the bark of the trunk, branches and, mostly, in the roots. Generally, alkaloid concentration is higher in trunk– than in branch-bark.[262] There is great variation in the alkaloid of wild trees of apparently the same species growing in different locations and also much variation among the various clones in cultivation.[451]

In *C. calisaya*, quinine ($C_{20}H_{24}N_2O_2$) predominates over the other major alkaloids, cinchonine ($C_{19}H_{22}N_2O$), cinchonidine ($C_{19}H_{22}N_2O$) and quinidine ($C_{20}H_{24}N_2O_2$).[451, 552] The alkaloids occur in combination with a tannin called cinchotannic acid. Quinine is bacteriostatic, highly active against protozoa and inhibits decay of meat and fermentation of yeast.[423] The alkaloid content of cinchonas presently or formerly of commercial importance have been reported as follows:

- *C. calisaya*; CALISAYA BARK, YELLOW BARK (Bolivia and Peru): 4.5 to 7.14% total alkaloids; 3.3 to 5.3% dry weight quinine.[262]
- *C. ledgeriana*; LEDGER BARK, BROWN BARK (prominent in India): average quinine content, 7%; selected clones, 14 to 16%.
- *C. officinalis*; CROWN BARK, LOXA BARK (Colombia and Ecuador): 2.4 to 4.8% total alkaloids; 0.1 to 1.58% quinine.
- *C. succirubra*; RED BARK (Ecuador, Guatemala, etc.): 5 to 7% total alkaloids;[423] 2 to 3% quinine, difficult to extract.[451, 458]
- *C. pitayensis* Wedd.; COLOMBIAN BARK, PITAYO BARK (Colombia and Ecuador): total alkaloids, 3 to 6%; 1.68% quinine.[262]
- *C. micrantha* Ruiz & Pavon; GRAY BARK (Ecuador and Peru): total alkaloids, 3.13 to 6.02%; higher alkaloid concentration in limb– than in trunk-bark;[262] little or no quinine; has been an important source of *Totaquina*, developed by the Philippine Bureau of Science in the early 1940s, which contains all of the alkaloids and has served as a substitute for quinine during shortages or when the latter is too expensive.[95]

Propagation, Cultivation and Harvesting:

Cinchona trees need acid soil;[394] are naturally adapted to mountainous rain forests without a dry season and with annual rainfall of no less than 85 in. (216 cm), but they require good drainage.[262] Seeds are of short viability; they are not embedded in soil but scattered on the sur-

Figure 89. *Cinchona officinalis* L. Reproduction from a color painting by David Blair in *Medicinal Plants*, R. Bentley and H. Trimen (J. & A. Churchill, London; 1880).

Figure 90. *Cinchona succirubra* Pav. ex Klotzsch. Reproduction from a color painting by David Blair in *Medicinal Plants*, R. Bentley and H. Trimen (J. & A. Churchill, London; 1880).

Figure 91. *Cinchona pubescens* Vahl. Photographed in a greenhouse at Longwood Gardens, Kennett Square, Pennsylvania.

face of prepared soil or sphagnum moss at the rate of 1 to 2 g seed per sq. yd. (.9 m).[249] If kept shaded and uniformly moist, they germinate in eleven to twenty-one days.[621] The seedlings require careful control of temperature, light and ventilation to avoid damping-off and other hazards.[450]

Early efforts at vegetative propagation other than by cuttings were discouraging, and seeds were commonly employed for planting until 1942 when a technique of side-veneer grafting proved practical.[451] Cleft grafting is employed with 90 percent success in India.[117] *C. succirubra* is usually used as a rootstock.

Seedlings are first transplanted when 2 to 3 in. (5 to 7.5 cm) high and are again spaced out when six to seven months old.[394, 450] They are set out in the field when about two years old and 1 to 2 ft. (.3 to .9 m) in height and placed at 4 x 4 or 5 x 5 ft. (1.2 x 1.2 or 1.5 x 1.5 m).[451]

At this time, those in rapid growth are pruned back about one-third, to mature wood. Others are defoliated by 50 percent. A palm-leaf shelter is arranged over each plant as protection until it has become well established,[249] or shade trees may be installed 20 ft. (6 m) apart throughout the plantation.[117]

Beginning in the third year after field planting, the least thrifty of the trees are progressively uprooted for total stripping of bark until the remaining trees are eight to twelve or even twenty years old and considered to be at the peak of alkaloid production.[450, 458] Then all are simultaneously uprooted. In the meantime, replacement plantings are made as the plantation is thinned out to assure a continuous supply.

In the Western Hemisphere, the trees are usually chopped down and the bark sliced from the trunk and branches by machete, usually in strips 1 to 2 ft. (.3 to .9 m) long. Sometimes the worker strips the bark from the trunks of standing trees, ignoring the overhead branches, but this is a wasteful practice. The roots are not dug up. They are left to produce new sprouts, which grow vigorously.

In the Orient, the entire tree is cut into sections, which are beaten with mallets to loosen the bark. The bark is then stripped and dried in sun, shade or in ovens[262] to reduce the moisture content from about 70 to 10%.[458] Thin strips of bark roll into "quills" during drying.

Yield of dried bark per acre/hectare may range from 9,000 to 16,000 lbs. per acre (9,000 to 16,000 kg per hectare). In the past, bark from wild trees had to be carried on the backs of humans, donkeys or mules to rafts navigating treacherous mountain streams. The construction of local airfields dramatically facilitated transport to warehouses or processing centers.[151]

The dried bark is nearly always ground to powder before shipment to pharmaceutical companies which extract the alkaloids.[450]

Medicinal uses:

Cinchona bark (*quina quina*) was valued by the Indians as a febrifuge. In 1633, in a Peruvian monastery, Padre Calancha wrote, "In the country of Loja [southwestern Ecuador] there is a tree which they call the fever-tree, the bark of which, ground to a powder and administered in the form of a potion, cures fevers; there have been marvelous examples of this in Lima."[450] The Corregidor of Loja, in 1638, sent cinchona bark to Lima to relieve the malarial fever of the Count of Chinchón, who was Viceroy of Peru from 1629 to 1639. Thereafter the praises of Spanish physicians as well as of the Jesuit priests created a great demand for cinchona in Spain. For 200 years, the wild trees were progressively decimated to meet the need for this drug which played an indispensable role in making possible man's exploration and colonization of the subtropical and tropical world.

Exports of cinchona bark from South America (mainly Colombia) to Europe totaled 7,000,000 lbs. (3,175 metric tons) in 1880.[458] In 1942, the United States' annual requirements of extracted quinine for medicinal, cosmetic, culinary and beverage purposes were reported to be between 3,000,000 and 4,000,000 oz. (85,058 to 113,411 kg).

Until World War II, quinine was esteemed as the most valuable malaria remedy available. The usual therapeutic oral dose for adults has ranged from 167 to 333 mg three times per day (after meals) for two days, followed by 600 mg three times per day for four days. As a prophylactic before exposure to malaria-carrying mosquitoes, 600 mg may be taken once daily after the main meal.[423]

Quinine sulfate, containing 83% quinine, is the most common salt of quinine given orally. For those who cannot tolerate bitterness, there are quinine-containing capsules or sirups, or those patients may be given quinine dihydrochloride (81% quinine) or hydrochloride (74% quinine) by injection, though this is a much less desirable mode of administration.

Cinchonidine and cinchonine are also utilized as antimalarial drugs but are only half as effective as quinine, and dosage must be doubled.

Among other medicinal uses of quinine, quinine bisulfate (59% quinine) is readily soluble in water and has been employed in ophthalmia because of its astringent, bactericidal and anesthetic action.

Quinine and urea hydrochloride (59% quinine) is injected as a sclerosing agent in treating internal hemorrhoids, hydrocele, varicose veins and pleural cavities after thoracoplasty.

The discovery, in 1912, that malaria patients given quinine were relieved of cardiac arrhythmias, led to the use of quinidine (twice as effective as quinine) as therapy for certain cardiac conditions: premature contractions or extrasystoles of atrial, nodal or ventricular origin; paroxysmal atrial tachycardia; nodal tachycardia; atrial flutter and fibrillation; ventricular flutter and fibrillation. It is usually given orally in the form of quinidine sulfate. In some cases, where the drug must be given by injection, solutions of quinidine hydrochloride or quinidine gluconate are employed as being easier to prepare than solutions of quinidine sulfate. A solution of quinidine lactate has been found effective but risky.[423]

Quinidine gluconate is available also in regular or sustained-action tablets.[423] A minor use of quinidine is in treating hiccups.[552]

Cinchona tincture has been used as a bitter to stimulate appetite and digestion.[423] Quinine is a common ingredient in non-prescription headache and cold remedies and gargles.[190, 584] It is added to rubbing alcohol as a rubefacient, to hair tonics as a rubefacient and antiseptic and is an element in sun-screen preparations.[190]

During World War II, the synthetic quinacrine (Atabrine®) was the dominant antimalarial and was notorious for yellowing the skin. It has been superseded in malaria therapy by more effective, less toxic synthetics including chloroquine, hydroxychloroquine, primaquine, pyrimethamine and ammodiaquine. These may in some cases give rise to side effects similar to those induced by quinine, but they have largely replaced quinine as antimalarials except where certain types of malaria are resistant to synthetics and when circumstances favor the less expensive natural product.[122, 423, 552]

Toxicity:

Contact dermatitis and/or asthma occasionally occur in factory workers where cinchona bark is ground[190] or quinine tablets are made.[423] Quinidine sulfate solution may also cause an eczematous reaction.[190] Various toilet products (hair tonics, sun-screens, rubbing alcohol, etc.) containing quinine may produce topical sensitization. Non-prescription cold and headache remedies containing quinine are also recognized as sources of dermatitis.[190]

Therapeutic doses of quinine have resulted in acute hemolytic anemia ("blackwater fever"). This condition and fatality from uremia have followed the taking of quinine as an abortifacient. Persons sensitive to quinine display "cinchonism" after taking 600 mg daily. The symptoms include a feeling of congestion in the head, ringing in the ears, slight deafness and skin eruptions; with increased dosage, these effects intensify and may be accompanied by facial flushing, dizziness, staggering and blurred vision.[423]

Large (toxic) doses of quinine may cause severe gastric and rectal irritation, nausea, vomiting and diarrhea, headache, dilated pupils, delirium, stupor, loss of sight and hearing, difficult respiration, weakness, convulsions, paralysis and collapse.[116] Some persons have recovered after ingesting 30 g, though single doses of 8 g have proved fatal.

Individuals who react adversely to small amounts of quinine may be sensitive also to cinchonidine but rarely to quinidine or cinchonine.

Subcutaneous and intravenous administration of quinine (as soluble salts) causes serious local necrosis and is avoided except when it cannot be otherwise tolerated or in emergencies where quick action is necessary.

Quinidine is rapidly absorbed from the gastrointestinal tract and thus presents a real hazard of overdosage which can result in diastolic arrest of the heart. Normal therapeutic dosage may produce mild adverse symptoms like those associated with quinine in sensitive persons (about half the patients receiving this drug).[423]

Other uses:

Cinchona bark has long been used in Europe for flavoring wine.[199] Quinine sulfate is the bitter principle in carbonated quinine water (tonic water) and also enters into vermouth, gin, vodka[190] and bitters.[210]

In the food industry, extracts of red cinchona bark are employed in ice cream, baked goods, condiments and soft drinks; yellow cinchona bark is used in soft drinks and candy.[210]

Cinchona alkaloids are ingredients in insecticides and moth repellants for protection of feathers, furs and clothing. Cinchona bark, after extraction of the alkaloids, is still useful in the tanning industry.[277]

Coffee

Botanical name:

Coffea spp., primarily *C. arabica* L., Arabian, or Arabica, Coffee; and *C. canephora* Pierre, Robusta Coffee and Nganda Coffee, also called Congo Coffee.

Other names:

none, in English.

Family:

Rubiaceae, the Madder Family.

Description:

Small trees—*C. arabica* to 16.5 ft. (5 m), *C. canephora* to 33 ft. (10 m). Leaves evergreen, opposite, oblong-elliptic, pointed at the tip, glossy, with conspicuous horizontal veins; to 6 in. (15 cm) long in *C. arabica*, to 1 ft. (30 cm) in *C. canephora*. Flowers fragrant, white, tubular, five-lobed, clustered in the leaf axils, abundant and showy. Fruit ("cherry" or "coffee berry") nearly round, ½ to ¾ in. (1.3 to 1.9 cm) long, red when ripe; skin smooth, glossy, tough; flesh soft, mucilaginous, yellowish. Seeds —normally 2—to ½ in. (1.3 cm) long, oval but flattened and grooved on inner side, gray-green or gray-blue, enclosed in a silvery "skin" (testa). The peeled, dried seeds are the so-called "coffee beans" on the market.[342, 458]

Origin and Distribution:

C. arabica, despite its name, originated in Ethiopia, where it flourishes at elevations between 4,500 and 6,000 ft. (1,373 to 1,830 m), and it is believed to have been introduced into Arabia prior to the 15th century. In 1690, it was first planted in Java and in the early 18th century was carried to Surinam, Martinique and Jamaica. Cultivation soon spread throughout the West Indies and Central America and favorable regions of South America. Later, it reached India and Ceylon. Today, nearly 90 percent of the world's coffee comes from this species. Coffee is the most

←

Figure 92. Coffee tree (*Coffea arabica* L.) in full bloom. Photographed in the Gifford Arboretum, University of Miami, Coral Gables, Florida. (*See also* color illustration 15.)

important crop in many tropical American countries, and Brazil is the world's leading producer.[395, 458]

C. canephora is found wild in equatorial Africa from Gabon to Uganda and is naturally adapted to lower elevations than *C. arabica*. In 1900, it was taken to Java and found resistant to a leaf rust which was destroying plantations of *C. arabica*. It has since been introduced into other regions where the disease, caused by *Hemileia vastatrix*, is a serious problem. It is the leading species cultivated in Africa and Asia but still little grown in the western hemisphere.

Constituents:

Among the numerous species of *Coffea*, there are many (such as *C. humboltiana*) in Madagascar and the Mascarene Islands which are nearly or wholly caffeine-free. However, caffeine (1,3,7-trimethylxanthine) ($C_8H_{10}N_4O_2$) is the main active principle in commercial coffee seeds. It varies from 1 to 3% in *C. arabica* (usually 1.5 to 2.5%) but may constitute as much as 3.21% in some clones of *C. canephora*.[572] Minor alkaloids in raw green coffee include theobromine, theophylline and trigonelline.[201, 617] A cup of coffee ordinarily contains about 100 mg caffeine. Caffeine acts primarily as a central nervous system stimulant. It increases the capacity for muscular activity, slightly amplifies the heart rate and flow of blood through the coronary artery and, in quantity, is a potent diuretic. It also promotes the formation of pepsin and hydrochloric acid in gastric juice.[423]

Green coffee beans possess 8% non-volatile acids, of which chlorogenic acid (formerly called caffetannic acid) represents about 7%. Roasting causes a 40% decomposition of chlorogenic acid, resulting in 5% caffeic acid and 0.5% quinic acid. Roasting results in substantial losses of tartaric, citric, malic and oxalic acid, and produces the following homologous volatile acids: acetic, 0.35%; propionic, 0.10%, butyric, 0.20%, and valeric, 0.20%.[535]

Propagation, Cultivation, Harvesting and Processing:

Coffee trees can be propagated vegetatively, but in commercial plantings they are usually grown from seeds which germinate in four to eight weeks. In Brazil, seeds are sown directly in the field on newly cleared land. Elsewhere, seedlings are raised in nurseries and transplanted when six to ten—or as much as twelve to twenty-four—months old, depending on the system employed. Spacing is variable. The plants may be 5 to 10 ft. (1.5 to 3 m) apart or set close to form "hedges." In the past, it was common to interplant with shrubs or trees to shade the coffee, but this practice is being abandoned in many areas, unshaded trees giving higher yields. Coffee trees may live one hundred years or more. Productivity declines after forty to fifty years or much sooner if conditions are unfavorable. Coffee plantations require much care in the form of weeding, mulching, pruning, fertilization, and disease and insect control. The trees begin to bear in three to four years, the fruits maturing seven to nine months after flowering.[458] Until recently, harvesting has been done selectively by hand, with repeated pickings over the ripening season of several weeks. Because of labor shortages and rising labor costs, other methods have been introduced, including one-time harvesting by stripping all the fruits onto sheets of plastic netting. This requires the later separation of green fruits but is still more efficient and saves the 12 percent of the ripe fruits which fall and are lost during the selective method of harvesting.[527] Removal of foreign material (trash) is accomplished by passing the fruits through an air tunnel. Thereafter, they may be washed by machine, de-pulped, fermented, then dried, the adhering "parchment" and testa mechanically removed and the beans polished; or drying and hulling may be the only post-harvest treatment. Currently, there is a trend away from sun-drying in yards or on raised concrete platforms and toward the use of artificial heat in silo-like towers. Humidity is reduced to about 85%, and then the coating of dehydrated skin and pulp must be mechanically removed.[395]

Intensively managed coffee plantations yield up to 2,000 lbs. of dried coffee per acre (approximately 1 metric ton per hectare). The cleaned coffee beans are shipped in sacks to warehouses for storage or to factories which blend and roast. Roasting enlarges the beans to nearly double their original size and changes the color to dark-brown.[20] Sometimes coffee beans several years old are mixed with recent harvests for processing. There is much variation in the chemistry and flavor of coffee depending on the raw material and the methods of curing and roasting. Because of the lower labor costs in the production of Robusta coffee, most of the instant coffee becoming increasingly popular is derived from this species.

Decaffeination of coffee, which is also on the increase, is the major source of natural caffeine for pharmaceutical use. As an industrial by-product, natural caffeine will continue to be available, even though much caffeine is also produced synthetically, and some is derived from tea waste.

Medicinal uses:

Caffeine is given mainly as a cerebral and respiratory stimulant. It is commonly employed in combination with aspirin, phenacetin or ergotamine tartrate to relieve migraine and other forms of headache. Caffeine with sodium benzoate may be injected as an antidote for poisoning, as an emergency treatment for respiratory failure, or to relieve headache induced by postspinal fluid puncture.[423]

Toxicity:

In some individuals, caffeine causes nervousness, restlessness, excitement and insomnia. Patients with peptic ulcers, hypertension, and other cardiovascular and nervous disorders are usually advised by their physicians to refrain from drinking coffee.[404] Chlorogenic acid may induce rhinitis and dermatitis in workers engaged in roasting, sorting or grinding coffee.[68]

Other uses:

In addition to its role as the source of one of the world's leading beverages, coffee is extensively used as a flavoring for carbonated beverages, liqueurs, baked goods, ice cream and confections.[210] The dehydrated pulp of coffee berries is being employed as cattle feed. In the East Indies, a "tea" is sometimes made from coffee leaves. The abundant nectar of the flowers is gathered by honeybees and yields a clear, dense honey. Coffee wood is compact and suitable for small cabinetwork. The roots are valued for carving.

1. A giant pineapple *(Ananas sativa)* and a four-headed stalk, photographed at the progressive fruit farm, "Multifrutas," San Felipe, Venezuela.

2. American hellebore *(Veratrum viride)* in bloom. Courtesy Doctor Donald G. Huttleston, Taxonomist, Longwood Gardens, Kennett Square, Pennsylvania.

3. May apple *(Podophyllum peltatum)* in bloom. Courtesy Doctor Hans R. Schmidt, Ormond Beach, Florida.

4. Great scarlet poppy *(Papaver bracteatum)* photographed in a garden in Shiraz, Iran, by Doctor Frank D. Venning, Consultant in Tropical Agricultural Development, Miami, Florida.

5. Great scarlet poppy color is actually more crimson than scarlet, having no orange cast. Courtesy USDA, Agricultural Research Service, Northeastern Region, Beltsville, Maryland.

6. The thorny *Acacia seyal*, with conspicuous red-brown bark, is a minor source of gum arabic. Photographed in the Gifford Arboretum, University of Miami.

7. Senegal senna *(Cassia obovata)*; formerly cultivated, now naturalized in the West Indies. Photographed in Aruba, Netherlands Antilles.

8. Coca *(Erythroxylum coca)* in fruit. Photographed at the Fairchild Tropical Garden, Coral Gables, Florida.

9. Scarlet-fruited strain of castor bean *(Ricinus communis)*. Immature fruits are more often a pale bluish-green.

10. Papaya *(Carica papaya)* plot at The University of the West Indies, St. Augustine, Trinidad.

11. A single bushy plant of the red-eyed form of periwinkle *(Catharanthus roseus)*. Photographed on the island of Curaçao, Netherlands Antilles.

12. Belladonna *(Atropa belladonna)* with flowers and unripe fruit. Photographed in the Beal-Garfield Botanic Garden, Michigan State University, East Lansing, Michigan.

13. Henbane *(Hyoscyamus niger)* in flower and fruit. Photographed at the Drug and Horticultural Experimental Station of the University of Illinois, Downer's Grove, Illinois.

14. Towering flower spike of foxglove *(Digitalis purpurea)*. Courtesy Doctor Hans R. Schmidt, Ormond Beach, Florida.

15. Coffee *(Coffea arabica)* in fruit. Photographed in the garden of the late Mr. Frank Rimoldi, Coral Gables, Florida.

16. Arrow-poison tree *(Acokanthera schimperi* Benth.), painted especially for this book by James Kahurananga, East African Herbarium, Nairobi, Kenya.

Appendix

TABLE I

MEDICINAL PLANTS NO LONGER OFFICIAL IN THE UNITED STATES OF AMERICA BUT STILL MENTIONED IN THE U. S. DISPENSATORY AND/OR AMERICAN TEXTBOOKS ON PHARMACOGNOSY AND STILL IN USE ABROAD

Family, Botanical Name and Common Name	Geographical Origin & Distribution	Plant Part	Constituents	Former Medicinal Use
LYCOPODIACEAE *Lycopodium clavatum* L. STAGSHORN CLUBMOSS	Temperate and colder regions of Old and New World.	Spores; called "vegetable sulphur".	50% fatty oil, of which 55% is oleic acid, 30% is hexadec-9-enoic acid; minor amounts of palmitic and linoleic.[122]	Absorbent on abrasions; coating for suppositories and pills to avoid adhesion; taken orally for urinary and digestive complaints in Europe. Abandoned in U. S.; external use may cause granulomas.[122, 423, 552, 584]
ASPIDIACEAE *Dryopteris filix-mas* Schott. (*Aspidium filix-mas* Sw.) MALE FERN	Europe, Asia.	Rhizome and stipes.	6.5% oleoresin, of which the main constituent is filicin (chiefly margaspicin); also filicic and flavaspidic acids; filmaron, albaspidin, etc.[552]	Anthelmintic, against tapeworm. Toxic and hazardous.[122, 423, 552, 584]
CUPRESSACEAE *Juniperus communis* L. JUNIPER	Europe, northern Africa, northern Asia, North America.	Fruit.	Volatile oil. Constituents vary with region. May contain 1-terpinen-4-ol, α-pinene, camphene, and the sesquiterpene cadinene;[122, 584] also *d*-limonene, cymene, borneol, myrcene.[210]	Urinary antiseptic; diuretic. Irritant action has proved injurious to kidneys[122, 423, 552, 584]
PALMAE *Areca catechu* L. BETEL NUT	Tropical Asia, Oceania.	Seed kernel.	Alkaloids: arecoline, arecaine; also arecolidine, guvacine, isoguvacine, guvacoline, arecaidine.[617]	Arecoline formerly used in glaucoma; also against tapeworm, roundworm, and as antidote for abrin poisoning. Abandoned on account of systemic toxicity. Still used in veterinary practice as cathartic and anthelmintic.[122, 423, 466, 552, 584]

TABLE I (Cont.)

Family, Botanical Name and Common Name	Geographical Origin & Distribution	Plant Part	Constituents	Former Medicinal Use
LILIACEAE *Urginea maritima* Baker RED SQUILL; SEA ONION	Mediterranean Europe and Algeria.	Dried bulb.	Glycosides related to digitalis: scillaroside (or scillaren) A and scillaroside (or scillaren) B; also a phytosterol and calcium oxalate.[87, 419]	Diuretic. Small doses expectorant; large doses emetic. More recent use as cardiotonic. Abandoned. Main use is as rat poison.[122, 138, 423, 466, 552, 584]
ZINGIBERACEAE *Zingiber officinale* Roscoe GINGER; JAMAICA GINGER	Probably from Pacific Islands; widely cultivated.	Dried rhizome.	Essential oil contains the sesquiterpene zingiberene; the terpenes, camphene and phellandrene; also methyl heptenone, farnesene, cineol, borneol, geraniol, linalool. Oleoresin contains aromatic ketones (zingerone, shogaol) and the pungent principle gingerol.[122, 210, 584]	Stimulant, carminative, in dyspepsia, mild diarrhea and colic. Still used in veterinary practice.[122, 423, 466, 552, 584]
PIPERACEAE *Piper cubeba* L. CUBEB; TAILED PEPPER	East Indies, Ceylon. Cultivated in West Indies.	Dried seeds of unripe fruit.	10 to 18% volatile oil which contains cubebin: 5-hydroxy-3, 4-bis (3, 4-methylene-dioxybenzyl) tetrahydrofuran; 1% cubebic acid,[552] 1, 4 cineol; terpene and sesquiterpene alcohols.[210]	Was given in chronic bronchitis, and as urinary antiseptic; also to promote healing of mucous membranes.[122, 423, 552]
SALICACEAE *Populus nigra* L. BLACK POPLAR and other *Populus* species	Algiers to Afghanistan. Widely cultivated; commercially in Hungary.	Dried bark.	Essential oil contains glycosides: salicin, chrysin, populin, salicortin, nigracin; also gallic acid and tannin.[210, 578]	Salicin has been much used in treatment of rheumatism. Now derived synthetically;[122, 466, 552] little or none from natural sources.
Salix fragilis L. CRACK WILLOW	Europe, northern Asia.	Dried bark.	Phenolic glycosides (.23%): salicin, populin, fragilin, glycosmin, grandidentatin, picein, salicyloyl tremuloidin, etc.[20]	
Salix purpurea L. PURPLE OSIER	Europe, northern Africa, northern and central Asia.			
	Both cultivated in Belgium and Poland.			

BETULACEAE *Betula lenta* L. BLACK BIRCH; SWEET BIRCH	Eastern U. S. (Maine to Alabama and Ohio)	Dried bark.	Distilled oil contains 97 to 99% methyl salicylate.	Counterirritant, applied topically to relieve pain of rheumatism, sciatica and lumbago. Some is absorbed through skin, but most evaporates unless covered. Very toxic orally. Now almost entirely synthetic.[122, 423, 466]
FAGACEAE *Quercus infectoria* Oliv. DYER'S OAK; GALL OAK	Greece to Iran.	Galls, resulting from attack by the insect *Adleria gallae-tinctoriae* Oliv.[20]	50 to 70% gallotannic acid.	Tannic acid has been used as an astringent and styptic in cases of anal fissures and hemorrhoids; also to halt diarrhea and dysentery, in gargles and as antidote for poisons. When used in burn therapy and in rectal enemas, has been absorbed and has caused severe liver necrosis.[423] Tannin increasingly recognized as carcinogenic and carcinostatic.[122, 439, 584, 552, 584]
ULMACEAE *Ulmus fulva* Michx. (*U. rubra* Muhl.) SLIPPERY ELM	Eastern Canada and eastern and central U. S.	Inner bark, dried.	Mucilage.	Demulcent and emollient. Powdered bark was used as a soothing poultice; infusion was given to relieve gastric and intestinal inflammation. Powder made into lozenges for throat irritation.[122, 466, 584]
POLYGONACEAE *Fagopyrum esculentum* Moench BUCKWHEAT	Asia; widely cultivated in temperate climates.	Leaves.	Contain 3% rutin, the 3-rhamno-glucoside of 5, 7, 3', 4'-tetrahydroxyflavonol.[122]	Vasoconstrictor. Allegedly halts capillary bleeding; reduces capillary fragility. Not of proven efficacy.

TABLE I (Cont.)

Family, Botanical Name and Common Name	Geographical Origin & Distribution	Plant Part	Constituents	Former Medicinal Use
Rheum emodi Wallich INDIAN RHUBARB	India, Pakistan.	Dried rhizome and root.	Emodin; chrysophanic acid; tannin.	Has been used as a substitute for *R. officinale* and *R. palmatum*; see below.[122,466]
Rheum officinale Baill. MEDICINAL RHUBARB	Tibet and western China; cultivated in Europe and U.K.	Dried rhizome and root.	Many 1, 8-dihydroxy-anthracene derivatives including chrysophanol, emodin, rhein, erythroeretin, rhabarberon; also rheotannic acid, methylchrysophanic acid, catechin, gallic acid, calcium oxalate.[296, 552, 584, 635]	Purgative in large doses; antidiarrheal in small doses.[122, 423, 466, 513, 552, 584, 624]
Rheum palmatum L. CHINESE RHUBARB	Tibet and western China.			No longer recognized.
CHENOPODIACEAE *Chenopodium ambrosioides* L. (*C. ambrosioides* var. *anthelminticum* Gray) AMERICAN WORMSEED	West Indies and tropical America; widely naturalized in U. S. and Old World.	Plant tops.	Distilled oil contains ascaridol; *p*-cymene, *l*-limonene, *d*-camphor, methyl salicylate.	Anthelmintic. Abandoned in favor of less toxic drugs.[122, 423, 466, 552, 584]
RANUNCULACEAE *Aconitum napellus* L. MONK'S-HOOD	Mountains of Europe and British Isles. Widely grown as garden flower in temperate climates.	Root, dried.	Alkaloids, including aconitine, aconine, napelline, picraconitine (benzaconine).[617]	Stimulates, then depresses central and peripheral nervous system. Was used for irritant and anesthetic effect in neuralgia and gastralgia. Abandoned in favor of less toxic drugs.[122, 423, 446, 552, 584]
HYDRASTIDACEAE (formerly in BERBERIDACEAE) *Hydrastis canadensis* L. GOLDENSEAL	Eastern U. S. Cultivated in Northwest.	Dried rhizomes and roots.	Alkaloids: hydrastine, berberastine, berberine, canadine, reticuline.[616]	Alterative and bitter tonic for gastric disorders;[618] diuretic; also used in eye lotions. Hemostat in uterine hemorrhage.[584] In 40 products sold in Canada.[218]

Appendix 363

LAURACEAE *Cinnamomum loureirii* Nees SAIGON CINNAMON	Southeast Asia.	Inner bark, powdered.	Volatile oil; cinnamic aldehyde; tannin.	Carminative, astringent; was given to relieve nausea, flatulence, diarrhea.[122, 423, 466, 582]
Cinnamomum zeylanicum Nees CEYLON CINNAMON	India, Malaya and widely cultivated in tropics.	Oil distilled from leaves and twigs.	Oil contains 90% cinnamaldehyde; small amounts of cinnamic acid, cinnamyl acetate, and eugenol.[423]	Was employed as stomachic, carminative and germicide[122, 423, 466, 552, 584]
Sassafras albidum Nees (*S. variifolium* Kuntze) SASSAFRAS	Eastern U. S.	Volatile oil from root.	Contains 80% safrole (*p*-allylmethylenedioxybenzene); some eugenol, pinene, phellandrene, sesquiterpene, *d*-camphor.[552]	Oil of sassafras has been used topically as an antiseptic, and to destroy lice, allay insect stings and bites. also has been taken internally as carminative and stimulant.[122, 522] Safrole has produced liver tumors in rats.[385]
PAPAVERACEAE *Sanguinaria canadensis* L. BLOODROOT	Canada to Nebraska, Arkansas and Georgia.	Dried rhizome and roots.	Alkaloids (isoquinoline derivatives), chiefly sanguinarine (pseudochelerythrine); also protopine, oxysanguinarine, chelerythrine, α– and β-allocryptopine,[617] berberine, chelilutine, chelirubine, coptisine, sanguilutine, sanguirubine.[616]	Emetic and expectorant.[122] Externally for eczema and skin cancers.[552] Toxic;discontinued. Sanguinarine is used to induce glaucoma in experimental animals.[552]

TABLE I (*Cont.*)

Family, Botanical Name and Common Name	Geographical Origin & Distribution	Plant Part	Constituents	Former Medicinal Use
CRUCIFERAE				
Brassica nigra Koch BLACK MUSTARD	Europe; universally cultivated and naturalized.	Seeds.	Glucoside sinigrin and the enzyme myrosin. When macerated in water release volatile oil containing .7 to 1.3% allyl isothiocyanate.[584]	Emetic and rubefacient; vesicant. Orally toxic in large doses.[122, 423, 552, 584]
Brassica juncea Coss. INDIAN MUSTARD	China, India; cultivated in Africa and Russia.			
HAMAMELIDACEAE				
Hamamelis virginiana L. WITCH HAZEL	Eastern Canada and U. S., west to Texas.	Bark leaves and twigs.	Bark contains gallitannins, mainly β-hamamelitannin; also α- and γ-hamamelitannin and elagitannins. Also condensed tannins— (+)-catechin, (+)-gallocatechin, (−)-epicatechin gallate and (−)-epigallocatechin gallate; plus proanthocyanidins of cyanidin and delphinidin.[202] Leaves contain no hamamelitannin but a mixture of gallotannins and some condensed catechins and proanthocyanidins.[203]	Astringent and hemostat; was used on wounds, sprains, bruises, hemorrhoids, and in eye lotions.[122, 423, 466, 552, 584]
LEGUMINOSAE				
Andira araroba Aguiar ARAROBA	Brazil	Goa Powder from trunk of tree.	Chrysarobin (containing chrysophanol, anthranol, etc.),[584]	Applied on skin diseases. Irritant to eyes, nose, etc.[122, 423, 466, 552, 584]
Cytisus scoparius Link SCOTCH BROOM	Europe and western Asia; cultivated in western U. S.	Plant tops, dried.	Alkaloid sparteine (to 1.5%); and the flavone scoparin[384]; genisteine, sarothamine, volatile oil, tannin.[582]	Sparteine was used as diuretic, cathartic, cardiac depressant and as oxytocic to induce and stimulate labor.[122, 423, 552, 584]
Sophora japonica L. JAPANESE PAGODA TREE	Japan, Korea, China. Widely cultivated as an ornamental.	Flower bud (called "yellow berry" in China)	Contains 4 to 5 times as much rutin as buckwheat. The dried plant yields 17.5%.[605]	Hemostat in China. Commercial source of rutin, which was being given to overcome capillary fragility but no longer officially utilized in U. S.[122, 423, 466]

Appendix 365

KRAMERIACEAE (formerly in LEGUMINOSAE) *Krameria triandra* Ruiz & Pavon PERUVIAN RHATANY	Peru, Bolivia	Root.	8 to 20% catechin tannin; also leucoanthocyanadins, krameric acid and calcium oxalate.	Tincture has been employed as oral and mucosal astringent; also in treating diarrhea and dysentery and on poultices for hemorrhoids; used for cleaning wounds. Dermatologically it is a primary sensitizer.[122, 190, 466, 552, 556, 584]
RUTACEAE *Agathosma betulina* Pillans (*Barosma betulina* Bartl. & Wendl.) BUCHU; ROUND BUCHU	South Africa	Leaves.	Diosmin, calcium oxalate, diosphenol, d-limonene, dipentene, l-menthone.	Diuretic and urinary antiseptic.[584] Dried leaves now sold through "health food" outlets for "tea."
Agathosma crenulata Pillans (*Barosma crenulata* Hook.; *B. serratifolia* Willd.) OVAL BUCHU (includes former "LONG BUCHU")				
SIMAROUBACEAE *Picrasma excelsa* Planch. (*Picraena excelsa* Lindl.) BITTERWOOD; JAMAICA QUASSIA	West Indies.	Wood.	Picrasmin, neoquassin and thiamine.[210]	Bitter tonic; enema to eliminate thread worms.[122, 552, 584]
BURSERACEAE *Commiphora abyssinica* Engl. MYRRH	Northeastern Africa; southern Arabia.	Gum resin from tree.	7 to 17% volatile oil containing terpenes, sesquiterpenes, esters, cuminic aldehyde and eugenol. 25 to 40% resin (myrrhin). 57 to 61% gum and bitter principles.[584]	Stimulant, astringent, antiseptic. An ingredient in mouthwash.[122, 466, 552, 584]
Commiphora molmol Engl.				

TABLE I (Cont.)

Family, Botanical Name and Common Name	Geographical Origin & Distribution	Plant Part	Constituents	Former Medicinal Use
POLYGALACEAE *Polygala senega* L. SENECA SNAKEROOT; RATTLESNAKE ROOT	Across southern Canada and northern U. S.	Root.	Triterpinoid saponins. On hydrolysis, the crude saponin senegin (polygalin) yields glucose, presenegenin, senegenin, senegenic acid, polygalic acid (hydroxysenegenin). Also present is polygalitol (1, 5-anhydro-sorbitol).[552]	Emetic and expectorant in chronic bronchitis. Still used in veterinary medicine.[466, 552, 584]
EUPHORBIACEAE *Croton tiglium* L. PURGING CROTON	From China and India to New Guinea.	Seed oil ("Croton oil").	55% oleic acid; 30% linoleic; 1.5% palmitic; 0.5% stearic;[122] croton oil factor A (12-0-tetradecanoyl-phorbol-13-acetate, TPA),[316] also there has been isolated phorbol 12-tiglate 13-decanoate, an antileukemic principle.[327]	Violent purgative; abandoned; is a powerful irritant and vesicant, and well-recognized co-carcinogen.[309]
GUTTIFERAE *Garcinia hanburyi* Hook.f. GAMBOGE	Southeast Asia.	Gum resin from trunk bark.	70 to 80% resin containing chiefly gambogic acid; 15 to 25% gum.	Powerful hydragogue cathartic. 4 g may be fatal.[122, 423, 552, 584]
FLACOURTIACEAE *Hydnocarpus laurifolia* Sleumer (*H. wightiana* Blume) CHAULMOOGRA	China, southeast Asia, western India; introduced into Africa.	Seed oil ("Hydnocarpus oil").	Oil rich in cyclopentenyl monocarboxylic acids;[291] 27% chaulmoogric acid; 48.7% hydnocarpic; 12.2% gorlic; 3.4% lower homologues of chaulmoogric acid; 6.5% oleic; 1.8% palmitic acid.[20]	Replaced oil of *H. kurzii* in treating leprosy, chronic skin diseases, rheumatic pains, etc. Toxic internally. Now replaced by synthetically produced ethyl esters and salts of hydnocarpic and chaulmoogric acid.[122, 423, 466, 552, 584]
PUNICACEAE *Punica granatum* L. POMEGRANATE	Southwestern Asia; widely cultivated in tropics.	Bark of trunk or roots.	20 to 22% tannin; 0.5 to 1% alkaloids, chiefly pelletierine; also isopelletierine, methylisopelletierine, methylpelletierine.[617] Isopelletierine has been shown to be the most active.[611]	Anthelmintic, especially to expel tapeworm.[122, 466, 552, 584]

MYRTACEAE				
Eucalyptus macrorryncha F. v. M. RED STRINGYBARK	Victoria and New South Wales, Australia.	Leaves.	When mature, leaves yield 12 to 20% rutin.[606]	This tree was being exploited as a commercial source of rutin when interest in this product was high;[456, 606] it was used for pulmonary hemorrhage and capillary hemorrhage from hypertension.[122, 466]
Melaleuca cajuputi Powell CAJEPUT (erroneously *M. leucadendron*)	Southeast Asia, Malaysia, to Queensland.	Leaves.	Cajeput oil distilled from leaves contains chiefly cineole; also terpineol and its esters, pinene, a little benzaldehyde.[85]	Internally, for coughs, colds. Externally applied on neuralgia and rheumatism. Used as an anthelmintic against roundworm.[85]
Melaleuca quinquenervia S. T. Blake	New Guinea, New Caledonia, New South Wales and Queensland.		Niaouli oil, very similar to cajeput oil.[85]	
Syzygium aromaticum Merr. & Perry (*Eugenia caryophyllus* Bullock & Harrison) CLOVE	East Africa, Madagascar, India, Ceylon, Indonesia, Haiti.	Flower buds, dried.	Distilled oil contains 70 to 90% eugenol (4-allyl-2-methoxyphenol); also caryophyllene (a sesquiterpene), acetyleugenol, vanillin, furfural, and methyl-*n*-amyl ketone.[122]	Eugenol formerly used as anodyne and antiseptic in dentistry; also as expectorant and antiemetic.[122, 423, 466, 552, 584]
UMBELLIFERAE				
Ferula assa-foetida L. ASAFETIDA	Iran to Kashmir.	Oleo-gum-resin from rhizomes and roots of living plant.	25% gum; 4 to 20% volatile oil, mainly isobutyl-propanyl disulfide, with other disulfides and terpenes. 40 to 65% resin,[122] containing asaresinol ferulate and free ferulic acid.[584]	Stimulant, antispasmodic, laxative, carminative, expectorant. Was prescribed in infantile convulsions, croup, flatulent colic and bronchitis. Still used in veterinary practice and as a dog and cat repellent.[122, 423, 466, 552, 584]
Ferula narthex Boiss.	Kashmir, Tibet, Afghanistan.			
Ferula rubricaulis Boiss.	Iran and western Afghanistan.			
Ferula foetida Regel	Iran and western Afghanistan.			

TABLE I (Cont.)

Family, Botanical Name and Common Name	Geographical Origin & Distribution	Plant Part	Constituents	Former Medicinal Use
ERICACEAE *Arctostaphylos uva-ursi* Spreng. BEARBERRY	Northern hemisphere: Asia to western U. S.	Leaves.	Arbutin (a hydroquinone derivative), 6.8 to 8%;[189, 265, 433] also corilagin pyroside, quercitin, gallic acid, elagic acid, ursolic acid.[122] Tannin content varies from 18 to 39.9%.[365, 473]	Diuretic, antiseptic, astringent; has been employed in urethritis, cystitis, and diarrhea.[122, 552, 584] Dried leaves sold for "tea" in "health food" outlets.
Gaultheria procumbens L. WINTERGREEN	Eastern U. S. and Canada.	Leaves.	Volatile oil contains 99% methyl salicylate; also paraffin, an aldehyde or ketone, an ester and a secondary alcohol.[117]	Methyl salicylate has been employed in liniments and ointments for pain relief in rheumatism, sciatica, lumbago.[423] Now almost entirely synthetic.[122, 466]
LOGANIACEAE *Gelsemium sempervirens* Ait.f. CAROLINA YELLOW JESSAMINE	Southeastern U. S., west to Texas and Mexico. Widely cultivated as an ornamental.	Rhizome and roots.	Alkaloids: gelsemine, gelsemicine, sempervirine, gelsedine, gelsemidine, gelseminine, gelsevirine,[656] also scopoletin.[584] Wichtl et al. have identified Gelsemium alkaloid A as 1-methoxygelsemine and Gelsemium alkaloid C as 14-hydroxygelsemicine.[613, 614]	Sedative, anodyne in neuralgia and migraine. Antispasmodic. Toxic and dangerous.[122, 423, 466, 552, 584]
STRYCHNACEAE (formerly in LOGANIACEAE) *Strychnos nux-vomica* L. POISON NUT	India and Malaya, southeast Asia. Cultivated in Africa and Hawaii.	Seeds dried.	Alkaloids: 1.23% strychnine and 1.55% less active brucine (dimethoxystrychnine); also struxine, vomicine, α- and β-colubrine;[617] the glycoside, loganin,[423] and chlorogenic acid.[584]	Strychnine has been given to aid appetite and digestion. Nux-vomica extract formerly used as a circulatory and respiratory stimulant, but effective doses are nearly equivalent to toxic doses. Abandoned.[122, 423, 466, 552, 584]

368   Major Medicinal Plants

GENTIANACEAE *Gentiana lutea* L. YELLOW GENTIAN	Mountains of southern and central Europe.	Rhizome and roots, dried.	Glucosides: 2% gentiopicrine; also gentiamarine, amarogentine[513] the glycoside gentiin; the tannin-like principle gentiamarin; gentisin (gentianin or gentianic acid), gentisic acid and the trisaccharide gentianose.[423]	Bitter tonic in dyspepsia and anorexia. Gentiopicrine was used as an antimalarial.[122, 423, 466, 552, 584]
APOCYNACEAE *Holarrhena antidysenterica* Wall. KURCHI	India.	Bark, seeds.	Alkaloids; chiefly conessine; also concuressine, conessimine, isoconessimine, holarrhimine, holonamine, kurchessine, kurcholessine, etc.;[616] also kurchicine, a protoplasmic poison like emetine.[500]	Conessine has been used in amebic dysentery and vaginitis. Toxic; oral doses may cause psychological disorders.[423, 552, 584]
CONVOLVULACEAE *Exogonium purga* Lindl. (*Ipomoea purga* Hayne) JALAP	Mexico, Central America; cultivated in India.	Dried root and resin.	Root contains 9 to 18% resin of which the main constituent is convolvulin; also present is scammonin.[571]	Drastic hydragogue cathartic.[584]
Ipomoea orizabensis Ledenois. MEXICAN SCAMMONY	Mexico.	Root and resin.	6 to 18% resin containing jalapin, ipuranol, ipurganol;[122] also sitosterol.[584]	Hydragogue cathartic.[122, 423, 466, 552, 584]
SOLANACEAE *Capsicum frutescens* L. (*C. minimum* Roxb.) RED PEPPER *Capsicum annuum* L.	Tropical America; universally cultivated.	Fruits, dried.	The alkaloid capsaicin (N-vanillyl-8-methyl-6-nonenamide) (I), 69%; dihydrocapsaicin (II), 22%; *nor*dihydrocapsaicin, 7%; homocapsaicin, 1%; homodihydrocapsaicin, 1%;[380] ascorbic acid, thiamine, red carotenoids (capsanthin, capsorubin).[584] An average of .45% capsaicin on a dry weight basis has been reported in Mexican peppers.[585]	Topically, a counterirritant; orally, an intestinal stimulant.[552]

TABLE I (Cont.)

Family, Botanical Name and Common Name	Geographical Origin & Distribution	Plant Part	Constituents	Former Medicinal Use
Datura metel L. (incl. *D. metel* var. *fastuosa* L.) DEVIL'S TRUMPET; BLACK DATURA	Asia, Africa. Widely cultivated and naturalized in tropics.	Leaves and seeds.	Alkaloids: mainly hyoscine. Leaves, 0.25 to .055%; Seeds, 0.12 to 0.5%; small amounts of hyoscyamine and atropine.[605] The alkaloid content varies with altitude and season of the year.[297]	Formerly source of atropine; now obtained from *Duboisia* (q.v.).[466]
Datura stramonium L. JIMSONWEED	Probably from South America; universally distributed.	Seeds; leaves (dried and powdered).	Leaves, total alkaloids: 0.25% to 0.45%;[552] Seeds, total alkaloids: 0.47 to 0.65%. Hyoscine content: 0.1% in leaves, 0.05% in stems, 0.1% in roots. Hyoscyamine content: 0.4% in leaves, 0.2% in stems, 0.1% in roots.[45]	Antispasmodic in asthma and Parkinson's disease. Atropine used for pupil dilation and as anticholinergic.[122, 423, 466, 552, 584]
NAUCLEACEAE (formerly in RUBIACEAE) *Uncaria gambier* Roxb. GAMBIER; GAMBIR; PALE CATECHU	Malaysia.	Extract of leaves and twigs.	7 to 30% catechins; 20 to 50% catechutannic acid (a phlobotannin); quercetin, and catechu red.[552]	Astringent employed against diarrhea, sorethroat, etc.[122, 466, 552, 584]
VALERIANACEAE *Valeriana officinalis* L. GARDEN HELIOTROPE; COMMON VALERIAN	Europe, Asia, British Isles.	Roots, dried.	.7 to .8% valepotriate and .5% ethereal oil.[600] Alkaloids: chatinine and valerine.[617]	Sedative, hypotensive;[491] antispasmodic, stomachic.[466, 552, 584]
CUCURBITACEAE *Citrullus colocynthis* Schrader COLOCYNTH	India.	Fruit pulp, dried and powdered.	"Colocynthin," which is a combination of an unnamed alkaloid[616] and citrullol; also α-elaterin, α-phytosterol, and hentriacontane.[617]	Powerful hydragogue cathartic; toxic in large doses.[116, 552]

Appendix

CAMPANULACEAE *Lobelia inflata* L. INDIAN TOBACCO	Eastern U. S. and Canada; cultivated in Europe.	Leaves and flowering tops, dried.	.36 to 2.25% lobeline (*l*-lobeline or α-lobeline); lobeliline; lobelanine, lobelanidine, isolobinine, etc.[220, 321, 584, 617]	Used as anti-tobacco drug;[319] is emetic, stimulant; served as antiasthmatic and as expectorant in cases of laryngitis and bronchitis. Toxic, in large doses may cause medullary paralysis,[408] can be fatal.[211]
COMPOSITAE *Anthemis nobilis* L. ROMAN CHAMOMILE	England, Belgium, France, Germany.	Flowers, dried.	Azulene, dihydrocinnamic acid, apigenin,[384] angelic acid.[406]	Infusion has been given in cases of dyspepsia, fever, nervousness, hysteria. Lotion applied to toothache, earache, neuralgia.[554]
Arnica montana L. ARNICA *Arnica fulgens* Pursh. *Arnica sororia* Greene *Arnica cordifolia* Hook	Europe. Western Canada and U. S.	Flowers and root.	Arnicin, arnisterol (arnidiol); anthoxanine; tannin.[552]	Externally as liniment on bruises, sprains, swellings.[423, 552, 618] Still used in homeopathy, though when given internally has caused fatal poisoning[624]
Artemisia cina Berg. *Artemisia maritima* L. WORMSEED	Western Asia.	Leaves and/or flowerheads.	Santonin; also cineole, thujone, camphene, terrisin.	Santonin has been widely used as an anthelmintic. Discontinued because of toxicity in effective doses.[122, 423, 466, 552]

TABLE II
PLANTS WHICH SERVE AS PHARMACEUTICAL AIDS OR ADJUNCTS

Family, Botanical Name and Common Name	Geographical Origin & Distribution	Plant Part or Product	Constituents	Pharmaceutical Uses
ALGAE *Gelidium cartilaginium* Gaillon (a red alga) RED LACE, and more than a dozen other species of this genus including: *G. corneum*	Off coasts of South Africa, Mexico, U. S. South Africa, Portugal, Spain, Morocco.	Agar.	A polysaccharide; a colorless, odorless, tasteless, transparent mucilaginous substance.	Suspending agent; ingredient in capsules, suppositories, emulsions, lubricants; a carrier for medication; a base and disintegrator in tablets. Also serves as a bulk laxative.[122, 456, 519, 552]
G. amansii, G. liatulum, G. lingulatum, G. pacificum	Japan.			
G. pristoides	South Africa			
G. sesquipedale	Portugal; Morocco.			
Gracilaria confervoides Grev. (a red alga) SEA STRING	Coasts of southern hemisphere. Harvested commercially in South Africa.	Agar.	See above.	See above.
Pterocladia capillacea J. Agardh (a red alga)	Egypt, Japan, New Zealand, Australia.	Agar.	See above.	See above.
Pterocladia lucida J. Agardh (a red alga)	New Zealand.	Agar.	See above.	See above.

Ascophyllum nodosum (a brown alga) KNOBBED WRACK; YELLOW TANG; SEA WHISTLE	Off coasts of eastern Canada, United Kingdom; Norway.	Algin (sodium alginate).	Algin and alginates serve as suspending agents, protective colloids, emollients. They give body and spreadability to ointments, stability to emulsions; they accelerate disintegration of tablets.[122, 307, 584]
Laminaria digitata Lamour. (a brown alga) SEA GIRDLES; HORSETAIL KELP; TANGLE	France; Norway, Atlantic coast of U.S.	Algin.	A carbohydrate material, soluble in water, insoluble in alcohol, chloroform or ether.[552]
Laminaria hyperborea (a brown alga)	United Kingdom.	Algin.	
Laminaria saccharina Lamour. (a brown alga) SUGAR WRACK	Atlantic coast of U.S.	Algin.	
Macrocystis pyrifera C. Agardh (a brown alga) CALIFORNIA GIANT KELP	Mainly off southern California; also Australia.	Algin (world's leading source).	
Chondrus crispus Stackh. (a red alga) IRISH MOSS; CARRAGEEN	Coasts of North Atlantic Ocean. Collected from Norway to Gibraltar and New England.	Carrageenan.	A complex of sulphated polysaccharides, mainly galactose and anhydrogalactose.[455]	Emollient, emulsifying and suspending agent, demulcent. Often in mineral oil and anti-cough preparations.[122, 466, 552, 564, 584]
Gigartina mammillosa J. G. Agardh (a red alga)	United Kingdom.	Carrageenan.	See above.	Degraded carrageenan, administered orally in quantity has produced ulcerations in the large intestine in guinea pigs and rabbits and in the colon of female rhesus monkeys.[70, 231]
Gigartina stellata Batt. (a red alga)	United Kingdom (Chief source of carrageenan since early 1940s).	Carrageenan.	See above.	
Furcellaria fastigiata Lamour. (a red alga) DANISH AGAR	Arctic Sea and North Atlantic Ocean, Mediterranean and Black Seas. Collected mainly in Denmark and Canada.	Furcellaran, a gum much like κ-carrageenan.[82]	A sulphated lactan, chemically midway between agar and carrageenan.[82, 584]	Came into use in World War II as substitute for agar. Now much in demand for suspensions, emulsions, foams and disintegrating tablets.[512]

Appendix 373

TABLE II (*Cont.*)

Family, Botanical Name and Common Name	Geographical Origin & Distribution	Plant Part or Product	Constituents	Pharmaceutical Uses
SMILACACEAE (formerly in LILIACEAE)				
Smilax aristolochiaefolia Mill. MEXICAN SARSAPARILLA	Southern Mexico, Guatemala, British Honduras.	Roots.	Steroidal sapogenins: mainly sarsasapogenin and smilagenin; also sitosterol and stigmasterol.[122, 552]	Extract used as a flavor.[122, 466, 552] *Smilax* spp. have been employed in partial synthesis of cortisone.[584]
Smilax febrifuga Kunth ECUADOREAN SARSAPARILLA	Ecuador and Peru.	Roots.	See above.	
Smilax regelii Killip & Morton (*S. ornata* Hook.) HONDURAS SARSAPARILLA	Guatemala, British Honduras and Honduras; cultivated in Jamaica.	Roots.	See above.	
ZINGIBERACEAE				
Elettaria cardamomum Maton CARDAMOM	India, Ceylon, Guatemala, El Salvador.	Seeds, for essential oil.	Essential oil contains eucalyptol (cineol); sabinene, d, α-terpineol and acetate, borneol, limonene, terpinene, 1-terpinen-4-ol with its formate and acetate.[552]	Used as a flavor.
ILLICIACEAE (formerly in MAGNOLIACEAE)				
Illicium verum Hook. f. STAR ANISE	Southern China, southeast Asia.	Fruits, for essential oil (2.5 to 5%).	Essential oil contains 80 to 90% anethole; also chavicol methyl ether, *p*-methoxyphenylacetone and safrole.[584]	Used as a flavor; carminative.

Appendix

LAURACEAE			
Cinnamomum burmannii Blume MALAY CINNAMON	India to China and Malaysia, commercialized mainly in Sumatra.	⎫ Bark, for essential oil (0.8 to 1.4%).	Essential oil contains approximately 60 to 75% cinnamic aldehyde, 4 to 10% phenols (primarily eugenol), pinene, phellandrine, caryophyllene[584]	Used as a flavor; carminative.[122, 423, 466, 552, 584]
Cinnamomum loureiri Nees SAIGON CINNAMON	Vietnam.			
Cinnamomum zeylanicum Nees CEYLON CINNAMON	Ceylon, Seychelles, Malagasy Republic.			
Cinnamomum cassia Nees CASSIA CINNAMON	China; has not been available for many years. ⎭			
MYRISTICACEAE			
Myristica fragrans Houtt. NUTMEG; MACE	India, Ceylon, Malaysia, Grenada.	Seed (nutmeg) for essential oil (5 to 15%). (Mace is the aril clasping the seed.)	Essential oil of nutmeg contains d-and l-α-pinene, sabinene; 60 to 80% d-camphene, 8% dipentene, 2% elemicin and isoelemicin; 0.6% safrole; eugenol; geraniol; d-borneol, l-terpineol; also 4% myristicin, which is toxic and narcotic.[122, 584]	Used as a flavor; carminative.[122, 466, 552, 584]

TABLE II (Cont.)

Family, Botanical Name and Common Name	Geographical Origin & Distribution	Plant Part or Product	Constituents	Pharmaceutical Uses
ROSACEAE *Prunus dulcis* D. A. Webb (*P. amygdalus* Batsch; *Amygdalus communis* L.) ALMOND and var. *amara* BITTER ALMOND	Iran, Italy, Spain, Morocco, Portugal, France, California.	Seed kernel, for fixed oil (40 to 55%).	Expressed almond oil (obtained mainly from the bitter almond because of the demand for the sweet almond kernel as a nut) contains these fatty acids: .2 to 1% myristic, 5 to 8.9% palmitic, 62 to 80% oleic, 14 to 30% linoleic, 4% stearic.[20, 242]	Used as a vehicle for injections; demulcent; lubricant.[20, 268]
			Essential oil, containing 2 to 4% HCN, is derived from bitter almond presscake.	Sometimes in cough remedies.
Prunus armeniaca L. APRICOT	California, Washington, Utah, Ecuador, France, Spain, Italy, Turkey, Syria, Palestine, Queensland, New Zealand.	Seed kernel, for fixed oil (40 to 45%), called "Persic oil" or Apricot Kernel Oil.	Oil resembles almond oil in fatty acid content.[242] Kernels contain the cyanogenetic glycoside, amygdalin.	Used as a substitute for almond oil, q.v. Amygdalin from apricot kernels, marketed as "Laetrile" is a controversial agent claimed to be effective in controlling cancer.[142]
Prunus persica Stokes PEACH	Temperate Asia, Australia, New Zealand, all but coldest and warmest areas of U. S.	Seed kernel, for fixed oil, "Persic oil."	See almond oil. In almond, apricot and peach oil β-sitosterol represents 95% of the sterol, and all three oils contain about 0.02% squalene.[242]	See almond oil.
Prunus serotina Ehrh. WILD BLACK CHERRY	Nova Scotia to North Dakota, south to Texas and northern Florida.	Bark.	Bark contains the cyanogenetic glycoside prunasin and the enzyme prunase (or emulsin), which yield on hydrolysis HCN.	Powdered bark is prepared as a syrup, which serves as a vehicle for cough remedies.[423]

LEGUMINOSAE				
Arachis hypogaea L. PEANUT; GROUNDNUT	Native to Brazil; grown throughout the subtropics and tropics.	Seeds for fixed oil (40 to 50%).	Oil contains glycerides of arachidic, stearic, lignoceric, oleic acids, etc.[584]	Oil is used as a substitute for olive oil. It serves as a vehicle for injections; also as an ingredient in liniments and ointments.[423]
Ceratonia siliqua L. CAROB	Middle East, Greece, Cyprus, Sicily, Italy, Spain, California.	Seed, a source of carob gum.	Carob gum (manogalactan) is a polysaccharide; contains 20% D-galactose and 80% D-mannose.[552]	The gum serves as a suspending agent, absorbent, demulcent, lubricant.[122, 126, 552]
Cyamopsis tetragonoloba Taub. (*C. psoralioides* DC.) GUAR; CLUSTER BEAN	India, Pakistan; widely cultivated.	Seed, a source of guar gum.	Guar gum (manogalactan) contains 32% anhydrogalactose and 59% anhydromannose.[126]	Used as a thickener in emulsions and suspensions; also as a binder and disintegrator in tablets.[81, 103, 122, 227, 552]
LINACEAE				
Linum usitatissimum L. LINSEED; FLAX	India, Iraq, Russia, eastern Africa, northern Africa, Europe, western Australia, north-central U. S., Argentina.	Seed, a source of oil.	Oil contains unsaturated fatty acids, chiefly 39 to 41% linolenic acid;[6] also linoleic.	Serves as a demulcent and emollient.[122, 423, 466, 552, 584]
RUTACEAE				
Citrus aurantium L. var. *amara* L. SOUR ORANGE; BITTER ORANGE; SEVILLE ORANGE	Spain, Italy, Paraguay.	Peel and flowers, for essential oils.	Peel oil contains .07% methoxyflavones, chiefly nobiletin.[515] Essential oil distilled from flowers contains pinene, camphene, dipentene, linalool, *l*-linalyl, acetate, α-terpineol, geraniol, nerol, nerolidol, methyl anthranilate, indole, farnesol and phenylacetic acid esters.[210]	These essential oils are employed as flavors.[122, 423, 584]
Citrus limon L. LEMON	Italy, Sicily, France, Spain, Portugal, India, Australia, New Zealand, Mexico, California, Florida.	Peel, for essential oil.	Essential oil contains 64.3% *d*-limonene, 0.78% *d*-α pinene, 0.43% camphene, 1.8% terpinene, 3.8% linalool, 5.2% hendecanal, 3.9% terpineol, 3.8% linalyl acetate, 1.2% cadinene.[20]	Essential oil is used as a flavor; carminative.[122, 423, 552, 584]

378 Major Medicinal Plants

TABLE II (*Cont.*)

Family, Botanical Name and Common Name	Geographical Origin & Distribution	Plant Part or Product	Constituents	Pharmaceutical Uses
Citrus sinensis Osbeck SWEET ORANGE	California, Florida, Israel, Mediterranean Europe, subtropical Africa and Asia, Australia, subtropical Latin America.	Peel, for essential oil.	Essential oil contains over 90% *d*-limonene; also decylic, octylic, monylic and dodecylic aldehydes; citral,[210] linalool; *d*-*l*-terpineol, methyl anthranilate.[20]	Essential oil serves as a flavor.[122, 552, 584]
MALVACEAE *Gossypium hirsutum* L. SHORT-STAPLE COTTON	South America; cultivated and naturalized throughout tropics and subtropics.	Seeds, for oil.	Cottonseed oil is a semi-drying oil with the following fatty acid composition: 1.4 to 3.3% myristic, 1.1 to 2.7% stearic, 0.6 to 1.3% arachidic, 22.9 to 29.6% oleic, 45.3 to 50.4% linoleic. There are also present some glycerides and small amounts of phospholipins and phytosterols.[20]	The oil is used as a vehicle for injections, as an emollient and as a lubricant.[423] It is decolored and treated to avoid rancidity.[147, 265]
STERCULIACEAE *Theobroma cacao* L. CACAO; CHOCOLATE	Southern Mexico, tropical South America. Widely cultivated in tropics. 70% of world supply from West Africa.	Seeds (nibs) as a source of fat; also powdered cocoa.	Expressed fat (called cocoa butter or Theobroma oil) is hard and brittle at ordinary temperatures. Contains 2.5% fully saturated glycerides, 73.0% monooleodi-saturated glycerides, 24.5% dioleomono-saturated glycerides. Cocoa, the powdered product of the partially defatted seeds, contains 0.85 to 2.7% theobromine; 0.09 to 0.19% caffeine.[107]	Cocoa butter is employed mainly as a suppository base; also as an emollient. Powdered cocoa is prepared as "cocoa syrup," a flavored vehicle.[122, 127, 423, 466, 497, 552, 584]

Appendix 379

MYRTACEAE *Eucalyptus dives* Schau. BROAD-LEAVED PEPPERMINT	New South Wales.	Leaves, for essential oil (4%).	Essential oil contains 40 to 50% (—)-piperitone, 15 to 25% α– and β-phellandrene, minor amounts of *p*-cymene, α– and γ-terpinene, terpinolene, etc.[238]	Essential oil employed currently only as a flavor.[423] These three species provide the preferred medicinal eucalyptus oils "free from the objectionable low boiling aldehydes so pronounced in the oils of *E. globulus* Labill., and *E. smithii* R. T. Baker."[238] *E. globulus* averages only 0.8% oil.
Eucalyptus radiata Sieb. ex DC. NARROW-LEAVED PEPPERMINT	Southern New South Wales.	Leaves, for essential oil (3 to 3.5%). ("Australiana Oil").	Essential oil contains 65 to 70% cineole; also geraniol, α-terpineol; 3.5% citral; methyl cinnamate.[238]	
Eucalyptus fruticetorum F. Muell. (*E. polybractea* R. T. Baker) BLUE MALLEE	Western New South Wales and Victoria.	Leaves, for essential oil (1.5 to 2.5%).	Essential oil contains 80 to 85% cineole; cuminal; phellandral; cryptone; macropone (4-isopropyl salicylaldehyde).[238]	
Pimenta officinalis Lindl. ALLSPICE	Mexico to the West Indies. Cultivated mainly in Jamaica and Guatemala.	Unripe (green) fruit, dried. Leaves.	Essential oil of fruits contains eugenol, *l*-α-phellandrene, caryophyllene, methyleugenol, cineole.[210] Essential oil of leaves contains 70% eugenol; also caryophyllene, acids and aldehydes.[210]	Essential oil used as a flavor; carminative.[122, 466, 552, 584]
UMBELLIFERAE *Carum carvi* L. CARAWAY	Europe, temperate Asia. Widely cultivated in cool climates of both hemispheres.	Fruits, for essential oil.	Essential oil contains 50 to 76% carvone, *d*-limonene, carveol, diacetyl, furfural, methyl alcohol, acetic aldehyde, etc.[210]	Employed as a flavor and acts as a carminative.[122, 423, 552, 584]

TABLE II (Cont.)

Family, Botanical Name and Common Name	Geographical Origin & Distribution	Plant Part or Product	Constituents	Pharmaceutical Uses
Coriandrum sativum L. CORIANDER	Mediterranean Europe. Widely grown in temperate and subtropical regions.	Fruits, for essential oil (0.3 to 2.2%).	Essential oil contains d- and dl-α-pinene, β-pinene, dipentene, P-cymene, d-linalool, aldehyde C_{10}, geraniol, acetic acid.[210]	Essential oil serves as a flavor; carminative.[122, 552, 584]
Foeniculum vulgare Mill. FENNEL	Mediterranean Europe, western Asia; universally cultivated.	Fruits, for essential oil (2.5 to 6.5%).	Essential oil contains d-pinene, camphene, d-α-phellandrene, dipentene, 50 to 65% anethole; also fenchone, methyl chavicol, aldehydes, anisic acid; sometimes 1, 3-dimethyl butadiene.[210]	Used as a flavor; carminative.[122, 423, 466, 552, 584]
Pimpinella anisum L. ANISE	Egypt and Asia Minor; cultivated in southern Europe, southern Russia, Iran, India.	Fruits, for essential oil (2 to 3.5%).	Essential oil contains 90% anethole; also methylchavicol; p-methoxyphenyl-acetone; and acetic aldehyde.[210]	This oil has been largely replaced by star anise oil, q.v.
OLEACEAE *Olea europaea* L. OLIVE	Mediterranean Europe to the Caucasus. Cultivated mainly in Spain, Italy and Greece; also California, to a lesser extent in Mexico, southern South America, Australia, etc.	Ripe fruits, for expressed oil (15 to 40%).	Olive oil is a non-drying oil with the following fatty acids: 0.2% myristic, 9.5% palmitic, 1.4% stearic, 0.2% arachidic, 81.6% oleic, 7.0% linoleic. Squalene, phytosterol and tocopherol are also present.[20]	Emollient, demulcent, and vehicle for injections.[122, 423, 466, 552, 584]
HYDROPHYLLACEAE *Eriodictyon californicum* Greene YERBA SANTA	Southern Oregon to central California.	Leaves.	Leaf extract contains flavone derivatives, eriodictyol, eriodictyonone, tannins, resin and volatile oil.[210]	Employed as a flavor to mask bitterness, especially of medications containing quinine.[122, 423, 552, 584]

LABIATAE (also known as LAMIACEAE) *Rosmarinus officinalis* L. ROSEMARY	Mediterranean region. Widely cultivated in temperate and subtropical climates.	Leaves and tender stems, for essential oil (0.5 to 1.2%).	Essential oil contains mainly α-pinene, also camphene, cineole, camphor, bornyl alcohol.[210]	Employed as a flavor; carminative.[122, 466, 552]
Mentha piperita L. PEPPERMINT	Central and southern Europe; widely cultivated in temperate regions of North and South America and Japan.	Flowering tops, for essential oil (0.3 to 0.7%).	Essential oil contains 56% free menthol; α- and β-pinene, limonene, cineole, ethyl amylcarbinol, menthone, isomenthone menthofuran, neomenthol, isomenthol menthyl acetate, and piperitone; also carvacrol and thymol.[210] Constituents fluctuate greatly with the stage of growth.[496]	Employed as a flavor; carminative.[122, 423, 552, 584]
Mentha spicata L. (*M. viridis* L.) SPEARMINT	Europe, North America; widely introduced and naturalized.	Flowering tops, for essential oil (.25 to .50%).	Essential oil contains α-pinene, α-phellandrene, *l*-limonene, octyl alcohol, dihydrocarveol, 55.8% carvone and, in some types, dipentene, cineole, or a terpenic glyoxal. Constituents vary with the type of plant grown and the region.[20, 210]	Employed as a flavor; carminative.[20, 122, 423, 552, 584]
Mentha cardiaca Gerard ex Baker SCOTCH MINT (believed to be a hybrid of *M. arvensis* (q.v.) and *M. spicata*).	U. S. (Indiana, Michigan, Washington).	Flowering tops, for essential oil.		Employed as a flavor; carminative.[20, 122]
PEDALIACEAE *Sesamum indicum* L. (*S. orientale* L.) SESAME	Probably originated in Africa; widely distributed. Cultivated mainly in India, Burma, China, Mexico, northern South America.	Seeds, for oil (50 to 56%).	Sesame oil is a semi-drying, polyunsaturated oil having the following fatty acids: 48% linoleic, 38% oleic, 9% palmitic, 5% stearic. It possesses antioxidant and synergistic properties (sesamolin, 0.3 to 5%, and sesamin, 0.5 to 1.0%).[407]	Sesame oil is employed as a vehicle for injections; also as a demulcent and emollient.[122, 423, 552]

Appendix 381

Bibliography

1. Abelin, I. and H. Pfister. 1953. Determination and activation of papain. *J. Physiol. Chem.* 295:232-332.
2. Abrol, B. K., L. D. Kapoor and K. S. Jamwal. 1955. Cultivation of Tinnevelly senna in Jammu Province. *J. Sci. Indus. Res.* 14A (9):432-433.
3. Adams, C. D. 1972. *Flowering Plants of Jamaica*. Univ. of the West Indies, Mona, Jamaica. 848 pp.
4. Adamson, A. D. 1971. *Oleoresins—production and markets with particular reference to the United Kingdom*. Trop. Prod. Inst., London. 46 pp.
5. Adzet Porredon, T. 1972. *Thymus vulgaris* of the Catalonian region. *Rev. Real Acad. Farm. Barcelona* (5):3-15; (6):29-47. [Span.] CA 79:35041x.
6. Agrawal, P. K. 1971. Effect of photoperiod on oil content, fatty acid composition and protein content of rape (*Brassica napus* L.) and flax (*Linum usitatissimum* L.) seeds. *Indian J. Exp. Biol.* 9 (2):252-254.
7. Akisue, G. 1971. Secretions of *Myroxylon peruiferum* L. f. I. Histological morphology of the secretory organs, and obtaining the balsam. *Rev. Farm. Bioquim. Univ. São Paulo* 9 (1):115-133. BA 56:15806.
8. ——— 1972. Secretions of *Myroxylon peruiferum* L. f. II. Physical and chemical characterization of the balsam and qualitative analysis of some components. *Rev. Farm. Bioquim. Univ. São Paulo.* 10 (1):73-96. BA 56:38846.
9. ——— 1972. Secretions of *Myroxylon peruiferum*. III. Physical and chemical characteristics of essential balsam oil and quantitative analysis of some components. *Rev. Farm. Bioquim. Univ. São Paulo* 10 (2):115-165. CA 79:57558y.
10. Aldrich, W. W. and H. Y. Nakasone. 1975. Day versus night application of calcium carbide for flower induction in pineapple. *J. Amer. Soc. Hort. Sci.* 100 (4):410-413.
11. Allegrini, J., S. DeBuochberg and A. Boillot. 1972. Essential oil antibacterial power. *Prod. Probl. Pharm.* 29 (9):819-827. [Fr.] CA 78:75830q.
12. Allen, P. H. 1943. Poisonous and injurious plants of Panama. Supp. to *Amer. J. Trop. Med.* 23 (1):1-76.
13. Al-Rawi, A. 1966. *Poisonous Plants of Iraq*. Tech. Bull. 145. Min. of Agric., Baghdad. 148 pp.
14. Alston, A. H. G. 1938. *The Kandy Flora*. Ceylon Gov't Press, Colombo. 109 pp. & pls.
15. Anderson, D. M. W., I. C. M. Dea, K. A. Daramalla and J. F. Smith. 1968. Studies on uronic acid materials: XXIV. An analytical study of different forms of the gum from *Acacia senegal* Willd. *Carbohyd. Res.* 6 (1):97-103. BA 50:15973.
16. Andrews, F. W. 1950, 1952, 1956. *The Flowering Plants of the Sudan*. Vols. I-III. T. Buncle & Co., Ltd., Arbroath, Scotland, pp. 237; 485; 579.
17. Anonymous. 1936. Camphor. *Fortune Mag.* 14 (4):18.
18. ——— 1936. *Coca, A Plant of the Andes*. Commod. of Comm. Ser. #20. Union of Pan Amer. Rep., Washington, D.C. 21 pp.

19. ———— 1942. *Quinine*. Commod. Ser. #24. Pan Amer. Union, Washington, D.C. 15 pp.
20. ———— 1948-1972. The Wealth of India. Vols. I-IX. Coun. Sci. & Indus. Res., Delhi.
21. ———— "D. E. R." 1951. Ephedra. *Col. Pl. & Anim. Prod.* 11 (2):119-124.
22. ———— 1951? *Vegetable Oils of Brazil*. Investment Opportunity Ser. Brazilian Gov't Trade Bur., New York, N.Y. Pp. 29-30.
23. ———— 1952. *Culinary and Medicinal Herbs*. Bull. 76. 2nd ed. Gt. Brit. Min. Agr. & Fish., London. 46 pp.
24. ———— 1954. Reserpoid: Antihypertensive. *Scope* (House Organ of the Upjohn Co., Kalamazoo, Mich.) 4 (3):10-12.
25. ———— 1957. The abolition of opium smoking in India. *Bull. on Narc.* 9(3):1-7.
26. ———— 1961. Transvaal lowveld farmers now "milk" their pawpaws. *The S. A. Sugar J.* 45 (10):886.
27. ———— 1962. Ten years of the coca monopoly in Peru. *Bull. on Narc.* 14 (1): 9-17.
28. ———— 1963. Anatomy of a project—"The Prohibited Plant." United Nations Tech. *Assist. Newsletter* 2 (15):1-8.
29. ———— 1963. Pectin and papain from raw papaya. Pp. 66-67. In: Fruit Technology, section of *Annual Report, 1961-62*. Cent. Food Tech. Res. Inst., Mysore, India.
30. ———— 1963. *Production of Mentha arvensis (Japanese mint) oil*. Trop. Prod. Inst. Rept. 17, pp. 1-7.
31. ———— 1965. Inter-American Consultative Group on coca leaf problems, Lima; 14-21 Dec. 1964. *Bull. on Narc.* 17 (4):37-41.
32. ———— 1965. Silvics of forest trees of the United States. *Agr. Hdbk.* 271. USDA. For. Serv. pp. 384-389; 458-463.
33. ———— 1966. Twenty years of narcotics control under the United Nations. *Bull. on Narc.* 18 (1):1-60.
34. ———— 1968. Review of the 22nd session of the Commission on Narcotic Drugs and the 44th session of the Economic and Social Council. *Bull. on Narc.* 20 (2):37–41.
35. ———— 1972. *Brazil castor situation*. For. Agr. Cir. FF09-72. USDA For. Agr. Serv., Washington, D.C. 14 pp.
36. ———— 1974. Report of the International Narcotics Control Board on its Work in 1973. *Bull. on Narc.* 26 (3):31-39.
37. ———— 1974. *Scientific literature reviews on generally recognized as safe (gras) food ingredients—papain*. U.S. Food & Drug Admin. Distrib. by Nat'l Tech. Info. Serv., U.S. Dept. Comm., Washington, D.C. 61 pp.
38. ———— 1975. Development and prospects of pineapple products industry. *J. Phil. Statist.* 26 (2):ix-xxix.
39. Antonelli, G. 1950. *Le piante che ridanno la salvte*. 4th ed. Azienda Libraria Cattolica Italiana, Rome. 491 pp.
40. Antoun, M. D. and M. F. Roberts. 1975. Enzymic studies with *Papaver somniferum*. *Pl. Med.* 28:6-11.
41. Arctander, S. 1960. *Perfume and flavor materials of natural origin*. Steffen Arctander, Elizabeth, N.J. 736 pp.
42. Arnold, H. L. 1944. *Poisonous Plants of Hawaii*. Tongg Pub'g Co., Honolulu. 71 pp.

43. Asahina, H. T., Kawatani, M. Ono and S. Fujita. 1957. Studies of poppies and opium. *Bull. on Narc.* 9 (2):20-33.
44. Audas, J. W. 1952? *Native trees of Australia*. Whitcombe & Tombs Pty, Ltd., Melbourne, Aust. 396 pp.
45. Avery, A. G., S. Satina and J. Rietsema. 1959. *Blakeslee: the genus Datura*. The Ronald Press Co., New York. 289 pp.
46. Aynehchi, Y. and S. Jaffarian. 1973. Determination of thebaine in various parts of *Papaver bracteatum* Lindl. during the growing season. *Lloydia* 36 (4): 427-429.
47. Aynilian, G. H., J. A. Duke, W. A. Gentner and N. R. Farnsworth. 1974. Cocaine content of *Erythroxylum* species. *J. Pharm. Sci.* 63 (12):1938-1941.
48. ———, S. G. Weiss, G. A. Cordell, D. J. Abraham, F. H. Crane and N. R. Farnsworth. 1974. *Catharanthus* alkaloids. XXIX. Isolation and structure elucidation of vincoline. *J. Pharm. Sci.* 63 (4):536-538.
49. Backer, C. A. and R. C. Bakhuizen van den Brink, Jr. 1963. *Flora of Java*. 3 vols. N. V. P. Noordhoff, Groningen, The Netherlands. Pp. 648; 641; 761.
50. Bailey, F. M. 1901. *The Queensland flora*. Pt. IV. The Queensland Gov't, Brisbane, Aust. Pp. 1031-1372.
51. Bailey, L. H. 1949. *Manual of cultivated plants*. Rev'd ed. The Macmillan Co., New York. 1116 pp.
52. Bala Narsaiah, B. and L. G. Kulkarni. 1969. Castor NPH-1 scores over HC-6. *Indian Farming* 18 (10): 11-13.
53. Balbaa, S. I. 1974. Effect of the use of different methods of drying *Digitalis lanata* leaves on their quality and glycosidal content. *Pl. Med.* 26 (1):20-25.
54. Balbaa, S. I., S. H. Hilal and M. Y. Haggag. 1971. A study of the effect of some factors on the growth and glycosidal content of *Digitalis lanata* Ehrh. grown in Egypt. *Quart. J. Crude Drug Res.* 11 (1):1689-1696.
55. ———, M. S. Karawya and M. S. Afifi. 1971. Pharmacognostical study of the seeds of certain *Plantago* species growing in Egypt. *U. A. R. J. Pharm. Sci.* 12 (1):35-52. CA 77:156311c.
56. Balint, G. A. 1973. Examination of the inflammatory effect of ricin with special reference to the endogenous corticosteroid mobilisation. *Toxicology* 1:329-336.
57. Balls, A. K., R. R. Thompson and M. W. Kies. 1941. Bromelin: properties and commercial production. *Indus. & Engrg. Chem.* 33:950-953.
58. Bampton, S. S. 1961. Yams and diosgenin. *Trop. Sci.* 3 (4):150-153.
59. Bandyopadhyay, P. C. 1960. Electrochemistry of high polymeric acids—acid polysaccharide from *Plantago ovata*. *J. Sci. Indus. Res.* 19B (10):378-381.
60. ——— 1961. Viscometric investigations on high polymeric acids: acid polysaccharide from *Plantago ovata*. *J. Sci. Indus. Res.* 20B (3):101-104.
61. Ban'kovskaya, A. N., V. I. Sheichenko, A. I. Ban'kovskii, L. D. Vechkanova, and V. S. Kabanov. 1973. Ergovalide, a new alkaloid from ergot. *Khim. Prir. Soedin.* 9 (1):134.
62. Barnard, C. 1952. The Duboisias of Australia. *Econ. Bot.* 6 (1):3-17.
63. Barton, B. H. and T. Castle. 1877. *The British Flora Medica*. Chatto & Windus, London. 447 pp. (Rev'd by J. R. Jackson).
64. Battiscombe, E. 1936. *Trees and shrubs of Kenya Colony*. Gov't Ptr., Nairobi. 201 pp.

65. Baytop, A. and T. Gozler. 1971. Origin and composition of Turkish gum tragacanth. *Istanbul Univ. Eczacilik Fak. Mecm.* 7 (1): 56-65. CA 77:16583t.
66. Bechtel, A. 1972. Gas chromatographic identification and quantitative determination of morphine, codeine, thebaine, papaverine, and narcotine in opium extract. *Chromatographia* 5 (7):404-407. [Ger.] BA 55:21520.
67. Becker, S. 1958. The production of papain—an agricultural industry for tropical America. *Econ. Bot.* 12 (1):62-79.
68. Behl, P. N., R. M. Captain, B. M. S. Bedi and S. Gupta. 1966. *Skin-irritant and sensitizing plants found in India.* Dr. P. N. Behl, Dept. of Dermat., Irwin Hosp. & M. A. Med. Coll., New Delhi. 179 pp.
69. Bejarano, J. 1961. Present state of the coca-leaf habit in Colombia. *Bull. on Narc.* 13 (1):1-5.
70. Benitz, K-F., L. Goldberg and F. Coulston. 1973. Intestinal effects of carrageenans in the rhesus monkey (*Macaca mulatta*). *Fd. Cosmet. Toxicol.* 11: 565-575.
71. Benthall, A. P. 1946. *Trees of Calcutta and its neighborhood.* Thacker, Spink & Co. (1933) Ltd., Calcutta. 513 pp.
72. Bergman, N. 1967. Manatee health officer gives warning on periwinkle smoke. *Herald-Tribune*, Bradenton, Fla. June 1.
73. Bergwein, K. 1972. Essential oils with bactericidal and bacteriostatic effects. *Seifen, Oele, Fette, Wachse* 98 (16):514. [Ger.] CA 77:130487q.
74. Berrens, L. and E. Young. 1962. Studies on the allergen in ipecacuanha. I. *Int. Arch. Allergy* 21:335-346.
75. ——— 1963. ——— II. ——— 22:51-59.
76. Bertler, A., A. Gustafson and A. Redfors. 1973. Massive digoxin intoxication. Report of two cases with pharmacokinetic correlations. *Acta Med. Scand.* 294 (4):245-249. BA: 57:69871.
77. Betts, T. J. 1965. The conservation of digitalis. *Lloydia* 28 (3):181-190.
78. Bhakuni, D. S., P. P. Joshi, H. Uprety and R. S. Kapil. 1974. Roseoside-A C-13 glycoside from *Vinca rosea*. *Phytochem.* 13 (11):2541-2543.
79. ——— and R. S. Kapil. 1972. Monoterpene glycosides from *Vinca rosea* Linn. *Ind. J. Chem.* 10 (4):454.
80. Bhattacharji, S., M. M. Dhar and M. L. Dhar. 1955. Isolation of sarpagine from *Rauwolfia canescens* Linn. *J. Sci. & Indus. Res.* 14B (6):310.
81. Bhatti, M. B. and M. B. Sial. 1971. Guar—its utility in food and nonfood industries. *Pakistan J. of Sci.* 23 (1 & 2):1-5.
82. Bjerre-Petersen, E., J. Christensen and P. Hemmingsen. 1973. Furcellaran. Pp. 123-136. In: *Industrial Gums.* Edited by R. L. Whistler. 2nd ed. Academic Press, Inc., New York.
83. Black, J. M. (Rev'd by E. L. Robertson). 1957. *Flora of South Australia.* Pt. IV. 2nd ed. K. M. Stevenson, Gov't Ptr., Adelaide, So. Aust. 1008 pp.
84. Blake, S. F. 1943. The "divine plant" of the Incas. *Agr. in the Amer.* 3 (6):114-116.
85. Blake, S. T. 1968. *A revision of Melaleuca leucadendron and its allies (Myrtaceae).* Contrib. from the Queensland Herbarium, No. 1. Dept. of Primary Indus., Brisbane, Aust. 114 pp.
86. Blakeslee, A. 1968. Today's Health News. *Today's Health*, A. M. A. Sept. P. 9.
87. Blanchard, M. 1972. Contribution à l'étude de la biologie et de la culture de la scille maritime en Algérie. *Agron. Trop., France* 27 (11):1101-1114.

88. Blohm, H. 1962. *Poisonous plants of Venezuela.* Harvard Univ. Press, Cambridge, Mass. 136 pp.
89. Blunden, G. and Yi Yi. 1974. A reinvestigation of the steroidal sapogenins of Agave sisalana. *Lloydia* 37 (1):10-16.
90. Boza, R. 1971. Breve estudio sobre la raicilla. *El Café de Nicaragua* 238:25-26.
91. Brendel, W. D. and D. Schneider. 1974. Quantitative determination of sennosides in senna pods and senna leaves. I. Method. *Pl. Med.* 25 (1):63-67.
92. Brocklehurst, K., J. Carlsson, M. P. J. Kierstan and E. M. Crook. 1973. Covalent chromatography. Preparations of fully active papain from dried papaya latex. *Biochem. J.* 133 (3):573-584.
93. Brown, W. H. 1951. *Useful plants of the Philippines.* Vol. I. Tech. Bull. 10. Phil. Dept. Agr. & Nat. Res., Manila. 590 pp.
94. ——— 1954. ——— Vol. II. ——— ——— 513 pp.
95. ——— 1946. ——— Vol. III. ——— ——— 507 pp.
96. Buck, A. A., T. T. Sasaki, J. J. Hewitt and A. A. Macrae. 1970. Coca chewing and health: an epidemiologic study among residents of a Peruvian village. *Bull. on Narc.* 22 (4):23-32.
97. Bugnariu, O. and V. Filipas. 1972. Quality of *Digitalis purpurea* leaves, raw material for some pharmaceutical forms. *Farmacia* (Bucharest) 20 (5): 285-296. [Rom.] CA 77:105554b.
98. Burkill, I. H. 1935. *Dictionary of the economic products of the Malay Peninsula.* Crown Agents for the Colonies, London. 2,402 pp.
99. Burks, J. S., J. E. Walker, B. H. Rumack and J. E. Ott. 1974. Tricyclic antidepressant poisoning. Reversal of coma, choreoathetosis and myoclonus by physostigmine. *J.A.M.A.* 230 (10):1405-1407.
100. Butcher, R. W. 1961. *A new illustrated British Flora.* Vols. I & II. Leonard Hill [Books] Ltd., London. 1016 pp.; 1080 pp.
101. Cann, H. J. 1966. The papaw. *Agr. Gaz. New South Wales* 77 (6):347-350.
102. Cardenas, M. 1969. *Manual de plantas económicas de Bolivia.* Imprenta Icthus, Cochabamba, Bolivia. 421 pp.
103. Carlson, W. A. and E. M. Ziegenfuss. 1965. The effect of sugar on guar gum as a thickening agent. *Food Tech.* 19 (6):64-68.
104. Carr, A. R. 1974. Duboisia growing. *Queensland Agr. J.* 100 (10):495-505.
105. Chandorkar, A. G. 1971. Contributions to the pharmacological studies of *Vinca rosea* (Linn.) II. Evaluation of central nervous system depressant, analgesic and anticonvulsant actions. *J. Shivaji Univ.* 4 (7/8):129-132. BA 55:45483.
106. Chandrasekhar, N. and C. S. Vaidyanathan. 1961. Some pharmacological properties of the blood anticoagulant from *Carica papaya* Linn. *J. Sci. Indus. Res.* 20C (7):213-215.
107. Chatt, E. M. 1953. *Cocoa, cultivation, processing, analysis.* (Econ. Crops Vol. 3). Interscience Publishers, New York. 302 pp.
108. Chatterjee, A. and A. B. Ray. 1962. Recent developments in the chemistry of indole alkaloids from *Rauvolfia serpentina* Benth. *J. Sci. Indus. Res.* 21A (11):515-527.
109. ——— and S. K. Talapatra. 1955. Synthesis of reserpine analogue from Rauwolscine—the alkaloid of *Rauvolfia canescens* Linn. *J. Sci. Indus. Res.* 14C (12):237-239.
110. Chatterjee, B. K. and P. K. Dutta. 1961. A simple method of determining leaf area in *Mentha arvensis* L. *J. Sci. Indus. Res.* 20C (12):359-360.

111. Chaudhri, I. I. 1957. Pakistani Ephedra. *Econ. Bot.* 11 (3):257-262.
112. Cheng, P. and N. J. Doorenbos. 1973. Alkaloid analysis of *Papaver bracteatum*. *Lloydia* 36 (4):440.
113. Cheshire, P. C. 1966. *Papain: trade and markets.* TPI Rpt. No. G25. Trop. Prod. Inst., London. 54 pp.
114. Chestnut, V. K. and E. V. Wilcox. 1901. *The stock-poisoning plants of Montana.* Bull. 26. USDA, Div. of Botany, Washington, D.C., 150 pp.
115. Chittenden, F. J. (Editor). 1956. *The Royal Horticultural Society Dictionary of Gardening.* 2nd ed. 4 Vols. & Supp. Clarendon Press, Oxford, Eng. 2316 pp. Supp. 2nd ed. 1969; 554 pp.
116. Chopra, R. N., R. L. Badhwar and S. Ghosh. 1965. *Poisonous Plants of India.* Vol. II. Indian Coun. of Agr. Res., New Delhi. Pp. 634-972.
117. ———, I. C. Chopra, K. L. Handa and L. D. Kapur. 1958. *Chopra's indigenous drugs of India.* 2nd ed. U. N. Dhur & Sons, Private Ltd., Calcutta. 816 pp.
118. Choudhary, D. K. 1975. Causes of poor and erratic germination in *Atropa belladonna*. *Pl. Med.* 27:18-23.
119. Chu, B-L. W. 1971. *Thermoperiodicity and the growth, development and alkaloid content of Hyoscyamus niger.* (Dissertation). Univ. Microfilms, Ann Arbor, Mich. 185 pp.
120. Ciulei, J. 1970. *Extraction and determination of cytostatic Vinca rosea alkaloids.* Rom. 52, 178 (Cl. A 61k), 17 Mar. 1970, Appl. 16 Aug. 1967; 3 pp. CA 74 (12):57258x.
121. Clarke, E. G. C. and T. King. 1974. Stability of anti-ricin serum. *J. Pharm. Pharmac.* 26 (7):567.
122. Claus, E. P., V. E. Tyler and L. R. Brady. 1970. *Pharmacognosy.* 6th ed. Lea & Febiger, Philadelphia. 518 pp.
123. Coats, A. M. 1971. *Flowers and their histories.* 2nd ed. McGraw-Hill Book Co., New York, N.Y. P. 198.
124. Codd, L. E. W. 1951. *Trees and shrubs of the Kruger National Park.* Bot. Surv. Mem. #26. Union of So. Afr., Dept. of Agr., Pretoria. 192 pp.
125. Coffman, C. B., C. E. Bare and W. A. Gentner. 1975. Thebaine variations between germplasm sources within one collection of *Papaver bracteatum* Lindl. *Bull. on Narc.* 27 (3):41-46.
126. Coit, J. E. 1951. Carob or St. John's Bread. *Econ. Bot.* 5 (1):81-96.
127. Collingwood, C. A. 1972. Cocoa in West Africa: the economics of pest control. *Span* 15 (2):74-77.
128. Collins, C. E. and C. Collins. 1935. Roentgen dermatitis treated with fresh whole leaf of *Aloe vera*. *Am. J. Roentgenol.* 33:396-397.
129. Collins, J. L. 1949. History, taxonomy and culture of the pineapple. *Econ. Bot.* 3 (4):335-359.
130. Connor, H. E. and N. M. Adams. 1951. *Poisonous plants in New Zealand.* Dept. Sci. Indus. Res., Wellington. 141 pp.
131. Conover, Dr. R. 1952. Communication to the Miami Herald, Apr. 6.
132. Cook, M. K. and B. H. Gominger. 1974. Glycyrrhizin. Chapt. 19 (pp. 211-215). In: *Symposium: Sweeteners*, edited by G. E. Inglett. The AVI Pub'g Co., Inc., Westport, Conn. 240 pp.
133. Cooper, W. C., V. B. Perone, L. D. Scheel and R. G. Keenan. 1964. Occupational hazards from castor bean pomace: Tests for toxicity. *Indus. Hyg. J.* 25:431-438.

134. Corner, E. J. H. 1952. *Wayside trees of Malaya*. Vols. I & II. 2nd ed. Gov't Ptg. Off., Singapore. 772 pp. & 228 pls.
135. Cornett, G. B. R. 1965. Rauvolfia serpentina. *World Crops* 17 (2):33-37.
136. Coulsen, J. F. and W. J. Griffin. 1968. The alkaloids of *Duboisia myoporoides*. II. Roots. *Pl. Med.* 16 (2):174-181.
137. Crabie, P., J. Pollet and F. Pebay-Peyroula. 1970. Study of hemostasis during acute colchicine intoxication. *J. Eur. Toxicol.* 3 (6):373-385.
138. Crabtree, D. G. 1947. Red squill—most specific of the raticides. *Econ. Bot.* 1 (4):394-401.
139. Cross, J. K. 1973. Tar burning, a forgotten art? *Forests & People* 23 (2):21-23.
140. Cruzado, H. J., H. Delpin and F. W. Martin. 1965. Effects of fertilizers on the sapogenin yields of *Dioscorea composita* in Puerto Rico. *J. Agr. Univ. of P.R.* 49 (2):254-258.
141. Cruzado, H. J., H. Delpin and B. A. Roark. 1965. Sapogenin production in relation to age of tuber in two *Dioscorea* species. *Turrialba Rev. Interam. Ciencias Agr.* 15 (1):25-28.
142. Culliton, B. J. 1973. Sloan-Kettering: the trials of an apricot pit—1973. *Science* 182: 1000-1003.
143. Curtin, L. S. M. 1947. *Healing herbs of the Upper Rio Grande*. Laboratory of Anthropology, Santa Fé, New Mexico. 281 pp.
144. Dale, I. R. and P. J. Greenway. 1961. *Kenya trees and shrubs*. Buchanan's Kenya Estates, Ltd., Nairobi. 654 pp.
145. Dalev, D., L. Iliev and R. Ilieva. 1960. Poppy cultivation in Bulgaria and the production of opium. *Bull. on Narc.* 12 (1):25-36.
146. Dalziel, J. M. 1948. *Useful Plants of West Tropical Africa* (Appen. to the Flora of W. T. A.) Crown Agents for the Colonies, London. 612 pp.
147. Dandy, A. J. 1967. The bleaching of cottonseed oil by sepiolite. *E. Afr. Agr. & For. J.* 32 (3):256-264.
148. Dastur, J. F. 1952? *Medicinal Plants of India and Pakistan*. D. B. Taraporevala Sons & Co. Ltd., Bombay. 317 pp.
149. Datta, D. D., P. C. Bose and D. Ghosh. 1972. Spectrophotometric estimation of cinnamic acid in tolu balsam. *Indian J. Pharm.* 34 (1):15-17. BA 54:50840.
150. Dauguet, J. C. and R. R. Paris. 1974. Flavonoid glycosides of *Rhamnus frangula*. *Pl. Med. Phytother.* 8(11):32-43. [Fr.] CA 81:60855q.
151. Davis, W. C. 1944. Pearl Harbor sent quinine home. *Agr. in the Amer.* 4 (3): 43-45, 50.
152. DeConti, R. C. and W. A. Creasey. 1975. Clinical aspects of the dimeric *Catharanthus* alkaloids. Pp. 237-278. In: *The Catharanthus Alkaloids*. (W. I. Taylor and N. R. Farnsworth, Editors). Marcel Dekker, Inc., New York. 332 pp.
153. Degener, O. 1946. *Flora Hawaiiensis*. Books 1-4. 2nd ed. Author, Riverdale, New York City [not page numbered]
154. Delpin, H. and F. W. Martin. 1970. Establishing sapogenin-bearing *Dioscoreas* from seed. *J. Agr. Univ. P.R.* 54 (2):334-340.
155. Dewey, L. H. 1943. *Fiber production in the western hemisphere*. Misc. Pub. 518. USDA, Washington, D.C. 95 pp.
156. Ditmer, W. P. 1965. *Poisonous plants of Pennsylvania*. Penn. Dept. Agr., Harrisburg. 51 pp.

157. Dodge, C. R. 1897. *Descriptive catalogue of useful fiber plants of the world*. Rpt. #9. USDA, Off. of Fiber Invest., Wash., D.C. 361 pp.
158. Doggett, N. S. 1973. Possible involvement of a dopaminergic pathway in the depressant effects of ouabain on the central nervous system. *Neuropharmacology* 12 (3):213-220. CA 78:154842y.
159. Dombernowsky, P., N. Nissen and V. Larsen. 1972. Clinical investigation of a new podophyllum derivative, epipodophyllotoxin, 4',-demethy-9-(4,6-0-2-thenylidine-beta-D-gluco-pyranoside) (NSC-122819), in patients with malignant lymphomas and solid tumors. *Cancer Chemoth. Rep. Pt. 1*. 56 (1):71-82.
160. Drury, H. 1873. *The useful plants of India*. William H. Allen & Co., London. 512 pp.
161. DuFrane, E. 1944. Gifts of the Americas: ipecac. *Agr. in the Amer*. 4 (7):1.
162. Duke, J. A., C. R. Gunn, E. E. Leppik, C. F. Reed, M. L. Solt and E. E. Terrell. 1973. *Annotated Bibliography on opium and oriental poppies and related species*. ARS-NE-28. USDA, Washington, D.C. 349 pp.
163. Dupaigne, P. 1975. Effets biochimiques des bromélines. Leur utilisation en thérapeutique. *Fruits* 30 (9):545-567.
164. Dutta, P. K. 1971. Cultivation of *Mentha arvensis* in India. *The Flavour Indus*. 2 (4):233-245.
165. ———, I. C. Chopra and L. D. Kapoor. 1963. Cultivation of *Rauvolfia serpentina* in India. *Econ. Bot*. 17 (4):243-251.
166. Dyer, C. D. 1963. History of the gum naval stores industry. *The ATFA J*. Jan. pp. 5-8.
167. Eddy, N. B. and E. L. May. 1973. The search for a better analgesic. *Science* 181 (4098):407-414.
168. Edmonds, M. 1965. *The market for gum arabic*. Rpt. G16. Trop. Prod. Inst., London. 27 pp.
169. Eggeling, W. J. 1953. (rev'd and enl. by I. R. Dale) *The indigenous trees of the Uganda Protectorate*. 2nd ed. Gov't of the Uganda Protectorate, Entebbe. 491 pp.
170. Eichbaum, F. W. and W. J. Yasaka. 1974. Antiarrhythmic effect of oxyfedrin. *Eur. J. Pharmacol*. 26 (1):82-88. BA 58:21285.
171. Eigsti, O. J. and P. Dustin. 1955. *Colchicine—in agriculture, medicine, biology and chemistry*. Iowa State Coll. Press, Ames, Iowa. 470 pp.
172. El Baradi, T. A. 1969. Castor bean (*Ricinus communis* L.) [review article] *Trop. Abs*. 24 (9):567-572.
173. Elgamal, M. H. A. and B. A. H. El-Tawil. 1975. Constituents of local plants. XVIII. 28-hydroxyglycyrrhetic acid, a new triterpenoid isolated from the roots of *Glycyrrhiza glabra*. *Pl. Med*. 27:159-163.
174. ———, and M. B. E. Fayez. 1972. Isolation of formononetin from the roots of *Glycyrrhiza glabra* Linn. collected locally. *Indian J. Exper. Biol*. 10 (1):128.
175. ———, ———, and G. Snatzke. 1965. Constituents of local plants. VI. Liquoric acid, a new triterpenoid from the roots of *Glycyrrhiza glabra*. *Tetrahedron* 21 (8): 2109-2115.
176. Elkinawy, M. and T. Hemberg. 1974. Effect of temperature and kinetin on the germination and endogenous hormones of *Hyoscyamus muticus* seeds. *Physiol. Plant*. 31 (1):64-66. BA 58:45654.

177. El-Tayeb, O., M. Kucera, V. O. Marquis and H. Kucerova. 1974. Contribution to the knowledge of Nigerian medicinal plants. III. Study on *Carica papaya* seeds as a source of a reliable antibiotic, the BITC. *Pl Med.* 26:79-89.
178. Espinel Ovalle, G. and I. Guzman Parra. 1971. Chromatographic separation and determination of alkaloids of *Erythroxylon coca* var. *novagranatensis. Rev. Colomb. Cienc. Quim. Farm.* 1 (4):95-118. CA 76:89994s.
179. Fairbairn, J. W. 1976. New plant sources of opiates. *Pl. Med.* 29 (1):26-31.
180. ——— and F. Hakim. 1973. *Papaver bracteatum* Lindl.—a new plant source of opiates. *J. Pharm. Pharmacol.* 25 (5):353-358.
181. ———and A. B. Shrestha. 1967. The taxonomic validity of *Cassia acutifolia* and *Cassia angustifolia. Lloydia* 30 (1):67-72.
182. Feldberg, W. and S. V. Shaligram. 1972. Hyperglycemic effect of morphine. *Brit. J. Pharmacol.* 46 (4):602-618. CA 78:67090h.
183. Felkova, M. 1972. Mineral nutrition of *Papaver somniferum*. Effect of nitrates and ammonium salts. *Cesk. Farm.* 21 (9):405-410. CA 78:109761k.
184. Felton, G. 1971. *The use of pineapple by-products for livestock feeding.* Misc. Pub. #75. Pp. 40-42. Coop. Exten. Serv., Univ. of Hawaii, Honolulu.
185. Fenzi, E. O. 1915. *Frutti tropicali e semitropicali.* Inst. Agr. Col. Ital., Firenze. 261 pp.
186. Fernald, M. L. and A. C. Kinsey. *Edible Wild Plants of eastern North America.* Spec. Pub. Gray Herb., Harvard Univ. Idlewild Press, Cornwall-on-Hudson, New York. 452 pp.
187. Feuell, A. J. 1955. The genus *Rauwolfia*; some aspects of its botany, chemistry, and medicinal uses. *Col. Pl. & Anim. Prod.* 5 (1):33.
188. Filatov, V. P. and V. A. Biber. 1948. Biogenic stimulators in *Aloe vera. Doklady Akad. Nuak. USSR* 62:259-262. CA 43:726f.
189. Fischer, R., K. Gloris and H. Claus. 1971. Chemical composition of bear berry leaves. *Deut. Apoth.-Ztg.* 111 (33):1225-1229. CA 75:121327u.
190. Fisher, A. A. 1967. *Contact dermatitis.* Lea & Febiger, Philadelphia. 324 pp.
191. Fisher, A. E. 1968. Growers test plastic mulch in pineapples. *Queensland Agr. J.* 94 (2):104-108.
192. Fisher, H. H. 1973. Origin and uses of ipecac. *Econ. Bot.* 27:231-234.
193. Floss, H. G., D. Goreger and D. Erge. 1968. The conversion of chanoclavine-I into tetracyclic ergot alkaloids. *Lloydia* 31 (4):425. [abs.]
194. Fonin, V. S. and V. V. Sheberstov. 1972. Biological nature of *Digitalis purpurea* and the role of external factors in its accumulation of glycosides. *Farmatsiya* (Moscow) 21 (3):86-89. [Russ.] CA 77:72516k.
195. ——— and ——— 1973. Effect of growth conditions on the glycoside level in digitalis leaves. *Farmatsiya* (Moscow) 22 (3):83-86. CA 79:96872r.
196. Font Quer, P. 1973. *Plantas medicinales—el Dioscórides renovado.* Editorial Labor, S. A., Barcelona. 1033 pp.
197. Forsyth, A. A. 1956. *British poisonous plants.* Bull. 161. H. M. Staty. Off., London. 111 pp.
198. Forte, F. A. 1974. Vincristine neuropathy. *J.A.M.A.* 227 (3):325.
199. Fosberg, F. R. 1947. Cinchona plantation in the New World. *Econ. Bot.* 1 (3):330-333.
200. Frank, P. J. 1957. A feeding trial with sisal waste. *E. Afr. Agr. J.* 22 (4):165-167.

201. Franze, C., K. S. Grunert, U. Hildebrandt and H. Griehl. 1968. Theobromine and theophylline content of raw (green) coffee and tea. *Pharmazie* 23 (9):502-503. [Ger.] BA 50:83875.
202. Friedrich, H. and N. Krüger. 1974. New investigations on the tannin of *Hamamelis* I. The tannin of the bark of *H. virginiana*. *Pl. Med.* 25 (2):138-148. [Ger.]
203. ——— and ——— 1974. New investigations on the tannin of *Hamamelis*. II. The tannin of the leaves of *H. virginiana*. *Pl. Med.* 26:327-332 [Ger.]
204. Freytag, A. 1954. Wound healing hormone in aloes. *Pharmazie* 9:705-710. [Ger.] CA 50:16897.
205. Friese, F. W. 1934. *Plantas medicinaes Brasileiras*. Inst. Agronomico do Estado, São Paulo. 494 pp.
206. Fujita, Y., S. Fujita and Y. Akama. 1972. Biogenesis of the essential oils in camphor trees. XXIX. Comparison of the essential oils of the seedlings of *Cinnamomum camphora* cultivated in different daylight. *Nippon Nogei Kagaku Kaishi* 46 (1):17-20. CA 76:151083z.
207. ———, ——— and S. Nishida. 1973. Biogenesis of the essential oils in camphor trees. XXXI. Components of young and old shoot oils of *Cinnamomum camphora* var. *linaloolifera*. *Nippon Nogei Kagaku Kaishi* 47 (6):403-405. CA 79:113210t.
208. Funatsu, M., G. Funatsu, M. Ishiguro and K. Hara. 1971. Structure and toxic function of ricin. Physical and physiological properties of subunits. *Proc. Jap. Acad.* 47 (10):786-790. CA 77: 1561k.
209. ——— and M. Ishiguro. 1971. Chemical structures and toxicities of ricins; proteins isolated from *Ricinus communis*. *Kagaku To Seibutsu* 9 (8):490-497. CA 76:110323j.
210. Furia, T. E. and N. Bellanca. 1971. *Fenaroli's handbook of flavor ingredients*. The Chemical Rubber Co., Cleveland, Ohio. 803 pp.
211. Fyles, F. 1920. *Principal poisonous plants of Canada*. Bull. 39, 2nd Ser. Dom. of Canada, Dept. Agr., Ottawa. 112 pp.
212. Gaevskii, A. V. and P. M. Loshkarev. 1972. Method for determining morphine in poppy pods. *Khim.-Farm. Zl.* 6 (6):54-60. [Russ.] CA 77:79578r.
213. Galeffi, C. and E. M. Delle Monache. 1974. Separation of dihydroergotoxine (hydergine) into dihydroergocornine, dihydroergocryptine and dihydroergocristine by counter-current distribution. *J. Chromatog.* 88 (2):413-415.
214. Gandotra, K. L. and D. Ganguly. 1962. Experimental cultivation of ergot in Jammu and Kashmir. Pt. I. Possibility of cultivating and improving the yield of ergot. *J. Sci. Indus. Res.* 21D (12):460-463.
215. Garbuzova, V. M. and N. I. Libizov. 1970. Improvement of a method for the separate determination of lanatosides A, B and C in foxglove leaves. *Sb. Nauch. Rab., Vses. Nauch. Issled. Inst. Lek. Rast.* (1):195-199. [Russ.] CA 76:56203e.
216. Gardner, C. A. and H. W. Bennetts. 1956. *The toxic plants of Western Australia*. West Australia Newspapers, Ltd., Perth. 253 pp.
217. Geggil, A. S., H. S. Yalabik and M. J. Groves. 1975. A note on tragacanth of Turkish origin. *Pl. Med.* 27:284-286.
218. Genest, K. and D. W. Hughes. 1969. Natural products in Canadian pharmaceuticals: IV. *Hydrastis canadensis*. *Can. J. Pharm. Sci.* 4 (2):41-45.

219. George, K. 1964. Regeneration of *Rauvolfia serpentina*. *Indian J. Exper. Biol.* 2 (3):161-163.
220. Gersmeyer, E. F. and W. C. Holland. 1963. Effect of heart rate on action of ouabain on calcium exchange in guinea pig left atria. *Am. J. Physiol.* 205 (4):795-798. CA 60:3384.
221. Giffard, P-L. 1966. Les Gommiers: *Acacia senegal* Willd. *Acacia laeta* R. Br. *Bois Forets Trop.* 105:21-28.
222. ———— 1975. Les gommiers, essences de reboisement pour les régions Sahelièннes. *Revue Bois et Forets Trop.* 161:3-21.
223. Ginzel, K. H. 1973. Muscle relaxation by drugs which stimulate sensory nerve endings. I. Effect of *Veratrum* alkaloids, phenyldiguanide and 5-hydroxytryptamine. *Neuropharmacology* 12 (2):133-148. CA 78:154762x.
224. Gjerstad, G. 1965. Intermediary metabolism of ergot. XI. *Quart. J. Crude Drug Res.* 5 (3):731-741.
225. Goff, S. and I. Levenstein. 1964. Measuring the effects of topical preparations upon the healing of skin wounds. *J. Soc. Cos. Chem.* 15:509-518.
226. Goldblatt, P. 1974. Biosystematic studies in *Papaver* section Oxytona. *Ann. Missouri Bot. Gard.* 61 (2):264-296.
227. Goldstein, A. M., E. N. Alter and J. K. Seaman. 1973. Guar gum. Pp. 303-321. In: *Industrial gums*. Edited by R. L. Whistler. 2nd ed. Academic Press, New York. 807 pp.
228. Gothoni, G., W. Nyberg and S. G. Jokipii. 1973. The laxative effect of pure sennosides A and B. *Ann. Clin. Res.* 5 (1):46-48. BA 56:38905.
229. Graham, E. H. 1935. *Poisonous plants of Pennsylvania*. Carnegie Mus., Pittsburgh, Penna. 15 pp.
230. Granier-Doyeux, M. 1962. Some sociological aspects of the problem of cocaism. *Bull. on Narc.* 14 (4):1-16.
231. Grasso, P., M. Sharratt, F. M. B. Carpanini and S. D. Gangolli. 1973. Studies on carrageenan and large-bowel ulceration in mammals. *Fd. Cosmet. Toxicol.* 11:565-575.
232. Greenfield, H. 1968. A forty-year's chronicle of international narcotics control: the work of the Permanent Central Narcotics Board 1928-1968 and of the Drug Supervisory Body 1933-1968. *Bull. on Narc.* 20 (2):1-8.
233. Grieve, M. 1967. *A modern herbal*. Vols. I & II. Hafner Pub'g Co., New York, N.Y. 888 pp.
234. Griffin, W. J., Sr. Lecturer, Pharmacy Dept., Univ. of Queensland, St. Lucia, Queensland, Aust. Personal communication, Oct. 14, 1974.
235. ————, H. P. Brand and J. G. Dare. 1975. Analysis of *Duboisia myoporoides* R. Br. and *Duboisia leichhardtii* F. Muell. *J. Pharm. Sci.* 64 (11):1821-1825.
236. Groff, G. W. and G. W. Clark. 1928. The botany of *Ephedra* in relation to the yield of physiologically active substances. *Univ. of Calif. Publications in Botany* 14 (7):247-282.
237. Guillen, A. 1967. Guía técnica para el cultivo de la raicilla o ipecacuana. *Nuestra Tierra* (Min. of Agr. y Ganad., Nicaragua) 11 (1):27-34.
238. Gunther, E. 1968. Australian *Eucalyptus* oils. *Perfum. & Essen. Oil Rec.* 59 (9):634-641.
239. Gupta, R. 1971. Ipecac: a promising subsidiary crop for north-eastern plantation regions. *Indian Farming* 21 (4):19-21.

240. ——— 1972. Cultivation and distillation of Japanese mint in India. *Indian Farming* 22 (3):18-23.
241. ——— 1972. Raise *Rauvolfia* for export. *Indian Farming* 21 (11):9-12.
242. Gutfinger, T., S. Romano and A. Letan. 1972. Characterization of lipids from seeds of the Rosaceae family. *J. Food Sci.* 37 (6):938-940. CA 78:62048r.
243. Gutierrez-Noriega, C. and V. Wolfgang von Hagen. 1951. Coca—the mainstay of an arduous native life in the Andes. *Econ. Bot.* 5 (2):145-152.
244. Habib, M. S. and K. J. Harkins. 1970. Quantitative determination of emetine and cephaeline in ipecacuanha root and its preparations. *Pl. Med.* 18 (3):270-274.
245. Haendler, H. and R. Huet. 1965. La papaine. *Fruits* 20 (8):411-415.
246. Handa, K. L. and H. L. Abrol. 1954. Chemical investigation of *Hyocyamus muticus*. *J. Sci. Indus. Res.* 13B:221-222.
247. Hanna, J. M. 1971. Responses of Quecha Indians to coca ingestion after cold exposure. *Am. J. Phys. Anthrop.* 34 (2):273-278. BA 51:139640.
248. Hara, K., M. Ishiguro, G. Funatsu and M. Funatsu. 1974. An improved method of the purification of ricin D. *Agr. Biol. Chem.* 38 (1):65-70. BA 58:33849.
249. Harper, R. E. and H. F. Winters. 1946. Cinchona investigations in Puerto Rico. *Agr. in the Amer.* 6 (2):30-32, 37.
250. Harper, R. M. 1928. *Economic botany of Alabama*. Pt. 2. Monog. #9. Geol. Surv. Ala.; State Comm. of For., University, Ala. 357 pp.
251. Harrington, T. A. 1969. Production of Oleoresin from Southern Pine Trees. *For. Prod. J.* 19 (6):31-36.
252. Hassan, A., B. Trabolsi and Z. Farid. 1974. Colchicine for familial Mediterranean fever (periodic peritonitis): Concl. *N. Eng. J. Med.* 290 (17):973.
253. Hedayatullah, S. 1959. Culture and propagation of *Rauvolfia serpentina* Benth. in East Pakistan. *Pakistan. J. Sci. & Indus. Res.* 2 (2-3):118-122.
254. Heinicke, R. M. and W. A. Gortner. 1957. Stem bromelain—a new protease preparation from pineapple plants. *Econ. Bot.* 11 (3):225-234.
255. Hemsley, W. B. 1882-1886. *Biologia Central-Americana; or contributions to the knowledge of the fauna and flora of Mexico and Central America—botany.* Vol. III. R. H. Porter, London. (P. 354).
256. Herrmann, W. 1972. *Cymarin from fermentation of Strophanthus glycosides.* Ger. 2,050,457 (Cl. C 07c, A 61k), 25 May 1972; Appl. p. 20, 50, 457.8-42, 14 Oct. 1970. 2 pp. CA 77:73684a.
257. ———, F. Zimmermann and G. Satzinger. 1970. *Purification of cymarin.* Ger. Offen. 1,920,177 (Cl. C 07c, A 61k), 26 Nov. 1970; Appl. 21 Apr. 1969. 9 pp. CA 74:54152x.
258. Hess, H. J. E. and R. P. Nelson. 1973. *18-beta-glycyrrhetinic acid amides useful as antiulcer agents.* US 3,766,206 (Cl. 260-309; C07d) 16 Oct. 1973; Appl. 195, 475. 10 Nov. 1971. 8 pp. CA 80: 37330k.
259. Heumann, W. R. 1957. The manufacture of alkaloids from opium. *Bull. on Narc.* 9 (2):34-40.
260. Hills, K. L. 1948. *Duboisia* in Australia—a new source of hyoscine and hyoscyamine. *J. N. Y. Bot. Gard.* 49 (584):185-188.
261. Hiroi, M. and D. Takaoka. 1974. Nonvolatile sesquiterpenoids in the leaves of camphor tree (*Cinnamomum camphora*). *Nippon Kagaku Kaishi* 4:762-765. CA 81:60880u.
262. Hodge, W. H. 1948. Wartime cinchona procurement in Latin America, *Econ. Bot.* 2 (3):229-257.

263. Hodkova, J., Z. Vesely, Z. Koblicova, J. Holubek and J. Trojanek. 1972. On alkaloids XXV. Minor alkaloids of poppy capsules. *Lloydia* 35 (1):61-68.
264. Hoehne, F. C. 1939. *Plantas substancias vegetais toxicas e medicinais*. Dept. de Botanico do Estado, São Paulo, Brazil. 355 pp.
265. Hoffman, M. L. 1966. Oxygen removal from cottonseed oil by sparging with nitrogen. *Food Tech.* 20 (4):204-208.
266. Hooper, D. and H. Field. 1937. *Useful plants and drugs of Iran and Iraq*. Pub. 387. Bot. Ser. Vol. 9, No. 3. Field Mus. Nat. Hist., Chicago. 241 pp.
267. Hornemann, U., M. K. Wilson, K. M. Kelley and H. G. Floss. 1968. The biosynthesis of indoleisopropionic acid. *Lloydia* 31 (4):425. [abs.]
268. Hotellier, F. and P. Delaveau. 1972. Oils of pharmaceutical, dietetic, and cosmetic interest. III. Almond (*Prunus amygdalus dulcis*), Prunus kernel, and hazelnut (*Corylus avellana*). *Ann. Pharm. Fr.* 30 (7-8):495-502. CA 78: 62044m.
269. Howes, F. N. 1949. *Vegetable gums and resins*. Chronica Botanica Co., Waltham, Mass. 188 pp.
270. Hsiao, S. and F. W. Putnam. 1961. The cleavage of human γ-globulin by papain. *J. Biolog. Chem.* 236 (1):122-135.
271. Hu, Shiu-Ying. 1969. *Ephedra* (Ma-Huang) in the new Chinese materia medica. *Econ. Bot.* 23 (4):346-351.
272. Hurst, E. 1942. *Poison plants of New South Wales*. N. S. W. Poison Pls. Comm., Univ. of Sydney & N. S. W. Dept. Agr., Sydney. 498 pp.
273. Hwang, K. and A. C. Ivy. 1951. A review of the literature on the potential therapeutic significance of papain. pp. 161-207. In: *Papain*. Edited by M. L. Tainter. *Ann. of N.Y. Acad. Sci.* Vol. 54, Art. 2, pp. 143-296.
274. Ikonomovski, K. 1973. Preparation of codeine by methylation of morphine with phenyltrimethylammonium methoxide. *Acta Pharm. Jugoslav.* 23 (3):169-171. CA 79:137324j.
275. Irvine, F. R. 1961. *Woody plants of Ghana*. Oxford Univ. Press, London. 868 pp. & 34 pls.
276. Jager, H. H., O. Schindler, E. Weiss and T. Reichstein. 1965. Die cardelolide von *Strophanthus gratus*. *Helvet. Chim. Acta.* 48 (1):202-219.
277. Jain, S. K. 1968. *Medicinal plants*. National Book Trust, New Delhi. 176 pp.
278. Jakovljevic, I. V. 1974. Digitoxin. *Anal. Profiles Drug Subst.* 3:149-172. CA 81: 41307w.
279. Janaki Ammal, E. K. and H. P. Bezbaruah. 1963. Induced tetraploidy in *Catharanthus roseus* (L.) G. Don. *Proc. Ind. Acad. Sci.* 57:341.
280. Janowsky, D. S., M. K. El-Yousef and J. M. Davis. 1973. Antagonistic effects of physostigmine and methylphenidate in man. *Amer. J. Psychiat.* 130 (12): 1370-1376.
281. Javanovics, K., E. Bittner, E. Dezseri, J. Eles and K. Szasz. 1971. Selective manufacture of antitumorous vinblastine, vinleurosine, and vincristine from *Vinca rosea*. Ger. Offen. 2, 124,023 (Cl. C 07d), 09 Dec. 1971; Hung. Appl. 27 May 1970. 18 pp. CA 76: 44112m.
282. ———, K. Szasz, G. Fekete, E. Bittner, E. Dezseri and J. Eles. 1973. Increasing the yield of vincristine when separating vincristine from *Vinca rosea* drug. 72-08,534 (Cl. A 61k) 15 Aug. 1973. Appl. 72, 85, 34, 01 Dec. 1972. 17 pp. CA 81: 68541c.
283. Jeppe, B. 1969. *South African aloes*. Purnell, Cape Town, S. Afr. 144 pp.

284. Jex-Blake, A. J. 1950. *Gardening in East Africa.* 3rd ed. Longmans, Green & Co., London. 398 pp.
285. Johnson, C. H., Prof., Coll. of Pharmacy, Univ. of Florida, Gainesville, Fla. Personal communication. May 2, 1975.
286. Johnson, C. and C. P. Johnson. 1861. *British poisonous plants.* 2nd ed. John van Voorst, London. 76 pp. & 32 pls.
287. Johnson, I. S. 1973. *Psoriasis treatment with vinblastine.* U.S. 3,749,784 (Cl. 424-262; A 61k), 31 Jul 1973; Appl. 84, 199, 26 Oct. 1970. 2 pp.
288. Joubert, J. S. G. 1971. *The cost and income structure of the cultivation and decortication of sisal in Natal.* Econ. Ser. 75. Div. of Agr. Prod. Econ., Dept. Agr. Econ. & Marketing. 25 pp.
289. Joye, N. M., Jr., A. T. Proveaux and R. V. Laurence. 1972. Composition of pine needle oil. *J. Chromatogr. Sci.* 10 (9):590-592. CA 79:96830a.
290. Julian, D. V. 1972. *Sterols from tall oil pitch.* U.S. 3,691,211 (Cl. 260/397.25). 12 Sep 1972. Appl. 95,735,07. Dec. 1970. 9 pp. CA 78:26707w.
291. Kamath, K. M. and J. G. Kane. 1961. Hydrogenation of the oil of *Hydnocarpus laurifolia* (Dennst.) Sleumer syn. *H. wightiana* Blume. *J. Sci. Indus. Res.* 20D (6):244-246.
292. Kamedulski, V., B. Dimov, and I. Tonev. 1972. Extraction of morphine from poppy capsules [with] organic solvents. *Farmatsiya* (Sofia) 22 (3): 34-39 [Bulg.]. CA 77:130524z.
293. Kapoor, L. D., K. L. Handa, I. Chandra and B. K. Abrol. 1955. Cultivation of *Mentha arvensis* in Jammu and Kashmir. *J. Sci. Indus. Res.* 14A (8):374-378.
294. Karawya, M. S., S. I. Balbaa and M. S. Afifi. 1971. Alkaloids and glycosides in *Plantago psyllium* growing in Egypt. *UAR J. Pharm. Sci.* 12 (1):53-61. CA 78:13729e.
295. ——— and M. S. Hifnawy. 1974. Analytical study of the volatile oil of *Thymus vulgaris* L. growing in Egypt. *J. Assoc. Off. Anal. Chem.* 57 (4):997-1001. BA 58:57279.
296. Karmazin, M., K. Michal, and V. Suk. 1966. Fluctuation of anthraquinone content in the roots of the cultivated medical rhubarb. *Herba Hung.* 5 (2/3): 131-134. BA 48:93115.
297. Karnick, C. R. and M. D. Saxena. 1970. On the variability of alkaloid production in *Datura* species. *Pl. Med.* 18 (3):266.
298. Kartnig, Th. and R. Danhofernoehammer. 1972. Chemical determination of digitalis drug value (*Digitalis purpurea* and *D. lanata*). XI. *Sci. Pharm.* 40 (2):110-119. [Ger.] CA 77:68683p.
299. Kaul, B. K. and L. D. Kapoor. 1961. Effect of gibberellic acid on vegetative growth and total alkaloids of *Rauvolfia serpentina* Benth. *J. Sci. Indus. Res.* 20C (12):358-359.
300. Kaul, B. L. and D. K. Choudhary. 1975. The effect of chemical mutagens on quantitative characters in *Atropa belladonna. Pl. Med.* 27:337-342.
301. Kay, D. E. 1965. *The production and marketing of pineapples.* TPI Rpt. #G10. Trop. Prod. Inst., London. 49 pp. & 17 tables.
302. Kerbosch, M. 1940. Notes on the cultivation of *Cinchona* and the world supply of quinine. *Monthly Bull. Agr. Sci. & Practice* (Rome) 31 (1):14-24.
303. Khanna, K. R. and U. P. Singh. 1975. Correlation studies in *Papaver somniferum* and their bearing on yield improvement. *Pl. Med.* 28:92-96.
304. King, D. B., H. B. Wagner and G. H. Goldsborough. 1962. *The outlook for naval stores.* USDA, Washington, D.C. 62 pp.

305. King, J. 1855. *American eclectic dispensatory*. Moore, Wilstach, Keys & Co., Cincinnati, Ohio. 1390 pp.
306. Kingsbury, J. M. 1964. *Poisonous plants of the United States and Canada*. Prentice-Hall, Inc., Englewood Cliffs, N.J. 626 pp.
307. Kirby, R. H. 1950-1951. Seaweeds in Commerce. Pt. I. *Col. Pl. & Anim. Prod.* 1 (3):183-216. Pt. II, *ibid.* 1 (4):284-293. Pt. III, *ibid.* 2 (1):1-22.
308. Kirk, D. R. 1970 *Wild edible plants of the western United States*. Naturegraph Publishers, Healdsburg, Calif. 326 pp.
309. Klein, M. 1953. The influence of croton oil on skin tumorigenesis in strain C57 brown mice. *Quart. J. Fla. Acad. Sci.* 16 (4):249-253.
310. Knuth, R. 1917. Dioscoreaceae americanae novae. *Notizblatt Bot. Gart.* (Berlin) 7:204.
311. Koch, P. 1972. *Utilization of the southern pines*. 2 vols. U.S. Gov't Ptg. Off., Washington, D.C. 734 pp.; 1,663 pp.
312. Koerber, W. L. and J. H. Reinhart. 1966. Cultivation of *Rauvolfia serpentina* under controlled conditions. *Lloydia* 29 (3):260-269.
313. Kononikhina, N. F. 1970. Conditions for the extraction of phytoestrogens of *Glycyrrhiza glabra* (licorice). *Aktual. Vop. Farm.* (1968):112-114. [Russ.] CA 76:49862r.
314. Kosa, F. and E. Viragos-Kis. 1969. A case of mortal intoxication of a 3-year-old child after ingestion of unripe poppy capsules. *Zacchia* 5:604-610. [Ger.]
315. Kotalawala, J. 1971. Mass production of pineapple planting material. *Trop. Agriculturist* 127:199-201.
316. Kreibich, G. and E. Hecker. 1970. On the active principles of croton oil. *Z. Krebsforsch.* 74:448-456.
317. Krikorian, A. D. and M. C. Ledbetter. 1975. Some observations on the cultivation of opium poppy (*Papaver somniferum* L.) for its latex. *The Bot. Rev.* 41 (1):30-103.
318. Kritikos, P. G. and S. P. Papadaki. 1967. The history of the poppy and of opium and their expansion in antiquity in the eastern Mediterranean area. *Bull. on Narc.* 19 (3):17-38; 19 (4):5-10.
319. Krochmal, A. and L. Wilken. 1970. The culture of Indian tobacco (*Lobelia inflata* L.) For. Serv. Res. Paper NI181. USDA For. Serv., N.E. For. Exp. Sta., Upper Darby, Penna. 9 pp.
320. ———, ——— and M. Chien. 1970. *Lobeline content of Lobelia inflata: structural, environmental and developmental effects*. For. Serv. Res. Paper NE-178. USDA, For. Serv., N/E For. Exp. Sta., Upper Darby, Penna. 13 pp.
321. ———, ——— and M. Chien. 1972. Plant and lobeline harvest of *Lobelia inflata* L. *Econ. Bot.* 26 (3):216-220.
322. ———, ———, D. van Lear and M. Chien. 1974. Mayapple (*Podophyllum peltatum* L.). USDA For. Serv. Res. Paper NE-296. Northeast. For. Exp. Sta., Upper Darby, Penna. 8 pp.
323. Krupinski, V. E., J. E. Robbers and H. G. Floss. 1974. Physiological studies on ergot. *Lloydia* 37 (4):646. (abs.)
324. Kumar, S. A. 1961. Active fragment of papain. *J. Sci. Indus. Res.* 20A (2):103.
325. Kupchan, S. M., R. W. Britton, C. K. Chiang, Ningur NoyanAlpan and M. F. Ziegler. 1973. Tumor inhibitors 88: The antileukemic principles of *Colchicum speciosum*. *Lloydia* 36 (3):338-340.
326. ——— and A. Karim. 1976. Tumor inhibitors 114: Aloe Emodin: Antileukemic principle isolated from *Rhamnus frangula* L. *Lloydia* 39 (4): 223-224.

327. ———, I. Uchida, A. R. Branfman, R. G. Dailey, Jr., and B. Yu Fei. 1976. Antileukemic principles isolated from Euphorbiaceae plants. *Science* 191:571-572.
328. Kusevíc, V. 1960. Cultivation of the opium poppy and opium production in Yugoslavia. *Bull. on Narc.* 12 (2):5-13.
329. Küssner, W. 1961. Poppy straw: a problem of international narcotics control. *Bull. on Narc.* 13 (2):1-6.
330. Kwong, K-H., T-H. Yang and J-B. Chen. 1972. Effects of different harvest time upon the root weight and alkaloid content of *Rauwolfia serpentina* in two-year old plants. *Tai-wan Tang Yeh Shih Yen so Yen Chiu Hui Pao* 58:49-56. CA 79: 123678r.
331. Lambert, J. 1965. La prairie à colchiques: une des plus intéressantes formations végétales d'Ardènne, en voie de disparition. *Neth. J. Agr. Sci.* 13 (2):129-142.
332. Langdon, O. G. and R. P. Schultz. 1973. *Longleaf and slash pine.* Agr. Handbook 445. USDA For. Serv., Washington, D.C. Pp. 85-89.
333. Lee, S. 1935. *Forest botany of China.* Commercial Press, Ltd., Shanghai. 991 pp.
334. Lenkey, B., P. Nanasi and P. Tetenyi. Glycoside content of *Digitalis lanata*. *Herba Hung.* 10 (1):33-35. CA 77:58755.
335. Lerche, K. 1963. The cultivation of sisal. *World Crops* 15 (12):437-443.
336. Lerman, F. S. 1970. Content of glycyrrhizic acid in roots of licorice growing under different conditions. *Mater. Biol. Vidov. Roda Glycyrrhiza L.* 188-194. [Russ.] CA 77:98798d.
337. Lewis, T. and E. F. Woodward. 1950. Papain—the valuable latex of a delicious tropical fruit. *Econ. Bot.* 4 (2):192-194.
338. Leyel, C. F. 1952. *Green medicine.* Faber & Faber, Ltd., London. 324 pp.
339. Lilly, Eli & Co. 1961. *Velban.* Information release 00351. Issued Feb. 27.
340. Linley, P. A. and J. M. Rowson. 1973. The evaluation of *Digitalis purpurea*. *Pl. Med.* 24 (3):211.
341. Little, E. L. and K. W. Dorman. 1952. Slash pine (*Pinus elliottii*): its nomenclature and varieties. *J. Forestry* 50 (12):918-923.
342. ——— and F. W. Wadsworth. 1964. *Common trees of Puerto Rico and the Virgin Islands.* Agr. Handbook 249. USDA, Washington, D.C. 548 pp.
343. Lock, G. W. 1963. Factors affecting sisal growing. *World Crops* 15 (4):141-145.
344. Lokich, J. J. and A. T. Skarin. 1972. Five-drug combination chemotherapy for disseminated adenocarcinoma. *Cancer Chemother. Rpt.* 56 (Pt. 1):761-767.
345. Loöf, B. 1966. Poppy cultivation. *Field Crop Abs.* 19 (1):1-5.
346. Lorand, L. and K. Konishi. 1964. Papain-induced polymerization of fibrogen. *Biochemistry* 3:915-919.
347. Lorenzetti, L. J., R. Salisbury, J. L. Beal and J. N. Baldwin. 1964. Bacteriostatic property of *Aloe vera*. *J. Pharm. Sci.* 53 (10):1287.
348. Lorincz, K. and P. Tetenyi. 1970. Results of poppy breeding by distant hybridization. *Postep Dziedzinie Leku Rosl., Pr. Ref. Dosw. Wygloszone Symp.* (Pub. 1972): 190-195 CA 78:69264s.
349. Lovell, J. H. 1926. *Honey plants of North America (north of Mexico).* The A. I. Root Co., Medina, Ohio. 408 pp.
350. Lugt, Ch. and L. Noordhoek-Ananias. 1974. Quantitative fluorimetric determination of the main cardiac glycosides in *Digitalis purpurea* leaves. *Pl. Med.* 25:267-273.

351. Luk'yanov, A. V., Onoprienko, V. S. and Zasosov, V. A. 1972. Commercial methods for preparing papaverine. *Khim.-Farm. Zh.* 6 (5):14-25. CA 77: 101940b.
352. Lund, M. H. and R. R. Royer. 1969. *Carica papaya* in head and neck surgery. *Arch. Surg.* 98 (2):180-182.
353. Lushbaugh, C. C. and D. B. Hale. 1953. Experimental acute radiodermatitis following beta irradiation. *Cancer* 6:690-698.
354. Lutomski, J. and M. Turowska. 1973. Sorption of moisture in the leaves of *Atropa belladonna* and *Datura innoxia* and its effect on the stability of alkaloids. *Herba Pol.* 19 (3):202-206. [Pol.] CA 81-41326b.
355. Lynn, K. R. 1973. An isolation of chymopapain. *J. Chromatog.* 84 (2):423-425.
356. Macbride, J. F. 1937. *Flora of Peru*, Vol. 13, Pt. 6, No. 2. Bot. Ser. Pub. 393. Field Mus. Nat. Hist., Chicago. 491 pp.
357. ——— 1943. ——— Pt. 3, No. 1. ——— 531. ———. 510 pp.
358. ——— 1949. ——— Pt. 3, No. 2. ——— 622. ———. Pp. 511-777.
359. ——— 1959. ——— Pt. 5, No. 1. ——— 880. ———. 536 pp.
360. MacKenzie, A. L. and J. F. G. Pigott. 1971. Atropine overdose in three children. *Br. J. Anaesth.* 43 (11):1088-1090. BA 53:46532.
361. MacLeod, J. 1963. Ipecac intoxication—use of a cardiac pacemaker in management. *N. Eng. J. Med.* 263 (3):146-147.
362. Maheshwari, J. K. 1963. *The flora of Delhi*. Coun. Sci. & Indus. Res., New Delhi. 447 pp.
363. Maiden, J. H. 1889. *Useful native plants of Australia (including Tasmania)*. Tech. Mus. N. S. Wales, Sydney. 696 pp.
364. Makarevich, I. F. 1972. Cardenolides of *Strophanthus kombe*. II. *Khim. Prir. Soedin* 2:180-188. [Russ.] CA 77:58746d.
365. Makarov, A. A. 1971. Chemical evaluation of *Arctostaphylos uva ursi* (bearberries). *Uch. Zap. Yakutsk. Gos. Univ.* 18:41-44. [Russ.] CA 79:27240.
366. Malawista, S. E. and K. G. Bensch. 1967. Human polymorphonuclear leukocytes: demonstration of microtubes and effect of colchicine. *Science* 156 (3774): 521-522.
367. Malinina, V. M. and R. M. Ivanova. 1974. Accumulation of morphine in several varieties of poppy in different climatic zones. *Dokl. Vses. Akad. Sel'skokhoz. Nauk.* 2:19-20. [Russ.] CA 81:60831d.
368. Manfred, L. 1947. *7,000 recetas botánicas, a base de 1,300 plantas medicinales americanas*. Editorial Kier, Buenos Aires. 778 pp.
369. Manialawi, M. 1973. Colchicine for familial Mediterranean fever. *New Eng. J. Med.* 289 (14):752.
370. Manning, C. E. F. 1969. *The market for steroid drug precursors with particular reference to diosgenin*. Rpt. G41. Trop. Prod. Inst., London. 20 pp.
371. Martin, F. W. 1969. The species of *Dioscorea* containing sapogenin. *Econ. Bot.* 23 (4):373-379.
372. ——— 1972. Current status of the sapogenin-bearing yams. *Pl. Fds. Human Nutr.* 2:139-143.
373. ———, E. Cabanillas and M. H. Gaskins. 1966. Economics of the sapogenin-bearing yam as a crop plant in Puerto Rico. *J. Agr. Univ. P.R.* 50 (1):53-64.
374. ———, ——— and S. Ortiz. 1963. Natural pollination, hand pollination and crossability of some Mexican species of *Dioscorea*. *Trop. Agr., Trin.* 40 (2): 135-141.

375. ——— and H. Delpin. 1967. The influence of some soil and climatic factors on sapogenin yields of *Dioscorea*. *J. Agr. Univ. P.R.* 51 (3):260-265.
376. ——— and ——— 1969. Techniques and problems in the propagation of sapogenin-bearing yams from stem cuttings. *J. Agr. Univ. P.R.* 53 (3): 191-198.
377. ——— and M. H. Gaskins. 1968. *Cultivation of the sapogenin-bearing Dioscorea* spp. Production Res. Rpt. 103. USDA, ARS, Washington, D.C. 19 pp.
378. Martinez, M. 1959. *Las plantas medicinales de México*. 4th ed. Ediciones Botas, Mexico, D. F. 656 pp.
379. ——— 1959. *Plantas útiles de la flora Mexicana*. Ediciones Botas, Mexico, D. F. 621 pp.
380. Mathew, A. G., Y. S. Lewis, N. Krishnamurthy and E. S. Nambudiri. 1971. Capsaicin. *The Flavour Industry* (Dec.) 691-695.
381. Medora, R. S., J. M. Campbell and G. P. Mell. 1973. Proteolytic enzymes in papaya tissue cultures. *Lloydia* 36 (2):214-216.
382. Meer Corporation (undated). Technical and Application Information 60504 #51.
383. Meijer, W. 1974. *Podophyllum peltatum*—May apple, a potential new cash-crop of Eastern North America. *Econ. Bot.* 28 (1):68-72.
384. Merlin, H. E. 1972. Azoospermia caused by colchicine: a case report. *Fert. Steril.* 23 (3):180-181.
385. Miller, J. A. 1973. Naturally occurring substances that can induce tumors. Pp. 508-549. In: *Toxicants Occurring Naturally in Foods*. 2nd ed. Nat. Acad. Sci., Washington, D.C. 624 pp.
386. Miner, R. W. (editor). 1954. *Reserpine (Serpasil) and other alkaloids of Rauwolfia serpentina: chemistry, pharmacology and clinical applications*. Ann. N.Y. Acad. Sci. 59. 140 pp.
387. Mira, L. and E. Livini. 1970. Neurotoxicity of vincristine in childhood leukemia. *Riv. Patol. Nerv. Ment.* 91 (3):139-146. BA 54:23124.
388. Mishinskaya, S., Y. Shostenko and E. Vysotskaya. 1972. Use of anion exchange for isolating codeine and morphine from poppy seeds. *Khim.-Farm. Zh.* 6 (12):34-37 [Russ.] CA 78:75812k.
389. Miyazaki, Y. and H. Watanabe. 1962. Experimental cultivation of coca (*Erythroxylon* sp.) at Izu. *Bull. Nat. Inst. Hyg. Sci.* (Tokyo) 80:155-158.
390. Moertel, C. G., A. J. Schutt, R. G. Hahn and R. J. Reitemeier. 1974. Treatment of advanced gastrointestinal cancer with emetine (NSC-33669). *Cancer Chemother. Rpt.* Pt. I. 58 (2):229-232.
391. Monachino, J. 1954. *Rauvolfia serpentina*—its history, botany and medical use. *Econ. Bot.* 8 (4):349-365.
392. Montesinos A. F. 1965. Metabolism of cocaine. *Bull. on Narc.* 27 (2):11-17.
393. Morice, J. and J. Louarn. 1971. Morphine content in the oil poppy (*Papaver somniferum*). *Ann. Amelior. Plant.* 21 (4):465-484. CA 77:2764r.
394. Morrison, B. Y. 1943. Quinine from seed. *Agr. in the Amer.* 3 (7): 131-133.
395. Mors, W. B. and C. T. Rizzini. 1966. *Useful plants of Brazil*. Holden-Day, Inc., San Francisco. 166 pp.
395a. ——— and N. Sharapin. 1973. Obtenção de esteroides do sisal. Bol. 11. Centro de Tecnologia Agricola e Alimentar. Repr. *Rev. Brasileira de Tec.* 4: 153-165.
396. Morton, J. F. 1961. Folk uses and commercial exploitation of aloe leaf pulp. *Econ. Bot.* 15 (4):311-319.

397. ——— 1974. *Folk remedies of the Low Country.* E. A. Seemann Pub., Inc., Miami, Fla. 176 pp.
398. Mowry, H. and R. D. Dickey. 1950. *Ornamental hedges for Florida.* Bull. 443, Rev'd. Univ. of Fla. Agr. Exper. Sta., Gainesville. 36 pp.
399. Moza, B. K., D. K. Basu and U. P. Basu. 1969. Survey of Indian medicinal plants: Pt. I: Search for hyoscyamine and hyoscine. *Ind. J. Chem.* 7:414-415.
400. Muenscher, W. C. 1951. *Poisonous plants of the United States.* Rev'd ed. The Macmillan Co., New York. 277 pp.
401. Munz, P. A. and D. D. Keck. 1968. *A California flora.* Univ. of Calif. Press, Berkeley, Calif. 1681 pp.
402. Murav'ev, I. A. and N. F. Kononikhina. 1972. Estrogenic properties of *Glycyrrhiza glabra. Rast. Resur.* 8 (4):490-497. CA 78:75811j.
403. Murphy, H. B. M., O. Rios and J. C. Negrette. 1969. The effects of abstinence and of re-training on the chewer of coca-leaf. *Bull. on Narc.* 21 (2):41-47.
404. Nagy, M. AMA Dept. of Foods & Nutr., Chicago. 1972. Questions and Answers. *J.A.M.A.* 219 (5):626.
405. Naik, K. C. 1949. *South Indian fruits and their culture.* P. Varadachary & Co., Madras, India. 477 pp.
406. Nano, G. M., T. Sacco and C. Frattini. 1973. Anthemis oils. *Atti, Conv. Naz. Olii Essenziali Sui Deriv. Agrum.* 2nd: 50-57. [Ital.] CA 81:111400x.
407. Nayar, N. M. and K. L. Mehra. 1970. Sesame: its uses, botany, cytogenetics, and origin. *Econ. Bot.* 24 (1):20-31.
408. Nayar, S. L. and I. C. Chopra. 1951. *Distribution of British Pharmacopoeial drug plants and their substitutes growing in India.* Coun. Sci. & Indus. Res., Pharm. & Drugs Res. Comm., New Delhi. 56 pp.
409. Nazir, B. N. and K. L. Handa. 1959. Chemical investigation of *Lochnera rosea. J. Sci. & Indus. Res.* 18B (4):175.
410. Negrette, J. C. and H. B. M. Murphy. 1967. Psychological deficit in chewers of coca leaf. *Bull. on Narc.* 19 (4):11-18.
411. Nishi, H. and J. Morishita. 1971. Components of licorice root used for tobacco flavoring. I. Fractionation of the substances in licorice root effective in improving the tobacco smoking quality. *Nippon Nogei Kagaku Kaishi* 45 (11): 507-512. CA 76: 151016e.
412. North, P. M. 1967. *Poisonous plants and fungi in colour.* Blanford Press, London. 161 pp.
413. Northmore, J. M. 1963. Sisal waste—mulch. *Kenya Coffee* 28 (330):201, 203, 206.
414. O'Brien, M. and J. B. Smith. 1972. *Bio-engineering factors in mechanically harvesting pineapple.* Spec. Pub. 01-72. Conf. Papers Internat. Conf. on Trop. & Subtrop. Agr., Amer. Soc. Agr. Engr., St. Joseph, Mich.
415. Ochse, J. J. and Bakhuizen van den Brink. 1931. *Vegetables of the Dutch East Indies.* Dept. Agr., Indus. & Comm. of Neth. E. Indies, Buitenzorg, Java. 1,005 pp.
416. Ogilvie, R. L. and J. Ruedy. 1973. An educational program in *Digitalis* therapy. *J.A.M.A.* 222 (1):50-55.
417. Ohn, T., Y. Khaing, M. Gale and M. Thein. 1970. Preliminary chemical examination of *Cephaelis ipecacuanha* test-cultivated in Burma. *Union Burma J. Sci. Technol.* 3 (2):251-254.

418. Olesen, J. and E. Skinhoej. 1972. Effects of ergot alkaloids (hydergine) on cerebral hemodynamics in man. *Pharmacol. Toxicol.* 3(1-2):75-85. CA 77: 14350r.
419. Oliver, B. 1960. *Medicinal plants in Nigeria.* Nigerian Coll. of Arts, Sci. & Tech., Ibadan. 138 pp.
420. Olizar, M. 1970. *A guide to the Mexican markets.* 4th ed. Author. Sinaloa 9, Mexico 7, D. F. P. 13.
421. Orellana Segovia, M. D. 1972. A new method for determining the 18-beta-glycyrrhentinic acid content in the roots and extract of *Glycyrrhiza glabra. An. R. Acad. Farm.* 38 (1):167-169. BA 55:68209.
422. Orr, E. and R. Pope. 1962. The market for peppermint, arvensis and spearmint oils and menthol. *Trop. Sci.* 4 (1):38-47.
423. Osol, A., R. Pratt and M. D. Altschule. 1960. *The United States Dispensatory and physicians' pharmacology.* 26 ed. J. B. Lippincott Co., Phila. 1,277 pp.
424. Otieno, L. H. 1966. Observations on the action of sisal waste on freshwater pulmonate snails. *E. Afr. Agr. & For. J.* 32 (1):68-71.
425. Paa, N. F., J. G. Blando and F. D. Milan. 1968. *Rauwolfia serpentina* (Benth.) and its introduction in the Philippines. *Phil. Geog. J.* 12 (1-2): 12-15.
426. Pailer, M. and E. Haslinger. 1972. Isolation of (R)-(-) armepavine from *Rhamnus frangula. Monatsh. Chem.* 103 (5):1399-1405. [Ger.] CA 78:26461m.
427. Pakrashi, S. C. and B. Achari. 1968. Rauwolfia alkaloids in retrospect. *J. Sci. & Indus. Res.* 27 (2):58-69.
428. Paris, R. and S. Etchepare. 1967. The flavonoids of the leaves of *Rauwolfia vomitoria* Afzel. *Ann. Pharm. Fr.* 25 (12):779-782.
429. ———, H. Pourrat and P. Chancel. 1954. Essais de culture de *Colchicum autumnale. Soc. Bot. de France, B.* 101:228-231.
430. Parker, H. L., G. Blaschke and H. Rapoport. 1972. Biosynthetic conversion of thebaine to codeine. *J. Am. Chem. Soc.* 94 (4):1276-1282.
431. Parker, J. B. 1966. Gum arabic from Sudan exported to markets all over the world. *For. Agr.* 4 (25):7.
432. Patiño, V. M. 1967. *Plantas cultivadas y animales domésticos en América Equinoccial.* Tomo III: Fibras Medicinas, Misceláneas. Imprenta Departamental, Cali, Colombia. 567 pp.
433. Pecherskaya, L. G., G. A. Glumov and S. V. Teslov. 1971. Accumulation of arbutin in leaves of bearberry and red whortleberry grown in the Perm region. *Nauch. Tr., Perm. Farm. Inst.* 4:111-115. From *Ref. Zh., Biol. Khim.* 1972. Abstr. No. 23F 1207. CA 79 (1973):2783n.
434. Pekic, B. and S. Djordjevic. 1972. Extraction of glycosides with water from *Digitalis* leaves. *Arh. Farm.* 22 (4):201-205. [Serbo-Croat.] CA 79:149304r.
435. ———, S. Dordevic and K. Petrovic. 1973. Determination of ergotamine in ergot. *Farm. Glas.* 29 (1):1-4 [Croat.]. CA 79:50436y.
436. ———, and D. Stamenkovic. 1972. Extraction of lanatosides from the leaves of *Digitalis lanata. Acta Pharm. Jugoslav.* 22 (4):145-152. [Croat.] CA 78: 75814n.
437. Pendse, G. S., P. S. Dange and S. R. Surange. 1974. The present and future of senna cultivation in India. *J. Univ. Poona* 46:151-162.
438. Pennington, T. D. and J. Sarukhan. 1968. *Arboles tropicales de México.* FAO & Inst. Nac. de Invest. For., Mexico, D. F. 413 pp.
439. Perdue, R. and J. Hartwell. 1969. The search for plant sources of anticancer drugs. *Morris Arb. Bull.* 20 (3):35-53.

440. Perez-Arbelaez, E. 1956. *Plantas útiles de Colombia*. 3rd ed. Libreria Colombiana-Camacho Roldan (Cia. Ltda), Bogotá. 831 pp.
441. Petelot, A. 1953. *Plantes médicinales du Cambodge, du Laos et du Vietnam*. Vol. 2, #18. Centre de Rech. Sci. et Tech., Arch. des Rech. Agron. au Camb., au Laos et au Vietnam, Saigon. 284 pp.
442. Petrova, Y., T. Tomova and L. Filipova. 1972. Quantitative determination of ergotamine and ergotamine-ergotaminine in ergot. *Farmatsiya* (Sofia) 22 (1): 9-13. [Bulg.] CA 76-158394k.
443. Pfeifer, S. 1971. New *Papaver* alkaloids. *Pharmazie* 26 (6):328-341. [Ger.].
444. Pillay, G. R. 1965. Kerala farmers can try the Japanese mint. *Indian Farming* 14 (11):15-17.
445. Pillay, P. P. and T. N. Santha Kumari. 1961. The occurrence of alstonine in *Lochnera rosea* (L.) Reichb. *J. Sci. Indus. Res.* 20B:458-459.
446. Pillen, D. 1972. Modern treatment of cardiac muscle insufficiency of coronary sclerotic origin. *Med. Welt.* 23 (20):739-742. [Ger.] BA 55:68266.
447. Pitcher, S. 1970. Brazil's rising tide of castor oil sales on the world market. *For. Agr.* 8 (14):10-11.
448. Poffenbarger, P. L. and B. R. Brinkley. 1974. Colchicine for familial Mediterranean fever: possible adverse effects. *N. Eng. J. Med.* 290 (1):56.
449. Pope, W. T. 1968. *Manual of wayside plants of Hawaii*. Charles E. Tuttle Co., Rutland, Vt. 289 pp.
450. Popenoe, W. 1942. Quinine from the "fever-tree". *Agr. in the Amer.* 2 (3):43-47.
451. ——— 1949. Cinchona cultivation in Guatemala—a brief historical review up to 1943. *Econ. Bot.* 3 (2):150-157.
452. Popov, P., Y. Dimitrov and S. Georgiev. 1973. Indigenous and foreign poppy varieties characterized by the morphine content of their dry capsules. *Bull. on Narc.* 25 (3):51-56.
453. ———, I. Dimitrov., S. Georgiev, T. Deneva and L. S. Iliev. 1974. Contents of morphine in the dried capsules of poppy (*Papaver somniferum*) cultivated in Bulgaria. *Farmatsiya* (Sofia) 24 (2):20-25. [Bulg.] CA 81:166307c.
454. Post, G. E. (Rev'd by J. E. Dinsmore). 1932, 1933. *Flora of Syria, Palestine and Sinai*. Vols. I and II. 2nd ed. American Press, Beirut. 658 pp. 928 pp.
455. Poulsen, E. 1973. Short-term toxicity of undegraded carrageenan in pigs. *Fd. Cosmet. Toxicol.* 1:219-227.
456. Pousset, J. L. and J. P. Foucher. 1973. Extraction of natural drugs. I. Rutoside. *Prod. Probl. Pharm.* 28 (4):290-293. [Fr.] CA 79:70134b.
457. Pratt, A. (undated). *The poisonous, noxious and suspected plants of our fields and woods*. Soc. for Promoting Christian Knowledge, London. 208 pp.
458. Purseglove, J. W. 1968. *Tropical Crops—Dicotyledons*. 2. John Wiley & Sons, Inc., New York. 719 pp.
459. Py, C. and R. Naville. 1973. Economic production of quality pineapples. *Span* 16 (1):21-23.
460. Qazilbash, N. A. 1971. Pakistan Ephedra, II. *Pharm. Weekblad* 106:373-382.
461. Quisumbing, E. 1951. *Medicinal Plants of the Philippines*. Tech. Bul. 16. Phil. Dept. Agr. & Nat. Res., Manila. 1234 pp.
462. Radford, A. E., H. E. Ahles and C. R. Bell. 1968. *Manual of the vascular flora of the Carolinas*. Univ. of No. Carolina Press, Chapel Hill, N.C. 1183 pp.
463. Rajagopalan, S. 1954. Recent developments on *Rauwolfia serpentina* Benth. *J. Sci. Indus. Res.* 13B (1):77.

464. Ram, M., J. Suri, A. Anand and H. Singh. 1972. Preparation of ergot alkaloids. *Res. Ind.* 17 (1):8. CA 78:62090y.
465. Ramanathan, V. S. and C. Ramachandran. 1974. Effect of temperature on the estimation of morphine in opium by the method of British Pharmacopoeia. *Bull. on Narc.* 26 (2):69-82.
466. Ramstad, E. 1959. *Modern pharmacognosy.* McGraw-Hill Book Co., Inc., New York. 480 pp.
467. Raymond, W. D. 1961. Castor beans as food and fodder. *Trop. Sci.* 3 (1):19-23.
468. ———, E. F. J. Thorpe and J. B. Ward. 1955. Sisal wax from Kenya. *Col. Pl. & Anim. Prod.* 5 (1):58-61.
469. Read, B. E. 1929. *Ephedra. Chinese Medicinal Plants.* Flora Sinensis. Ser. B. Vol. XXIV. #1. Pt. 1. 15 pp.
470. Record, S. H. and R. W. Hess. 1948. *Timbers of the New World.* Yale Univ. Press, New Haven, Conn. 640 pp.
471. Reddy, V. V. S., J. N. Choudhry, S. Vadlamudi, V. S. Waravdeker and A. Goldin. 1973. Antileukemic and immunosuppressive properties of abrin and ricin. *Fed. Proc.* 32 (3, Pt. 1):735.
472. Refsnes, K., S. Olsnes and A. Pihl. 1974. On the toxic proteins abrin and ricin. Studies of their binding to and entry into Erlich ascites cells. *J. Biol. Chem.* 249 (11):3557-3562.
473. Regir, V. G. 1968. Accumulation of tannins in bearberry leaves dependent on the phases of vegetation in different habitats. *Tr. Leningrad Khim.-Farm. Inst.* 26:179-184. CA 73:63235h.
474. Rehder, A. 1940. *Manual of cultivated trees and shrubs.* 2nd ed. The Macmillan Co., New York. 996 pp.
475. Reynolds, G. W. 1950. *The aloes of South Africa.* The Aloes of South Africa Book Fund, Johannesburg. 520 pp.
476. Rhodes, D. G. 1962. Menispermaceae. Pp. 157-172. In: *Flora of Panama.* Ann. Missouri Bot. Gard. 49 (3-4).
477. Robb, G. L. 1957. The ordeal poisons of Madagascar and Africa. *Bot. Mus. Leaflets* (Harvard Univ.) 17 (10):307-311.
478. Robbers, J. E. and H. G. Floss. 1968. Biosynthesis of ergot alkaloids. *Lloydia* 31 (4):425 [abs.]
479. ———, L. G. Jones and V. M. Krupinski. 1974. Ergot physiology. The production of ergot alkaloids in saprophytic culture by a homokaryotic strain of ergot. *Lloydia* 37 (1):108-111.
480. Robinson, H. J., F. S. Harrison and J. T. L. Nicholson. 1971. Cardiac abnormalities due to licorice intoxication. *Penna. Med.* 74:51-54.
481. Robinson, W. D. 1965. The present status of colchicine and uricosuric agents in management of primary gout. Chap. VIII, pp. 865-882. In: *Arthritis and Rheumatism.* Vol. 8, No. 5, Pt. I.
482. Roboz, E. and A. J. Hagen-Smit. 1948. A mucilage from *Aloe vera. J. Amer. Chem. Soc.* 70:3248-3249.
483. Rock, J. F. 1920. *The leguminous plants of Hawaii.* Hawaiian Sugar Planter's Assn., Honolulu. 234 pp.
484. Rodriguez, A. A. 1965. Possibilities of crop substitution for the coca bush in Bolivia. *Bull. on Narc.* 17 (3):13-23.
485. Roig y Mesa, J. T. 1945. *Plantas medicinales, aromáticas o venenosas de Cuba.* Cultural, S. A., Havana. 870 pp.

486. —— and J. Acuña-Galé, A. Mesa-Esnard, and E. Ledón Ramos, E. 1958. *Apocináceas hipotensoras de Cuba*—I. Estud. Trab. Invest. Serie #5. Inst. Cubano de Invest. Tecn., Havana. 72 pp.
487. Rondina, R. V. D., A. L. Bandoni and J. D. Coussio. 1973. Quantitative determination of morphine in poppy capsules by different spectrophotometry. *J. Pharm. Sci.* 62 (3):502-504.
488. Rosca, M. and V. Cucu. 1972. Naphthalenic components from the cortex of *Rhamnus frangula*. *Farmacia* (Bucharest) 20 (11):695-699. [Rom.] CA 79: 15843g.
489. —— and —— 1975. About a monoglucoside of emodine from the bark of *Rhamnus frangula*. *Pl. Med.* 28:343-345.
490. —— and —— 1975. Naphthaline glycosides from the bark of *Rhamnus frangula*. *Pl. Med.* 28:178-181. [Rom.].
491. Rosecrans, J. A., J. J. Defeo and H. W. Youngken, Jr. 1961. Pharmacological investigation of certain *Valeriana officinalis* L. extracts. *J. Pharm. Sci.* 50 (3): 240-244.
492. Rosenberg, H. 1974. Physostigmine reversal of sedative drugs. *J.A.M.A.* 229 (9): 1168.
493. Rosengarten, F. 1969. *The book of spices*. Livingston Pub'g Co., Wynnewood, Penna. 489 pp.
494. Rosenthal, C. 1963. Vermehrungsmöglichkeiten bei *Colchicum* im hinblick auf die züchterische bearbeitung. *Archief fur. Gartbau* 11 (1):55-65.
495. Rothbaecher, H. and H. Heltmann. 1968. Zur dynamik der biosynthese einiger terpeninhaltsstoffe des oels von *Mentha piperita* L. *Pharmazie* 23 (7):387-388. BA 50:55309.
496. Roubicek, J., Ch. Geiger and K. Abt. 1972. An ergot alkaloid preparation (hydergine) in geriatric therapy. *J. Am. Geriatric Soc.* 20 (5):222-229. BA 54: 50930.
497. Rourke, B. E. 1971. Ghana's contribution to year to year changes in world production of cocoa beans. *Ghana J. Agr. Sci.* 4:3-6.
498. Rowe, T. D. 1940. Effect of fresh *Aloe vera* jell in the treatment of third-degree roentgen reactions on white rats. *J. Amer. Pharm. Assn.* 29:348-350.
499. Roxburgh, W. 1874. *Flora Indica*. Thacker, Spink & Co., Calcutta. 763 pp.
500. Roy, A. C. and B. Mukerji. 1958. Kurchi alkaloids: their isolation, constitution and biological activity. *J. Sci. Indus. Res.* 17A (4):158-164.
501. Rubio, J. F. 1967. El alambrillo, fuente de ingresos adicionales para la finca. *Rev. Cafetalera* 75:27-28.
502. Russell, G. F. and K. V. Olson. 1972. Volatile constituents of oil of thyme. *J. Food Sci.* 37 (3):405-407.
503. Saetre, H., G. Ahlmark and G. Ahlberg. 1974. Haemodynamic effects of ajmaline in man. *Eur. J. Clin. Pharmacol.* 7 (4):253-257. BA 58:62692.
504. Sagel, J. and R. Matisonn. 1972. Neuropsychiatric disturbance as the initial manifestation of digitalis toxicity. *S. Afr. Med. J.* 46 (17):512-514. BA 55:348-349.
505. Saihoo, P. 1963. The hill tribes of northern Thailand and the opium problem. *Bull. on Narc.* 15 (2):35-45.
506. Salkin, R. 1971. *Cathartic cascarosides from cascara bark*. U.S. 3,627,888 (Cl. 424/195; A 61k) 14 Dec. 1971; Appl. 757,100, 03 Sep. 1968. 3 pp. CA 76: 103749z.

507. Salser, J. K., Jr. 1970. Cubeo acculturation to coca and its social implications. *Econ. Bot.* 24 (2):182-186.
508. Sargent, C. S. 1933. *Manual of the trees of North America (exclusive of Mexico).* Houghton Mifflin Co., New York. 910 pp.
509. Saunders, C. F. 1934. *Useful wild plants of the United States and Canada.* Robert M. McBride & Co., New York. 283 pp.
510. Savonius, K. 1973. Isolation and identification of emodin glucoside B from *Frangula alnus. Farm. Aikak.* 82 (9-10):136-139. CA 81:41319b.
511. Sawer, J. Ch. 1894. *Odorographia.* Gurney and Jackson, London. 533 pp.
512. Schachat, R. E. and M. Glicksman. 1959. Furcellaran, a versatile seaweed extract. *Econ. Bot.* 13 (4):365-370.
513. Schauenberg, P. and F. Paris. 1969. *Guide des plantes médicinales.* Delachaux et Niestle S. A., Neuchatel, Switz. 355 pp.
514. Schenk, G. 1955. *The book of poisons.* (Trans. from the German by M. Bullock). Rinehart & Co., Inc., New York. 310 pp.
515. Schneider, G., G. Unkrich and P. Pfaender. 1968. Die methoxylierten flavone von *Citrus aurantium* L. ssp. *amara* L. *Arch. Pharm. Ber. Deut. Pharm. Ges.* 301 (10):785-792. BA 50:10621.
516. Schultes, R. E. 1957. A new method of coca preparation in the Colombian Amazon. *Harvard Univ. Bot. Mus. Leaflets* 17 (9):241-246.
517. Science Service. 1943. Southern pine stump and pine oil. *Science* 97 (2514): 10-11.
518. Segal, A., J. A. Taylor and J. C. Eoff, III. 1968. A re-investigation of the polysaccharide material from *Aloe vera* mucilage. *Lloydia* 31 (4):423-424 [abs.]
519. Selby, H. H. and W. H. Wynne. 1973. Agar. Pp. 29-48. In: *Industrial Gums.* Edited by R. L. Whistler. 2nd ed. Academic Press, New York.
520. Sethi, H., B. K. Sharma and R. M. Beri. 1965. Optimum stage of growth for harvesting the leaves of Indian belladonna cultivated in the Chakrata Hills. *Indian Forester* 91 (10):751-753.
521. Sharghi, N. and I. Lalezari. 1967. *Papaver bracteatum* Lindl., a highly rich source of thebaine. *Nature* 213:1,244.
522. Shibata, S. and T. Saitoh. 1968. Chemical studies on the oriental plant drugs. XIX. Some new constituents of licorice root: The structure of licoricidin. *Chem. Pharm. Bull. Tokyo* 16 (10):1932-1936.
523. Shivpuri, D. N. and K. L. Dua. 1963. Allergy to papaya tree (*Carica papaya* L.) *Ann. of Allergy* 21 (3):139-144.
524. Shostenko, Yu. V., S. Kh. Mushinskaya, E. Vysotskaya, E., Cherkashina and N. G. Bozhko. 1973. *Codeine from the pods of the opium poppy.* USSR (patent) 372,222 (Cl. C 07d, A 61k), 01 Mar. 1973; Appl. 23 Mar. 1970. CA 79: 63638r.
525. Shuljgin, G. 1969. Cultivation of the opium poppy and the oil poppy in the Soviet Union. *Bull. on Narc.* 21 (4):1-8.
526. Sievers, A. F. 1930. *American medicinal plants of commercial importance.* Misc. Pub. 77. USDA, Washington, D.C. 74 pp.
527. Silva, S., J. Vicente-Chandler and F. Abruña. 1969. Field losses of coffee and improved harvesting methods for intensively managed plantations. *J. Agr. Univ. Puerto Rico* 53 (4):268-273.
528. Simonin, C. 1960. Double fatal poisoning after the therapeutic ingestion of leaves of *Colchicum automnale* [sic] on the prescription of a 'curer.' *J. For. Med.* 7 (4):184-185.

529. Singh, C. 1969. Poppy cultivation for opium. *World Crops* 21:270.
530. Singh, L. B. and R. D. Tripathi. 1957. Studies in the preparation of papain. I & II. *Indian J. Hort.* 14 (2):77-82; (3):141-144 & 6 figs.
531. ——— and ——— 1961. A new method of papain production. Pp. 90-92. In: *Ann. Rpt. Year. 1960.* Hort. Res. Inst., Saharanpur, U. P., India.
532. ——— and ——— 1962. Studies in the preparation of papain. VI *Horticultural Adv.* 6 (Feb.):64-70.
533. Singh, P. 1967. Comparative diagnostic characters of leaves of medicinal *Digitalis* of USSR. *Quart. J. Crude Drug Res.* 7 (1):967-981.
534. Singleton, V. L. and F. H. Kratzer. 1969. Toxicity and related physiological activity of phenolic substances of plant origin. *J. Agr. Food Chem.* 17 (3):497-512.
535. Sivetz, M. 1972. How acidity affects coffee flavor. *Food Tech.* 26 (5):70-77.
536. Smirnova, N. D., N. I. Libizov, F. A. Vereshchako, N. N. Nesterov and N. F. Bezukladnikova. 1968. Biokhimicheskoe izuchenie maperstyanki sherstistor (*Digitalis lanata* Ehrb.) *Farmatsiya* 17 (6):33-39.
537. Smith, D. 1974. Insect and other pests of *Duboisia*. *Queensland Agr. J.* 100 (6):243-251.
538. Smith, T. W. and E. Haber. 1973. Digitalis. 3rd of 4 parts. *N. Eng. J. Med.* 289 (20):1063-1072.
539. ——— and ——— 1973. Digitalis. 4th of 4 parts. *N. Eng. J. Med.* 289 (21):1125-1129.
540. Snyder, B. D., L. Blonde and W. R. McWhirter. 1974. Reversal of amitriptyline intoxication by physostigmine. *J.A.M.A.* 230 (10):1433-1435.
541. Sobti, S. N., K. L. Handa and I. C. Chopra. 1957. Propagation of *Rauvolfia serpentina* in Jammu and Kashmir. *J. Sci. Indus. Res.* 16A (6):268-269.
542. Soderholm, P. K., M. H. Gaskins, V. E. Green, Jr., G. A. White, J. W. Garvin and C. C. Seale. 1968. Yield trials of steroid-producing *Dioscorea* on Florida's Everglades peat soils. *Econ. Bot.* 22 (1):80-83.
543. Srinivasulu, C. and S. N. Mahapatra. 1971. Modified process for the isolation of hecogenin from *Agave sisalana*. *Res. Ind.* 16 (3):183-184. CA 77:45207d.
544. Standley, P. C. 1922. *Trees and shrubs of Mexico.* Contrib. U.S. Nat'l Herb. V. 23, Pt. 2. Smithsonian Inst., Washington, D.C. Pp. 171-515.
545. ——— 1924. *Trees and shrubs of Mexico.* Contrib. U.S. Nat'l Herb. V. 23, Pt. 4. Smithsonian Inst., Washington, D.C. Pp. 849-1312.
546. ——— 1928. *Flora of the Panama Canal Zone.* Contrib. U.S. Nat'l Herb. Vol. 27, Smithsonian Inst., U.S. Nat'l Mus., Washington, D.C. 416 pp.
547. ——— 1930. *Flora of Yucatan.* Bot. Ser. Vol. 3, No. 3. Pub. 279. Field Mus. Nat. Hist., Chicago. 492 pp.
548. ——— 1938. *Flora of Costa Rica.* Pt. 3. Bot. Ser. V. 18, Pub. 420. Field Mus. Nat. Hist., Chicago. Pp. 783-1133.
549. ——— and J. A. Steyermark. 1946. *Flora of Guatemala.* Fieldiana: Botany, Vol. 24, Pt. 4 (Pub. 577). Chicago Nat. Hist. Mus., Chicago. 493 pp.
550. ——— and L. O. Williams. 1961. *Flora of Guatemala.* Fieldiana: Bot. Vol. 24, Pt. 7, #1. Chicago Nat. Hist. Mus., Chicago. 185 pp.
551. Stanev, D. 1974. *Thymus vulgaris* in Bulgaria. *Rastenievud* 11 (2):29-33. [Bulg.] CA 81:111395z.
552. Stecher, P. G., Editor. 1968. *The Merck index: an encyclopedia of chemicals and drugs.* 8th ed. Merck & Co., Rahway, N.J. 1,713 pp.

553. Stefanov, Zh., E. Mermerska and N. Stoyanov. 1972. Morphine content of some *Papaver somniferum* (poppy) varieties. *Tr. Nauchnoizsled. Khim.-Farm. Inst.* 8:157-161. [Bulg.] CA 78:156663q.
554. Stehle, H. and M. Stehle. 1962. *Flore médicinale illustrée.* Vol. IX. Flore Agron. des Antilles Francaises. Anibal Lautric, Pointe-a-Pitre, Guadeloupe. 184 pp.
555. Steinmetz, E. F. (undated) *Materia medica vegetabilis.* 3 vols. Dr. E. F. Steinmetz, Amsterdam. 479 pp.
556. ——— 1959. *Drug guide.* Dr. E. F. Steinmetz, Amsterdam. 212 pp.
557. ——— 1965. Ouabain. *Quart. J. Crude Drug Res.* 5 (1):683-685.
558. ——— 1967. Personal communication, Jan. 2nd.
559. Steyn, D. G. 1934. *The toxicology of plants in South Africa.* (S. Afr. Agr. Ser. Vol. XIII). Central News Agency, Ltd., Johannesburg. 631 pp.
560. Stockberger, W. W. 1935. *Drug plants under cultivation.* FB 663, rev'd. USDA, Washington, D.C. 37 pp.
561. Stockdale, F. A. 1926. *The cultivation of papaya and the preparation of papain.* Leaflet 44. Dept. of Agr., Colombo, Ceylon. 5 pp.
562. Stolinsky, D. C., D. L. Bogdon, R. P. Pugh, J. Braunwald, R. Hestorff and J. R. Bateman. 1973. Vinblastine (NSC-49842) plus vincristine (NSC-67574) given in intensive 3-day courses as therapy for lymphomas, sarcomas, and other neoplasms. *Cancer Chemother. Rpt.* Pt. I 57 (4):481-484.
563. ———, G. J. Hum, E. M. Jacobs, J. Solomon and J. R. Bateman. 1973. Clinical trial of weekly doses of vinblastine (NSC-49842) combined with vincristine (NSC-67474) in malignant lymphomas and other neoplasms. *Cancer Chemother. Rpt.* Pt. I 57 (4):477-480.
564. Stoloff, L. 1949. Irish moss—from an art to an industry. *Econ. Bot.* 3 (4):428-435.
565. Strauss, A. 1968. Castor bean allergy. *Rev. Inst. Med. Trop.* São Paulo. 10 (6): 342-348.
566. Stuart, G. A. 1911. *Chinese Materia Medica: Vegetable Kingdom.* American Presbyterian Mission Press, Shanghai. 558 pp.
567. Su, S. C. Y. and N. M. Ferguson. 1973. Extraction and separation of anthraquinone glycosides. *J. Pharm. Sci.* 62 (6):899-901.
568. Svoboda, G. H. 1962. The current status of research on the alkaloids of *Vinca rosea (Catharanthus roseus). Lloydia* 25 (4):334-335.
569. ——— 1969. The alkaloids of *Catharanthus roseus* G. Don in cancer chemotherapy. Pp. 303-335. In: *Current Topics in Plant Science.* Edited by J. E. Gunckel. Academic Press, New York. 461 pp.
570. ——— and D. A. Blake. 1965. The phytochemistry and pharmacology of *Catharanthus roseus* (L.) G. Don. Pp. 45-83. In: *The Catharanthus Alkaloids.* Edited by W. I. Taylor and N. R. Farnsworth. Marcel Dekker, Inc., New York. 323 pp.
571. Swamy, A. V., S. Kalyanasundaram and M. Balagopal. 1965. The plant that gives us jalap. *Indian Farming* 14 (12):30-31.
572. Sylvain, P. G. 1972. The problem of caffeine content in coffee. *Indian coffee* 36 (11-12):393-399.
573. Synge, P. M., Editor. 1969. *The Royal Horticultural Society Dictionary of Gardening.* 2nd ed. Supplement. Clarendon Press, Oxford, Eng. 554 pp.
574. Täckholm, V. 1956. *Students' flora of Egypt.* Anglo-Egyptian Bookshop, Cairo. 649 pp.

575. Takacs, E. A. and C. B. Davis. 1973. *Catalytic isomerization of terpenes*. Ger. Offen. 2,213,055 (Cl. CO7c), 20 Sep 1973. Appl. P22, 13, 055.8, 17 Mar. 1972. 22 pp.

576. Taylor, N. 1943. *Quinine: the story of cinchona*. Reprinted from Sci. Monthly for July 1943; published by Amer. Assn. Adv. Sci., Smithsonian Inst., Washington, D.C. 18 pp.

577. Taylor, W. I. and N. R. Farnsworth. 1975. The *Catharanthus alkaloids: Botany, Chemistry, Pharmacology and Clinical Use*. Marcel Dekker, Inc., New York. 322 pp.

578. Thieme, H. and R. Benecke. 1967. Isolierung eines neuen phenolglykosids aus *Populus nigra* L. *Pharmazie* 22 (1):58-59.

579. Thudichum, J. L. W. 1885. *The coca of Peru and its immediate principles: their strengthening and healing powers*. Bailliere, Tindall & Cox, London. 41 pp.

580. Toffoli, F. and U. Avico. 1965. Coca paste. Residues from the industrial extraction of cocaine, ecgonine and anhydroecgonine. *Bull. on Narc.* 17 (4):27-36.

581. Tookey, H. L., G. F. Spencer and M. D. Grove. 1975. Effects of maturity and plant spacing on the morphine content of two varieties of *Papaver somniferum* L. *Bull. on Narc.* 27 (4):49-57.

582. Tovo, S. 1967. Un caso de avvelenamento mortale da colchico. *Minerva Med.* 87 (6):283-287.

583. Townsend, C. C. and E. Guest. 1966. *Flora of Iraq*. Vol. II. Min. of Agr., Rep. of Iraq, Baghdad. 182 pp.

584. Trease, G. E. and W. C. Evans. 1972. *Pharmacognosy*. 10th ed. Williams and Wilkins Co., Baltimore. 795 pp.

585. Trejo-Gonzalez, A. and C. Wold-Altamirano. 1973. New method for the determination of capsaicin in *Capsicum* fruits. *J. Food Sci.* 38 (2):342-344.

586. Tucakov, J. 1971. The dynamics of alkaloid accumulation in the leaves of belladonna (*Atropa belladonna* L.) during various phases of vegetation. *Glas. Srp. Akad. Nauka Umet. Od. Med. Nauka.* 279 (23):55-59. [Yug.] BA 54:44901.

587. Uline, E. B. 1897. Dioscoreae Mexicanae. *Bot. Jahrb.* 22:424.

588. Upadhyay, S. N. and D. K. Chattoraj. 1972. Microelectrophoretic study of the conformational change of gum tragacanth on the solid-liquid interface. *Ind. J. Biochem. & Biophysics* 9:17-20.

589. Van Hulle, C., P. Braeckman and M. Vandewalle. 1971. Isolation of two new flavonoids from the root of *Glycyrrhiza glabra* var. *typica*. *Pl. Med.* 20 (3): 278-282.

590. Van Wauwe, J. P., F. G. Loontiens, and C. K. deBruyne. 1973. Interaction of *Ricinus communis* haemagglutinin with polysaccharides and low molecular weight carbohydrates. *Biochim. Biophys. Acta* 313 (1):99-105.

591. Van Wyk, J. J. 1966. Sisal—look before you leap. *Farming in S. Afr.* 42 (2): 49-50.

592. Varadarajan, P. D. 1963. Climate and soil conditions of *Rauwolfia serpentina* in India. *Econ. Bot.* 17 (2):133-138.

593. Vaughan, J. G. and E. I. Gordon. 1973. Taxonomic study of *Brassica juncea* using the techniques of electrophoresis, gas-liquid chromatography and serology. *Ann. Bot.* (London) 37 (149):167-184.

594. Vechkanova, L. D., A. I. Ban'kowskii and A. N. Ban'kovskaya. 1972. A method of quantitative determination of ergometrine in ergot sclerotia. *Khim. Prir. Soedin.* (Tashk.) 8 (4):483-487. BA 56:38806.

595. Verdcourt, B. and E. C. Trump. 1969. *Common poisonous plants of East Africa.* Collins, St. James' Place, London. 254 pp.
596. Verhulst, H. L. and L. A. Page. 1961. The small roadside emetic plant and accidental poisoning. Nat'l Clearing House for Poison Control Centers. [bulletin]. Mar.-Apr. 7 pp.
597. Verma, U. P. 1956. Properties and therapeutic value of papain. *Indian Food Packer* 10 (2):9-10, 26.
598. Vietti, T. J., K. Starling, J. R. Wilbur, D. Lonsdale and D. M. Lane. 1971. Vincristine, prednisone, and daunomycin in acute leukemia of childhood. *Cancer* (Philadelphia) 27 (3):602-607.
599. Vincent, P. G., C. E. Bare and W. A. Gentner. 1976. Rapid semi-quantitative spot test for determination of thebaine and differentiation of *Papaver bracteatum* from *P. orientale* and *P. pseudo-orientale. Lloydia* (39) 11:76-78.
600. Wagner, H., R. Schaette, L. Hörhammer and J. Hölzl. 1972. Valepotriat—und ätherischer ölgehalt bei *Valeriana officinalis* L. s. l. in abhängigkeit von verschiedenen exogenen und endogenen factoren. *Arzneim.-Forsch.* 22:1204-1209.
601. Wahlenberg, W. G. 1946. *Longleaf pine: its use, ecology, regeneration, protection, growth and management.* Chas. Lathrop Pack For. Found. and USDA For. Serv., Washington, D.C. 429 pp.
602. Walker, G. T. 1968. Balsam of Peru. *Perf. & Ess. Oil Rec.* 59 (10):705-707.
603. Walker, A. R. and R. Sillans. 1961. *Les plantes utiles de Gabon* (Encyc. Biologique LVI). Editions Paul LeChevalier, Paris. 614 pp.
604. Watt, G., Editor. 1899. *The use of Indian henbane as an intoxicant in Sind.* The Agricultural Ledger #5. 13 pp. (Med. & Chem. Ser. #13: Narcotics).
605. Watt, J. M. and M. G. Breyer-Brandwijk. 1962. *Medicinal and poisonous plants of southern and eastern Africa.* 2nd ed. E. & S. Livingstone, Ltd., Edinburgh. 1457 pp.
606. Webb, L. J. 1948. *Guide to the medicinal and poisonous plants of Queensland.* Bull. 232. Coun. for Sci. & Indus. Res., Melbourne, Aust. 202 pp.
607. Wei, C. H. 1973. Two phytotoxic anti-tumor proteins: ricin and abrin. Isolation, crystallization, and preliminary x-ray study. *J. Biol. Chem.* 248 (10):3745-3747.
608. Weiss, Y., R. E. Merceron, A. Sobel, M. Safar and P. Milliez. 1973. Comparison of the antihypertensive effects of alpha-methyl dopa, clonidine, guanethidine and reserpine. I. Clinical study. *J. Pharmacol. Clin.* 1 (1):24-29. BA 57: 68252.
609. Wellman, F. L. 1945. Balsam of Peru from El Salvador. *Agr. in the Amer.* 5 (5):86-88.
610. Whistler, R. L., Editor. 1973. *Industrial Gums: Polysaccharides and their derivatives.* 2nd ed. Academic Press, New York and London. 807 pp.
611. Wibaut, J. P. and U. Hollstein. 1957. Investigation of the alkaloids of *Punica granatum* L. *Arch. Biochem. & Biophys.* 69:27-32.
612. Wichlinski, L. 1971. *Separation of ergotamine from the ergotoxine of ergot crude alkaloids.* Pol. 64,158 (Cl. C 07d), 10 Nov. 1971. Appl. 11 Apr. 1969. 3 pp. CA 76:153988x.
613. Wichtl, M., A. Nikiforov and S. Sponer. 1973. Structure of *Gelsemium* alkaloid A. Mass spectrometry of *Gelsemium* alkaloids. *Monatsh. Chem.* 104 (1):87-89. CA 78:124787s.

614. ———, ———, G. Schulz and S. Sponer. 1973. Structure elucidation of *Gelsemium* alkaloid C (14-hydroxygelsemicin) by NMR and mass spectrometry. *Monatsh. Chem.* 104 (1):99-104. CA 78:124788t.
615. Wilcox, E. V. and W. McGeorge. 1912. *Sisal and the utilization of sisal waste.* Press Bull. 35. Hawaii Agr. Exper. Sta., Honolulu. 24 pp.
616. Willaman, J. J. and Hui-lin Li. 1970. Alkaloid-bearing plants and their contained alkaloids 1957-1968. *Lloydia* 33 (3A) [supplement]:1-286.
617. ——— and B. G. Schubert. 1961. *Alkaloid-bearing plants and their contained alkaloids.* Tech. Bull. 1234. USDA, ARS, Washington, D.C. 287 pp.
618. Williams, L. O. 1960. *Drug and condiment plants.* Agr. Handbook 172. USDA, ARS, Washington, D.C. 37 pp.
619. Williams, R. O. and R. O. Williams, Jr. 1951. *Useful and ornamental plants of Trinidad and Tobago.* 4th ed. Guardian Com'l Pty., Port-of-Spain, Trin. 335 pp.
620. Williamson, J. 1955. *Useful plants of Nyasaland.* Gov't Ptr., Zomba, Nyasaland. 168 pp.
621. Winters, H. F. 1950. *Cinchona propagation.* Bull. 47. USDA, Fed. Exper. Sta. in Puerto Rico, Mayaguez. 26 pp.
622. Wintersteiner, O. and J. D. Dutcher. 1943. Curare alkaloids from *Chondodendron tomentosum. Science* 97 (2525):467-470.
623. Woodward, R. B. 1965. A total synthesis of colchicine. *Harvey Lectures* 59:31-47.
624. Wren, R. W. 1970. *Potter's new cyclopaedia of botanical drugs and preparations.* 7th ed. Health Science Press, Rustington, England. 400 pp.
625. Wright, A. E. 1958. The battle against opium in Iran. *Bull. on Narc.* 10 (2):8-11.
626. Wright, D. G. and S. E. Malawista. 1973. Mobilization and extracellular release of granular enzymes from human leukocytes during phagocytosis. Inhibition by colchicine and cortisol but not by salicylate. *Arthritis Rheum.* 16 (6):749-758. BA 58:9542.
627. Wu, C-H., E. M. Zabawa and P. M. Townsley. 1974. Single cell suspension culture of the licorice plant, *Glycyrrhiza glabra. Can. Inst. Food Sci. Technol. J.* 7 (2):105-109. CA 81:166295x.
628. Yakobashvili, N. Z. 1971. Essential oil industry of the Georgian SSR and prospects for its development. *Mezhdunar. Kongr. Efirnym Maslam.* (Mater.) 4th (1):11-18. CA 79:83387m.
629. Yang, T.-H., K-T. Chen. L-H. Wu and K-H. Kwong. 1972. Alkaloid content in differently harvested biennial *Rauwolfia serpentina* trial planting in Taiwan. *Tai-wan Yao Hsueh Tsa Chih* 24 (2):82-85. [Ch.] CA 81:23070j.
630. Yee, W. *et al.* 1970. *Papayas in Hawaii.* Cir. 436. Univ. of Hawaii, Coop. Exten. Serv., Honolulu. 57 pp.
631. Youngken, H. W. 1947. Ergot—a blessing and a scourge. *Econ. Bot.* 1 (4):372-380.
632. ——— and J. K. Karas. 1964. *Common poisonous plants of New England.* PHS Pub. 1220. Rhode Isl. Health Dept. & US HEW, Pub. Health Serv. 23 pp.
633. Zbinden, G., M. Hegele and L. Grimm. 1972. Toxicological effects of vincristine sulfate on blood platelets. *Agents Actions* 2 (5):241-245. BA 56:34846.
634. Zinkel, D. F. 1975. Chemicals from trees. *Chemtech.* Apr.: 235-241; and letter of May 16, 1975.

635. Zwaving, J. H. 1965. Trennung und isolierung der anthrachinonglykoside von *Rheum palmatum*. Pl. Med. 13 (4):474-484.

ADDENDUM:

Anon. 1974. *Markets for selected medicinal plants and their derivatives.* [*Catharanthus roseus*; *Cinchona* bark; *Datura metel*; *Dioscorea* and diosgenin; Ipecacuanha root; Liquorice; Papain; *Rauwolfia*; Valerian root] International Trade Centre UNCTAD/GATT, Geneva. 192 pp.

Index

A

a-allocryptopine, 363
a, B-hexenic acid, 272
Abrin, 194
Abrus precatorius, 194
Acacia arabica, 140
Acacia laeta, 141
Acacia senegal, 136
Acacia seyal, 141
Acacia verek, 137
Acetic acid, 354, 380
acetic aldehyde, 379, 380
Acetylcholine betaine, 6
Acetyldiginatin, 320
Acetyldigitoxin, 320
Acetyldigoxin, 320
Acetyleugenol, 367
Acetylgitaloxin, 320
Acetylgitoxin, 320
Acetyltrimethylcolchicinic acid, 54
Acocantherin, 233
Acocanthin, 234
Acokanthera friesiorum, 233
Acokanthera schimperi, 233
Acolongifloroside G, 234
Acolongifloroside H, 234
Acolongifloroside K, 234, 260
Acolongifloroside K acetate, 234
a-colubrine, 368
Aconine, 362
Aconitine, 362
Aconitum napellus, 362
Acovenoside A, 234
Adenosine, 238
Aden senna, 151
a-elaterin, 370
Aframomum, 261
African rauwolfia, 255
African serpent-wood, 254
Agar, 372-373
Agar, Danish, 373
Agathosma betulina, 365
Agathosma crenulata, 365
Agavaceae, 65, 67
Agave family, 67
Agave fourcroydes, 68
Agave rigida var. *sisalana*, 67

Agave sisalana, 66, 67
Agmatine, 6
Ailanthus, 290
Ajmalicine, 238, 244-245, 252
Ajmaline, 244, 248, 252, 255
Ajmalinine, 244-245
Ajowan, 279
Alambrillo, 76
Albaspidin, 359
Aldehyde C_{10}, 380
Aldobiouronic acid, 212, 325-326
Alexandrian senna, 151
Alfavaca, 187
Algae, 372
Algin, 372, 373
Alkaloid A, 255
Alligator tree, 129
Alloyohimbine, 244
Allspice, 379
Allyl isothiocyanate, 364
4-allyl-2-methoxyphenol, 367
Almond, 376
Almond oil, 376
Aloe, 47-50
Aloe barbadensis, 47-48
Aloe-emodin, 48, 147, 202, 204
Aloe-emodin-anthrone-diglucoside, 147
Aloe-emodin-8-glucoside, 147
Aloe ferox, 47
Aloe perryi, 47
Aloe succotrina, 47
Aloe vera, 47
Aloin, 48
Aloinoside, 48
Alpinigenine, 124
Alstonia congensis, 261
Alstonine, 238, 252, 255
Althaea officinalis, 206-207
a-lumicolchicine, 54
Amarogentine, 369
Amatillo, 251
American false hellebore, 59
American hellebore, 58-61
American mandrake, 87
American serpent-wood, 250
American storax, 129
American veratrum, 59
American white hellebore, 59

American wormseed, 362
Ammocallis rosea, 237
Amsonine, 252
Amygdalin, 376
Amygdalus communis, 376
Anabasine, 293
Anamirta cocculus, 92-95
Anamirta paniculata, 93
Anamirtin, 94
Ananas comosus, 38-39
Ananas sativus, 39
Anatolian licorice, 155
Anatolian tragacanth, 143
Andira araroba, 364
Anethole, 21, 374, 380
Angelic acid, 371
Anhydrogalactose, 373, 377
Anhydromannose, 377
1,5-anhydro-sorbitol, 366
Anise, 380
Anise, star, 374
Anisic acid, 380
Anthemis nobilis, 371
Anthostyrax tonkinense, 217
Anthoxanine, 371
Anthranol, 364
Anthraquinones, 48
a-peltatin, 88
a-phellandrene, 381
a-phytosterol, 370
Apigenin, 371
a-pinene, 272, 278, 359, 381
Apocynaceae, 231, 233, 243, 251, 255, 259, 265, 369
Apohyoscine, 294
Apomorphine hydrochloride, 118
Aporeine, 113
Apricot, 376
Apricot kernel oil, 376
Arabian coffee, 353
Arabic acid, 138
Arabica coffee, 353
Arabin, 138
Arabinose, 143, 325
Arachidic acid, 377, 378, 380
Arachis hypogaea, 377
Araroba, 364
Arbutin, 368
Arctostaphylos uva-ursi, 368
Arecaidine, 359
Areca catechu, 359
Arecaine, 359
Arecolidine, 359
Arecoline, 359

Argassi, 233
Aricine, 252
Arnica, 371
Arnica cordifolia, 371
Arnica fulgens, 371
Arnica montana, 371
Arnica sororia, 371
Arnicin, 371
Arnidiol, 371
Arnisterol, 371
Arrow-poison tree, 233-234
Artemisia cina, 371
Artemisia maritima, 371
Arvensis oil, 274
Asafetida, 367
Asaresinol ferulate, 367
Ascaridol, 362
Ascophyllum nodosum, 373
Ascorbic acid, 369
Asiatic storax, 133
Asimina triloba, 223
Asparagine, 156, 208, 284
Aspidiaceae, 359
Aspidium filix-mas, 359
A-storesin, 133
Astragalus adscendens, 145
Astragalus brachycalyx, 145
Astragalus gummifer, 142-143
Astragalus kurdicus, 145
Astragalus leiocladus, 145
Astragalus microcephalus, 145
Astragalus strobiliferus, 145
Astragalus verus, 145
Atabrine, 349
a-terpinene, 278, 379
a-terpineol, 377, 379
Atropa acuminata, 288-289
Atropa belladonna, 282-283, 289
Atropa belladonna var. *acuminata*, 289
Atropine, 284, 286, 294, 299-300, 304
Atropine sulfate, 307-308
a-truxilline, 178
Aucubin, 326, 332
Autumn crocus, 52-57
a-yohimbine, 244
Azulene, 371

B

B-allocryptopine, 363
Balsam of Peru, 164-167
Balsam of tolu, 160-163
Barbados aloe, 47
Barbaloin, 48, 201
Barbasco, 75

Barberry family, 87
Barosma betulina, 365
Barosma crenulata, 365
Barosma serratifolia, 365
Bassorin, 143
B-barbaloin, 48
B-colubrine, 368
6-*B*-*d*-glucuronosido-*d*-galactose, 138
Bearberry, 201, 368
Bearcorn, 59
Bear's oil, 198
Bearwood, 201
Belladonna, 282-286
Benjamin tree, 217
Benzaconine, 362
Benzaldehyde, 218, 367
Benzoic acid, 161, 166, 217, 219
Benzoin, Compound tincture, 50, 130, 134, 219
Benzoin trees, 217-219
Benzyl benzoate, 162, 166
Benzyl cinnamate, 129, 162, 166
Berberastine, 362
Berberidaceae, 87, 362
Berberine, 362, 363
Betaine, 6
Betel nut, 359
Betulaceae, 361
Betula lenta, 361
Big hellebore, 59
Bilsted, 129
Bipindoside, 260
Birch, black, 361
Birch, sweet, 361
Bitter almond, 376
Bitter aloes, 49-50
Bitter bark, 201
Bitter orange, 377
Bitterwood, 365
Black balsam, 165
Black birch, 361
Black datura, 370
Black henbane, 303
Black mustard, 364
Black poplar, 360
Black psyllium, 330
Blight kernels, 5
Blonde psyllium, 325
Bloodroot, 363
Blue mallee, 379
B-lumicolchicine, 54
B-methylaesculetin, 284
Boboró, 251
2-bornanone, 103

Borneol, 18, 105, 359, 360, 374
Bornyl alcohol, 381
Bornyl chloride, 18
Borrachera, 251
B-peltatin, 88
B-phellandrene, 379
B-pinene, 278, 380, 381
Branched plantain, 332
Brassica juncea, 364
Brassica nigra, 364
Brazilian ipecac, 335
Broad-leaved peppermint, 379
Bromelain, 40, 43
Bromeliaceae, 37, 39
Bromelin, 40
Broom pine, 17
Brown bark, 343
Brown-berried cedar, 27
Brown-fruited juniper, 27
Brown strophanthus, 266
Brucine, 368
B-sitosterol, 147, 332, 376
B-storesin, 133
B-truxilline, 178
Buchu, 365
Buckthorn bark, 204
Buckthorn berries, 204
Buckthorn family, 201
Buckwheat, 361
Bugbane, 59
Bugwort, 59
Burseraceae, 365
Butyric acid, 331, 354
B,*y*-hexenyl phenyl acetate, 272
B-yohimbine, 252

C

Cacao, 378
Cadaverine, 6
Cade oil, 28
Cade-oil plant, 27
Cadinene, 18, 104, 359, 377
Cafetannic acid, 354
Caffeic acid, 278, 354
Caffeine, 354-356, 378
Cajeput, 367
Cajeput oil, 367
Calabar bean, 170-173
Calabarine, 171
Calcium oxalate, 148, 336, 362, 365
California giant kelp, 373
Calisaya bark, 343
Caltha palustris, 61
Campanulaceae, 371

Camphene, 18, 104, 272, 278, 359, 360, 371, 377, 380, 381
Camphor, 18, 102-107, 381
Camphora camphora, 103
Camphoracene, 104
Camphora officinarum, 103
Camphor laurel, 103
Camphor oil, 104-105, 107
Canadine, 362
Canescine, 252
Cape aloe, 47
Capsaicin, 369
Capsanthin, 369
Capsicum annuum, 369
Capsicum frutescens, 369
Capsicum minimum, 369
Capsorubin, 369
Carapanaubine, 255
Caraway, 379
Cardamom, 374
Cardinal's bark, 341
Caricaceae, 221, 223
Carica papaya, 223, 227
Carnation poppy, 111
Carob, 377
Carob gum, 377
Carolina yellow jessamine, 368
Carrageen, 373
Carrageenan, 373
Cartagena ipecac, 338
Carum carvi, 379
Carum copticum, 275, 279
Carvacrol, 278, 381
Carveol, 379
Carvone, 379, 381
Caryophyllene, 104, 272, 278, 367, 379
Cascara buckthorn, 201
Cascara sagrada, 200-204
Cascaroside A, 201
Cascaroside B, 201
Cascaroside C, 201
Cascaroside D, 201
Cassia acutifolia, 151
Cassia angustifolia, 146-147, 151-152
Cassia cinnamon, 375
Cassia italica, 152
Cassia lanceolata, 147
Cassia obovata, 152
Cassia obtusa, 152
Cassia senna, 151, 152
Castor bean, 192-198
Castor bean allergen (CBA), 194, 197
Castor oil, 195-196
Castor oil plant, 193

Castor pomace, 198
Catechin, 18, 362, 364, 365
Catechu red, 370
Catechutannic acid, 370
Catharanthus roseus, 236-237
Ceará jaborandi, 187
Cecropia, 181
Cephaëline, 336, 339
Cephaelis acuminata, 335, 338
Cephaelis ipecacuanha, 334-336, 339
Ceratonia siliqua, 377
Cevadine, 60
Ceylon cinnamon, 363
Chalchupa, 251
Chalchupine A, 252
Chalchupine B, 252
Chamomile, Roman, 371
Chandrine, 244-245
Chatinine, 370
Chaulmoogra, 366
Chaulmoogric acid, 366
Chavicol methyl ether, 374
Chelerythrine, 363
Chelilutine, 363
Chelirubine, 363
Chenopodiaceae, 362
Chenopodium ambrosioides, 362
Chenopodium ambrosioides var. *anthelminticum,* 362
Chenopodium quinoa, 181
Cherry, wild black, 376
Chinese rhubarb, 362
Chinese storax, 134
Chittem bark, 201
Chlorogenic acid, 354, 356, 368
Chocolate, 378
Chocolate family, 211
Choline, 6, 266
Chondocurine, 98
Chondodendron, 97
Chondodine, 98
Chondrodendron tomentosum, 96, 97, 99
Chondrus crispus, 373
Chrysaloin, 201
Chrysarobin, 364
Chrysin, 360
Chrysophanic acid, 148, 362
Chrysophanol, 202, 362, 364
Church-flower, 237
Chymopapain, 224
Cinchona, 340-350
Cinchona calisaya, 340-343
Cinchona ledgeriana, 341-343
Cinchona officinalis, 341, 343-344

Cinchona pubescens, 341, 346
Cinchona succirubra, 341, 343, 345-346
Cinchonidine, 343, 348
Cinchonine, 343, 348
Cinchotannic acid, 343
Cineole, 104, 360, 367, 371, 374, 379, 381
Cinnamaldehyde, 363
Cinnamein, 166
Cinnamic aldehyde, 363, 375
Cinnamic acid, 129, 130, 133, 161-162, 218, 363
Cinnamomum burmannii, 375
Cinnamomum camphora, 102-104
Cinnamomum camphora var. *linaloolifera,* 104
Cinnamomum cassia, 375
Cinnamomum loureirii, 363, 375
Cinnamomum zeylanicum, 363, 375
Cinnamom, Cassia, 375
Cinnamon, Ceylon, 375
Cinnamon, Malay, 375
Cinnamon, Saigon, 375
Cinnamyl acetate, 363
Cinnamyl alcohol, 130
Cinnamyl benzoate, 217
Cinnamyl cinnamate, 129, 166, 218
Cinnamylcocaine, 178
Cissampelos pareira, 99
Citral, 378, 379
Citronellol, 275
Citrullol, 370
Citrullus colocynthis, 370
Citrus aurantium var. *amara,* 377
Citrus limon, 377
Citrus sinensis, 378
Clammy psyllium, 331
Claviceps purpurea, 4, 5
Clavicepsin, 6
Clavicipitaceae, 3, 5
Clavine, 6
Clerodendron spp., 249
Clove, 367
Clubmoss, Stagshorn, 359
Cluster bean, 377
Coca, 176-183
Coca bush, 177
Coca citrin, 178
Coca family, 177
Cocaine, 178-183
Cocaine plant, 177
Cocaine tree, 177
Cocatannic acid, 178
Cocculin, 94
Cocculus fructus, 93

Cocculus indicus, 93
Cocculus lacunosus, 93
Cocculus suberosus, 93
Cochlospermum gossypium, 213
Cochlospermum religiosum, 213
Cocoa, 378
Cocoa butter, 378
Cocotombo, 251
Codamine, 113
Codeine, 113, 118-119, 124-125
Coffea arabica, 353-354
Coffea canephora, 353-354
Coffea humboltiana, 354
Coffee, 352-356
Coffeeberry, 201
Coffee-tree, 201
Colchamine, 54
Colchiceine, 54
Colchicerine, 54
Colchicine, 54, 57
Colchicum autumnale, 52, 53
Colchicum luteum, 56-57
Colchicum speciosum, 57
Colocasia, 261
Colocynth, 370
Colocynthin, 370
Colombian bark, 343
Colophony, 21
Comida de culebra, 251
Commiphora abyssinica, 365
Commiphora molmol, 365
Common althea, 207
Common thyme, 277
Common valerian, 370
Compositae, 371
Compound tincture benzoin, 50, 130, 134, 219
Concuressine, 369
Conessimine, 369
Conessine, 369
Congo coffee, 353
Coniferyl benzoate, 217
Convolvulaceae, 369
Convolvulin, 369
Coptisine, 363
Coriander, 380
Coriandrum sativum, 380
Corilagin pyroside, 368
Corkwood, 292
Corn mint, 271
Corrimiento, 76
Cortisone, 80
Corynanthine, 244, 252
Cotarnine, 118
Cotton, short-staple, 378

Cottonseed oil, 378
Coumarin, 166
Crack willow, 360
Creosol, 18
Cresol, 28
Croton oil factor A, 366
Croton tiglium, 366
Crow killer, 93
Crown bark, 343
Crow poison, 59
Cruciferae, 364
Cryptone, 379
Cryptopine, 113
Cubeb, 360
Cubebic acid, 360
Cubebin, 360
Cucurbitaceae, 370
Cuminal, 379
Cuminic aldehyde, 365
Cupressaceae, 25, 27, 359
Curacao aloe, 47
Curare, 98
Curine, 98
Cuscohygrine, 178
Cyamopsis psoralioides, 377
Cyamopsis tetragonoloba, 328, 377
Cyanidin, 364
Cyclopentenyl monocarboxylic acids, 366
Cymbopogon jwarancusa, 275
Cymbopogon nardus, 275
Cymarin, 266
Cymarose, 265
Cymene, 359
Cypress family, 27
Cytisus scoparius, 364

D

Danish agar, 373
d-a-phellandrene, 380
d-a-pinene, 104
D-arabinose, 48
Datura metel, 370
Datura metel var. *fastuosa*, 370
Datura stramonium, 370
d-cadinene, 28
d-camphene, 375
d-camphor, 362, 363
d-chondocurine, 97
D-digitoxose, 314
Deadly nightshade, 283
Dead men's bells, 313
Decylic aldehydes, 378
Dehydrostearic acid, 194
Delphinidin, 364

Delta 9(11)-hecogenin, 68
Demecoline, 57
2-demethylcolchicine, 54
Deoxybarbaloin, 201
6-deoxy-1-galactose, 143
6-deoxy-1-mannose, 212
Deoxyloganin, 238
Desacetyllanatoside C, 320
Desacetyllanatoside D, 320
Deserpidine, 244-245, 252, 253
Desert tea, 36
11-desmethoxyreserpine, 245
3-desmethylcolchicine, 57
Devil pepper, 251
Devil's apple, 87
Devil's bite, 59
Devil's trumpet, 370
d-galactopyranose, 138
d-galacturonic acid, 143, 212
d-galactose, 143, 212, 326, 377
3-*d*-galactoside-1-arabinose, 138
Diacetyl, 379
Dianhydrogitoxigenin, 314
Digicorin, 314
Digilanidase, 320
Diginatin, 320
Digitalin, 314
Digitalinum verum, 320
Digitalis ambigua, 321
Digitalis ferruginea, 321
Digitalis grandiflora, 321
Digitalis lanata, 319-320
Digitalis lutea, 321
Digitalis purpurea, 312-313
Digitonin, 314
Digitoxigenin, 314
Digitoxin, 314, 316, 320
Digoxin, 320
Dihydrocapsaicin, 369
Dihydrocarveol, 381
Dihydrocinnamic acid, 371
Dihydroergocornine, 8
Dihydroergocristine, 8
Dihydroergocryptine, 8
Dihydroergotamine, 8
Dihydroergotoxine, 8
22,23-dihydrostigmasterol, 156
3', 6-diisopentenyl-2', 4', 5-trihydroxy-7-methoxyisoflavan, 156
Dimethylnaphthalene, 28
Dimethoxystrychnine, 368
1,3-dimethyl butadiene, 380
Dimethylmorphine, 113

Dioleomonosaturated glycerides, 378
Dioscoreaceae, 73, 75, 76
Dioscorea composita, 75-82
Dioscorea deltoidea, 81
Dioscorea elephantipes, 82-83
Dioscorea floribunda, 76-82
Dioscorea friedrichsthalii, 82
Dioscorea mexicana, 81
Dioscorea spiculiflora, 81
Dioscorea sylvatica, 82
Dioscorea tepinapensis, 75
Diosgenin, 68, 80
Diosmin, 365
Diosphenol, 365
Dipentene, 104, 365, 375, 377, 380, 381
d-isochondodendrine, 97
d-isochondodendrine dimethylether, 97
dl-*a*-pinene, 380
d-limonene, 359, 365, 377, 378
d-linalool, 380
dl-menthol, 21
dl-sesquiterpene alcohol, 272
dl-terpineol, 378
d-mannose, 377
d-nerolidol, 166
d-3-octanol, 272
Dodecylic aldehydes, 378
Dogbane family, 233, 243, 251, 255, 259
d-pinene, 375, 377, 380
Dryopteris filix-mas, 359
d-tubocurarine, 97
Duboisia Leichhardtii, 294-295, 299-301
Duboisia myoporoides, 292-296
Duckretter, 59
Duck's foot, 87
Dwale, 283
d-xylose, 143, 325-326
Dyer's oak, 361

E

Earth gall, 59
Ecuadorean sarsaparilla, 374
Ecgonine, 178
Egyptian henbane, 306-309
Elagic acid, 368
Elagitannin, 364
Elemicin, 375
Elettaria cardamomum, 374
Elliptine, 252
Elm, 293
Emetamine, 336
Emetine, 336
Emodin, 201, 362
Emulsin, 331

English belladonna, 283
Ephedra, 33-36
Ephedraceae, 31, 33
Ephedra equisetina, 33-34
Ephedra family, 33
Ephedra gerardiana, 32-36
Ephedra intermedia, 33-36
Ephedra major, 33-34
Ephedra nebrodensis, 33
Ephedra sinica, 33-34
Ephedra viridis, 36
Ephedra vulgaris, 33
Ephedrine, 34-36
3-epi-*a*-yohimbine, 244
Epicatechin gallate, 364
Epigallocatechin gallate, 364
Epipodophyllotoxin, 90
Ergobasine, 5
Ergochrysin, 6
Ergocornine, 6
Ergocristine, 6
Ergocryptine, 6
Ergoflavin, 6
Ergometrine, 5, 7
Ergonovine, 5-7
Ergosine, 6
Ergosterol, 6
Ergostetrine, 5-6
Ergot, 4-9
Ergotamine, 6, 7
Ergotic acid, 6
Ergotocine, 5
Ergot of rye, 5
Ergovalide, 6
Ericaceae, 368
Eriodictyol, 380
Eriodictyon californicum, 380
Eriodictyonone, 380
Erysimoside, 266
Erythrocephaleine, 336
Erythroeretin, 362
Erythrophleum guineense, 261
Erythroxylaceae, 175, 177
Erythroxylum bolivianum, 177
Erythroxylum coca, 177
Erythroxylum coca var. *coca*, 177
Erythroxylum coca var. *novogranatense*, 177
Erythroxylum coca var. *Spruceanum*, 177
Erythroxylum novogranatense, 177
Erythroxylum truxillense, 177
Eseramine, 171
Eseridine, 171
Eserine, 171
Ethyl amylcarbinol, 381

Ethyl cinnamate, 129
Ethyl guiacol, 28
Etorphine, 125
Eucalyptol, 374
Eucalyptus citriodora, 275
Eucalyptus dives, 379
Eucalyptus fruticetorum, 379
Eucalyptus macrorryncha, 367
Eucalyptus polybractea, 379
Eucalyptus radiata, 379
Eugenia caryophyllus, 367
Eugenol, 363, 365, 367, 375, 379
Euphorbiaceae, 191, 193, 366
European white hellebore, 63
Exogonium purga, 369
Eye-plant, 293

F

Fagaceae, 361
Fagopyrum esculentum, 361
Fairy cap, 313
Fairy finger, 313
False hellebore, 59
False pareira, 99
Farnesene, 360
Farnesol, 166, 377
Fat pine, 17
Fenchone, 380
Fennel, 380
Fern, male, 359
Ferula assa-foetida, 367
Ferula foetida, 367
Ferula narthex, 367
Ferula rubricaulis, 367
Ferulic acid, 367
Fetid nightshade, 303
Fever tree, 341
Ficin, 40, 230
Ficus glabrata, 230
Figwort family, 313, 319
Filicic acid, 359
Filicin, 359
Filmaron, 359
Fish berry, 93
Fish killer, 93
Flacourtiaceae, 366
Flavaspidic acid, 359
Flax, 377
Flea seed, 331
Fleawort, 331
Foeniculum vulgare, 380
Formononetin, 156
Foxglove, 312-316
Fragilin, 360

Fragrant maple, 134
Frangula bark, 204
French psyllium, 331, 332
Friar's balsam, 219
Fruta bomba, 223
Fungisterol, 6
Furcellaran, 373
Furcellaria fastigiata, 373
Furfural, 238, 272, 367, 379

G

Galactose, 138, 325, 373
Galacturonic acid, 325
Gallic acid, 360, 362, 368
Gallitannin, 364
Gall oak, 361
Gallocatechin, 364
Gallotannic acid, 361
Gambier, 370
Gambir, 370
Gamboge, 366
Gambogic acid, 366
Garcinia hanburyi, 366
Garden heliotrope, 370
Garcinia kola, 261
Garden thyme, 277
Gaultheria procumbens, 368
Gelidium corneum, 372
Gelidium amansii, 372
Gelidium cartilaginium, 372
Gelidium liatulum, 372
Gelidium lingulatum, 372
Gelidium pacificum, 372
Gelidium pristoides, 372
Gelidium sesquipedale, 372
Gelsedine, 368
Gelsemicine, 368
Gelsemine, 368
Gelsemidine, 368
Gelseminine, 368
Gelsemium sempervirens, 368
Gelsevirine, 368
Geneserine, 171
Genisteine, 364
Gentiamarin, 369
Gentiamarine, 369
Gentianaceae, 369
Gentiana lutea, 369
Gentianic acid, 369
Gentianin, 369
Gentianose, 369
Gentian, yellow, 369
Gentiin, 369
Gentiopicrine, 369

Gentisic acid, 369
Gentisin, 369
Gentrogenin, 68
Georgia pine, 17
Geraniol, 360, 377, 379, 380
Germerine, 63
Germidine, 60
Germine, 60
Germitrine, 60
Gigartina mammillosa, 373
Gigartina stellata, 373
Ginger, 360
Gingerol, 360
Gitaloxigenin, 314
Gitaloxin, 314, 320
Gitonin, 314
Gitorin, 320
Gitoxigenin, 314
Gitoxin, 314, 320
Gloriogenin, 68
Gloriosa superba, 94
Glucogitaloxin, 314, 320
Glucotropaeolin benzyl isothiocyanate, 225
Glucuronic acid, 156
Glycerides, 378
Glycestrone, 156
Glycosmin, 360
Glycyramarin, 156
Glycyrrhetic acid, 156
Glycyrrhetinic acid, 156-157
Glycyrrhiza echinata, 158
Glycyrrhiza glabra, 154-155
Glycyrrhiza glabra var. *glandulifera*, 155
Glycyrrhiza glabra var. *typica*, 155
Glycyrrhiza glabra var. *violacea*, 155
Glycyrrhiza uralensis, 158
Glycyrrhizic acid, 155
Glycyrrhizin, 155
Gnoscopine, 113
Goat's thorn, 143
Golden root, 335
Goldenseal, 362
Gorlic acid, 366
Gossypium hirsutum, 378
Gracilaria confervoides, 372
Grandidentatin, 360
Gray bark, 343
Great scarlet poppy, 123
Grecian foxglove, 319-321
Green agave, 67
Green dragon, 143
Green hellebore, 59
Green strophanthus, 264-267
Green veratrum, 59

Ground lemon, 87
Groundnut, 377
G-strophanthin, 233, 259
Guaiacol, 18, 28
Guar, 328, 377
Guar gum, 377
Guataco colorado, 251
Gulu, 211
Gum acacia, 136-141
Gum arabic tree, 137
Gum Benjamin, 217
Gum camphor, 103
Gum dragon, 143
Gum tragacanth, 143
Guttiferae, 366
Guvacine, 359
Guvacoline, 359

H

Hamamelidaceae, 127, 129, 364
Hamamelis virginiana, 364
Hamamelitannin, 364
Hard pine, 17
Heart pine, 17
HCN (hydrocyanic acid), 376
Hecogenin, 68-69
Heliotropin, 105
Hellaridine, 284
Henbane, 302-305
Hendecanal, 377
Henequen, 68
Hentriacontane, 370
Herniarin, 156
Heroin, 113, 117-118
Heterophyllin, 252, 255
Hexadex-9-enoic acid, 359
Higuereta, 193
Higuerilla, 193
Himalayan may apple, 89
Histamine, 6
Histidine, 6
Hockle elderberry, 93
Hog apple, 87
Hog bean, 303
Hog gum, 213
Holarrhena antidysenterica, 369
Holarrhimine, 369
Holonamine, 369
Homocapsaicin, 369
Homodihydrocapsaicin, 369
Honduras sarsaparilla, 374
Horn seed, 5
Horsemint, 279
Horsetail kelp, 373

Hydergine, 8
Hydnocarpic acid, 366
Hydnocarpus kurzii, 366
Hydnocarpus laurifolia, 366
Hydnocarpus wightiana, 366
Hydrastidaceae, 362
Hydrastine, 362
Hydrastis canadensis, 362
Hydrocortisone, 80
Hydrocotarnine, 113
Hydrophyllaceae, 380
5-hydroxy-3, 4-bis (3,4-methylene-dioxybenzyl) tetrahydrofuran, 360
14-hydroxygelsemicine, 368
28-hydroxyglycyrrhetic acid, 156
6-hydroxyoleanolic acid, 218
19-hydroxyoleanolic acid, 217
Hydroxysenegenin, 366
Hygrine, 178
Hygroline, 178
Hyoscine, 284, 293-294, 300-301, 304, 308, 370
Hyoscyamine, 284, 289, 293-294, 304, 308, 370
Hyoscyamus muticus, 295, 307-308
Hyoscyamus niger, 302-303, 308
Hypocreaceae, 5

I

Illiciaceae, 374
Illicium verum, 374
Indian apple, 87
Indian atropa, 289
Indian balsam, 165
Indian belladonna, 288
Indian berry, 93
Indian coccles, 93
Indian henbane, 303, 307
Indian may apple, 89
Indian mustard, 364
Indian plantago, 325
Indian poke, 59
Indian rhubarb, 362
Indian senna, 146
Indian tobacco, 371
Indian tragacanth, 211
Indian uncus, 59
Indole, 377
Invertase, 331
Ipecac, 334-339
Ipecacuanic acid, 336
Ipecacuanhin, 336
Ipecoside, 336
Ipomoea orizabensis, 369

Ipomoea purga, 369
Ipuranol, 369
Ipurganol, 369
Irish moss, 373
Isha, 259
Isoajmaline, 244
Isoamylamine, 6
Isobarbaloin, 48
Isoborneol, 18
Isobornyl acetate, 18
Isobutylpropanyl disulfide, 367
Isoconessimine, 369
Isoelemicin, 375
Isoguvacine, 359
Isoliquiritigenin, 156
Isoliquiritin, 156
Isoliquiritoside, 156
Isolobinine, 371
Isomenthol menthyl acetate, 381
Isomenthone menthofuran, 381
Isopelletierine, 293, 366
Isophedrine, 34
Isophysostigmine, 172
Isopilocarpine, 187
Isopilosine, 187
Isoporoidine, 293
4-isopropyl salicylaldehyde, 379
Isorauhimbine, 244
Isoraunescine, 252
Isoreserpiline, 252, 255
Isoreserpinine, 252
Isorhamnetin, 148
Isoricinoleic acid, 194
Isoyohimbine, 244
Ispaghula, 325
Italian senna, 152
Itch weed, 59

J

Jaborandi, 187
Jaborandine, 187
Jalap, 369
Jalapin, 369
Jamaica ginger, 360
Jamaica quassia, 365
Jamaica senna, 152
Japanese camphor tree, 103
Japanese mint, 270-275
Japanese pagoda tree, 364
Japanese peppermint oil, 274
Jervine, 60, 63
Jesuit's bark, 341
Jesuit's powder, 341
Jimsonweed, 370

Joint fir, 33
Juniper, 359
Juniper berry oil, 29
Juniper tar oil, 28
Juniperus communis, 359
Juniperus oxycedrus, 26-29

K

Kadira gum, 211
Kaempferin, 148
Kaempferol, 148, 256
Karaya, 210-213
Karaya gum, 211-213
Katira gum, 213
Keliot, 233
Kelp, California giant, 373
Kelp, horsetail, 373
Kibai, 233
Knobbed wrack, 373
Kombé, 265
Kombé arrow poison plant, 265
Kordofan gum, 138
Krameriaceae, 365
Krameria triandra, 365
Krameric acid, 365
K-strophanthin, 265, 266
K-strophanthoside, 265
Kulu, 211
Kurchessine, 369
Kurchi, 369
Kurchicine, 369
Kurcholessine, 369

L

Labiatae, 269, 271, 277, 381
Lady's thimble, 313
Laetrile, 376
Lamiaceae, 271, 381
Laminaria digitata, 373
Laminaria hyperborea, 373
Laminaria saccharina, 373
Lanatoside A, 319
Lanatoside B, 319
Lanatoside C, 319-320
Lanatoside D, 320
Lanatoside E, 320
Langwort, 63
Lanthopine, 113
l-a-phellandrene, 379
l-a-pinene, 375
l-arabinose, 138, 143, 325-326
l-arabofuranose, 138
Large-fruited juniper, 27
Laudanidine, 113

Laudanine, 113
Laudanosine, 113
Lauraceae, 101, 103, 363, 375
Laurel family, 103
laurin, 104
Laurus camphora, 103
l-borneol, 278
l-cadinol, 28
l-8-cineol, 278
Lechosa, 223
Lecithin, 120
Ledger bark, 343
Leguminosae, 135, 137, 143, 147, 151, 155, 161, 165, 171, 364, 377
Leichhardt corkwood, 298-301
Lemon, 377
Leucoanthocyanadins, 18, 365
Leurocristine, 238
Levant berry, 92
Levant storax, 133-134
l-fucose, 143
l-hyoscine, 299
l-hyoscyamine, 284, 299, 308
Licorice, 155
Licoricidin, 156
Lignoceric acid, 326, 377
Liliaceae, 45, 47, 53, 63, 360
Lily family, 47, 53, 63
Limonene, 104, 278, 374, 381
Linaceae, 377
Linalool, 278, 360, 377, 378
Linalyl acetate, 377
Linoleic acid, 326, 359, 366, 376, 377, 378, 380, 381
Linolenic acid, 326, 377
Linseed, 377
Linum usitatissimum, 377
Lipase, 194
Liquidambar formosana, 134
Liquidambar orientalis, 132-3
Liquidambar styraciflua, 128-129
Liquid storax, 133
Liquiritigenin, 156
Liquiritin, 156
Liquiritoside, 156
Liquorice, 154
l-limonene, 272, 362, 381
l-linalyl, 377
l-menthone, 365
l-methoxygelsemine, 368
Lobelanidine, 371
Lobelanine, 371
Lobelia inflata, 371
Lobeline, 371

Lobelinine, 371
Loblolly pine, 23
Lochnera rosea, 237
Lochnerol, 238
Loganiaceae, 368
Loganin, 238, 368
Long buchu, 365
Longleaf pine, 16-23
Longstraw pine, 17
Louseberry, 93
Loxa bark, 343
l-rhamnopyranose, 138
l-rhamnose, 138
l-scopolamine, 304
Lycopodiaceae, 359
Lycopodium clavatum, 359
Lysergic acid diethylamide, 6

M

Mace, 375
Macrocystis pyrifera, 373
Macropone, 379
Madagascar periwinkle, 237
Madder family, 335, 341, 353
Magdalena, 237
Magnoliaceae, 374
Maguey, 67
Ma-huang, 33
Malay cinnamon, 375
Male fern, 359
Mallow family, 207
Malvaceae, 205, 207, 378
Manchurian licorice, 158
Mandrake, 87, 148
Mannitol, 156
Manogalactan, 377
Maranham jaborandi, 187
Margaspicin, 359
Marsh mallow, 206-208
Marsh marigold, 61
Matacoyote, 251
Maw seed, 120
May apple, 86
May apple family, 87
Mbolo, 265
Meadow poke, 59
Meadow saffron, 53
Mecca senna, 147
Meconic acid, 113
Meconidine, 113
Medicinal rhubarb, 362
Mediterranean aloe, 47
Melaleuca cajuputi, 367
Melaleuca leucadendron, 367

Melaleuca quinquenervia, 367
Menispermaceae, 91, 93
Menispermine, 94
Menispermum cocculus, 93
Menispermum lacunosum, 93
Mentha arvensis subsp. *haplocalyx* var. *piperascens*, 270-271
Mentha cardiaca, 381
Mentha piperita, 275, 381
Mentha pulegium, 275
Mentha spicata, 381
Mentha viridis, 381
Menthol, 272, 274-275, 381
Menthone, 272, 381
Menthyl acetate, 272
Mescal, 67
Methoxyl, 143
11-methoxy-δ-yohimbine, 244
Methyl alcohol, 379
Methyl anthranilate, 377, 378
Methyl chavicol, 380
Methylchrysophanic acid, 362
Methyl cinnamate, 379
3-methylcolchicine, 54
Methylcreosol, 18
Methyleugenol, 379
Methyl heptenone, 360
Methylisopelletierine, 366
Methylmorphine, 113
Methyl-n-amyl ketone, 367
Methylpelletierine, 366
Methyl reserpate, 244
Methyl salicylate, 362, 368
Methyltransferase enzymes, 113
Mexican sarsaparilla, 374
Mexican scammony, 369
Mexican tea, 36
Mexican yams, 75-82
Mint family, 271, 277
Mitoridine, 255
Monarda didyma, 279
Monarda punctata, 279
Monk's-hood, 362
Monooleodisaturated glycerides, 378
Monylic aldehydes, 378
Moonseed family, 93
Moraceae, 230
Mormon tea, 36
Morphine, 113, 117-119
Mother-of-rye, 5
Mountain hemp, 307
Mucic acid, 331
Murichu, 233
Mururu, 233

Mustard, black, 364
Mustard, Indian, 364
Muttercorn, 5
Myrcene, 278, 359
Myristicaceae, 375
Myristic acid, 376, 378, 380
Myristica fragrans, 375
Myristicin, 375
Myrosin, 364
Myroxylon balsamum, 160-161
Myroxylon balsamum var. *genuinum*, 161
Myroxylon balsamum var. *pereirae*, 164-165
Myroxylon pereirae, 165
Myroxylon peruiferum, 165
Myroxylon toluiferum, 161
Myrrh, 365
Myrrhin, 365
Myrtaceae, 367, 379
Myrtle, 237

N

Naked ladies, 53
Naloxone, 124, 125
Napelline, 362
Narceine, 113
Narcotoline, 113
Narcotine, 113
Narrow-leaved peppermint, 379
Naucleaceae, 370
N-deacetyl-N-methylcolchicine, 54
Neoajmaline, 244
Neogermitrine, 60
Neomenthol, 381
Neopine, 113
Neoprotoveratrine, 60
Neoquassin, 365
Neo-tigogenin, 68
Nerol, 377
Nerolidol, 166, 377
N-formyl-N-deacetylcolchicine, 54
Nganda coffee, 353
N-glycyrrhetinoyl amino acid, 157
Ngmoo, 293
Niaouli oil, 367
Nicaragua ipecac, 338
Nicotine, 293
Nightshade family, 283, 293, 299, 303, 307
Nigracin, 360
N-methyl-pyrroline, 284
N-8-norphysostigmine, 172
Nobiletin, 377
Nor-atropine, 299
Norcycleanine, 98
Nordihydrocapsaicin, 369

Norhyoscyamine, 293, 299
Nornicotine, 293
Noscapine, 113, 118
Nubian senna, 151
Nutmeg, 375
N-vanillyl-8-methyl-6-nonenamide, 369

O

4-O-(a-d-galactopyranosyluronic acid)-d-galactose, 212
2-O-(a-d-galactopyranosyluronic acid)-1-rhamnose, 212
Oak, Dyer's, 361
Oak, gall, 361
O-(B-d-glucopyranosyluronic acid)-1→3-(a-d-galactopyranosyluronic acid)-(1→2)-1-rhamnose, 212
Octyl alcohol, 381
Octylic aldehydes, 378
O-demethyl-N-methylcolchicine, 54
Oil nut, 193
Oil of origanum, 278
Oil of storax, 130
Old maid, 237
Oleaceae, 380
Olea europaea, 380
Oleic acid, 326, 331, 359, 366, 376, 377, 378, 380, 381
Olive, 380
Onaye, 259
Oncovin, 238
Ophiorrhiza mungos, 249
Opium, 112-120
Opium poppy, 111
Orange, sweet, 378
Ordeal bean, 171
Oriental berry, 93
Oriental sweet gum, 132
Orungurabie, 293
12-0-tetradecanoyl-phorbol-13-acetate (TPA), 366
Ouabain, 233-234, 259, 261
Oval buchu, 365
Oxalic acid, 331, 354
Oxanthrone, 202
Oxynarcotine, 113
Oxysanguinarine, 363

P

Pale catechu, 370
Palembang benzoin, 218
Palisota, 261
p-allylmethylenedioxybenzene, 363
Palma cristi, 193

Palmae, 359
Palmitic acid, 326, 359, 366, 376, 380, 381
Panama ipecac, 338
Papain, 40, 224
Papaveraceae, 109, 111, 363
Papaveramine, 113
Papaver bracteatum, 123-125
Papaverine, 113, 118, 244
Papaver orientale, 114
Papaver orientale var. *bracteatum*, 123
Papaver pseudo-orientale, 123
Papaver setigerum, 125
Papaver somniferum, 111, 114, 124
Papaw, 223
Papaw table, 255
Papaya, 222-230
Papaya family, 223
Paraguay jaborandi, 187
Paramenispermine, 94
Pareira, 96-98
Pareira brava, 97
p-coumaric acid, 48
p-cymene, 278, 362, 379, 380
Peach, 376
Pea family, 137, 143, 147, 151, 155, 161, 165, 171
Peanut, 377
Pedaliaceae, 381
Pelletierine, 366
Penpen, 255
Peppermint, 381
Peppermint, broad-leaved, 379
Peppermint, narrow-leaved, 379
Pepper, red, 369
Pepper, Tailed, 360
Periplocin, 266
Periplocymarin, 266
Periwinkle, 236-241
Pernambuco jaborandi, 187
Persian licorice, 155
Persic oil, 376
Peruvian balsam, 165
Peruvian bark, 341
Peruvian rhatany, 365
Phellandral, 379
Phellandrene, 104, 360, 363, 375
Phenol, 18, 278, 375
Phenylacetic acid, 377
Phenylethylene, 130, 218
Phenylpropyl cinnamate, 129, 133
Phlorol, 18
Phorbol 12-tiglate 13-decanoate, 366
Phospholipins, 378
P-hydroxy-cinnamic acid, 48

Physostigma venenosum, 170-171
Physostigmine, 171-173
Physovenine, 171
Phytolacca acinosa, 290
Phytolacca americana, 61
Phytosterol, 378, 380
Picein, 360
Picked turkey gum, 138
Picraena excelsa, 365
Picraconitine, 362
Picrasma excelsa, 365
Picrasmin, 365
Picrinine, 255
Picropodophyllin, 88
Picrotin, 94
Picrotoxin, 94-95
Picrotoxinin, 94
Pilocarpidine, 187
Pilocarpine, 187-189
Pilocarpus jaborandi, 187
Pilocarpus microphyllus, 187
Pilocarpus pennatifolius, 187-188
Pilocarpus trachylopus, 187
Pilosine, 187
Pimenta officinalis, 379
Pimpinella anisum, 380
Pinaceae, 11, 13
Pineapple, 38-43
Pineapple family, 39
Pinene, 18, 104, 363, 367, 375, 377
Pine straw, 21
Pine tar, 18, 21
Pinus elliottii, 12-15, 275
Pinus maritima, 105
Pinus palustris, 16-23, 275
Pinus pinaster, 105
Pinus taeda, 23
Piperaceae, 360
Piper cubeba, 360
Piperitone, 272, 275, 379
Pitayo bark, 343
Pitch pine, 17
Plantaginaceae, 323, 325, 331
Plantago amplexicaule, 327
Plantago arenaria, 332
Plantago isphagula, 325
Plantago ovata, 324-325, 331
Plantago psyllium, 330-332
Plantago ramosa, 332
Plantain family, 325, 331
p-methoxyphenylacetone, 374, 380
Podophyllaceae, 85, 87
Podophyllic acid, 88
Podophyllin, 88

Podophyllotoxin, 88
Podophyllum emodi, 89
Podophyllum hexandrum, 89-90
Podolphyllum peltatum, 86, 90
Podophyllum resin, 88
Poisonberry, 93
Poison black cherry, 283
Poison nut, 368
Poison tobacco, 303
Poison vine, 265
Poke root, 59
Pokeweed, 61
Polygalaceae, 366
Polygala senega, 366
Polygalic acid, 366
Polygalin, 366
Polygalitol, 366
Polygonaceae, 361
Pomegranate, 366
Poppy family, 111
Poppy straw, 116
Populin, 360
Populus spp., 360
Populus nigra, 360
Poroidine, 293
Porphyroxine, 113
Port Royal senna, 152
Potassium chloride, 308
Pourouma, 181
Presenegenin, 366
Prickly cedar, 27
Prickly juniper, 26-29
Proanthocyanidins, 364
Propionic acid, 354
Protopine, 113, 363
Protoveratrine, 60
Protoveratrine A, 63
Protoveratrine B, 63
Prunase, 376
Prunasin, 376
Prunus amygdalus, 376
Prunus armeniaca, 376
Prunus dulcis, 376
Prunus persica, 376
Prunus serotina, 376
Pseudochelerythrine, 363
Pseudoephedrine, 34
Pseudojervine, 60, 63
Psychotria ipecacuanha, 335
Psychotrine, 336
Psychotrine methyl ether, 336
Psyllium, 324-328
Pterocladia capillacea, 372
Pterocladia lucida, 372

Pulegone, 275
Punicaceae, 366
Punica granatum, 366
Puppet root, 59
Purging croton, 366
Purpeline, 255
Purple foxglove, 313
Purple osier, 360
Purpurea glycoside A, 314, 320
Purpurea glycoside B, 314, 320
Putrescine, 6
Pyridine, 284

Q

Quassia, Jamaica, 365
Quercitin, 88, 368
Quercus infectoria, 361
Quinacrine, 349
Quinic acid, 354
Quinidine, 343
Quinine, 343, 348-350
Quinoquino, 161

R

Ram-goat rose, 237
Ranunculaceae, 362
(R)-(—)-armepavine, 204
Rattlesnake root, 366
Rauhimbine, 244
Raujemidine, 245, 252
Raunatine, 244
Raunescine, 252
Raupine, 244, 252
Rauvanine, 255
Rauvolfia canescens, 251
Rauvolfia densiflora, 249
Rauvolfia heterophylla, 251
Rauvolfia hirsuta, 251
Rauvolfia micrantha, 249
Rauvolfia nitida, 251
Rauvolfia serpentina, 60, 64, 242-243, 247-248, 254, 256
Rauvolfia tetraphylla, 250-251, 253
Rauvolfia vomitoria, 254, 255-257
Rauvomitine, 255
Rauvoxine, 255
Rauvoxinine, 255
Rauwolfia serpentina, 243
Rauwolfine, 244
Rauwolfinine, 244-245
Rauwolscine, 244-245, 252
Recanescine, 252
Red bark, 343
Red gum, 129

Red lace, 372
Red pepper, 369
Red periwinkle, 237
Red squill, 360
Red stringybark, 367
Red thyme oil, 278, 280
Rescidine, 255
Rescinnamine, 244, 248, 255
Reserpiline, 244, 252, 255
Reserpine, 238, 244-245, 248, 253, 255
Reserpinine, 244-245, 252
Reserpoxidine, 244, 252
Reterophyllin, 252
Reticuline, 362
Rhabarberon, 362
Rhamnaceae, 199, 201
Rhamno-liquiritin, 156
Rhamno-isoliquiritin, 156
Rhamnose, 325
Rhamnus californica, 202
Rhamnus cathartica, 203-204
Rhamnus frangula, 204
Rhamnus purshiana, 200-201
Rhatany, Peruvian, 365
Rhein, 147, 362
Rhein-anthrone-8-glucoside, 147
Rhein-I-glucoside, 147
Rhein-8-diglucoside, 147
Rhein-8-glucoside, 147
Rheotannic acid, 362
Rheum emodi, 362
Rheum officinale, 362
Rheum palmatum, 362
Rhoaedine, 113
Rhubarb, Chinese, 362
Rhubarb, Indian, 362
Rhubarb, medicinal, 362
Ricin, 193
Ricinine, 193
Ricino, 193
Ricinoleic acid, 194
Ricinolein, 194
Ricinus communis, 192
Rio ipecac, 335
Robusta coffee, 353
Roman chamomile, 371
Rosaceae, 376
Roseoside, 238
Rosmarinus officinalis, 381
Rosemary, 381
Rosemary pine, 17
Rosin, 20
Round buchu, 365
Rubiaceae, 333, 335, 341, 353, 370

Rubijervine, 60
Rue family, 187
Russian licorice, 155
Rusty foxglove, 321
Rutaceae, 185, 187, 365, 377
Rutin, 361, 364, 367

S

Sabinene, 374, 375
Sacred bark, 201
Safrole, 104-105, 363, 374, 375
Sagangur, 289
Saigon cinnamon, 363
Salicaceae, 360
Salicin, 360
Salicortin, 360
Salicylic acid, 148
Salicyloyl tremuloidin, 360
Salix fragilis, 360
Salix purpurea, 360
Salvia aegyptiaca, 327
Sand plantain, 332
Sanguilutine, 363
Sanguinaria canadensis, 363
Sanguinarine, 363
Sanguirubine, 363
Santonin, 371
Sapogenin, 68-71, 76-82
Saponin, 266, 278, 320, 336, 366
Sarhamnoloside, 260
Sarmentoside A, 260
Sarmentoside D, 260
Sarmentoside E, 260
Sarna de perro, 251
Sarothamnine, 364
Sarpagine, 252, 255
Sarsaparilla, Ecuadorean, 374
Sarsaparilla, Honduras, 374
Sarsaparilla, Mexican, 374
Sarsasapogenin, 374
Sassafras, 363
Sassafras albidum, 363
Sassafras variifolium, 363
Savanilla ipecac, 338
Sawai, 259
Scammonin, 369
Scillaren, 360
Scillaroside A, 360
Scillaroside B, 360
Sclererythin, 6
Scoparin, 364
Scopolamine, 293, 308
Scopoletin, 284, 368
Scotch broom, 364
Scotch mint, 381

Scotch mercury, 313
Scrophulariaceae, 311, 313, 319
Sea girdles, 373
Sea onion, 360
Sea string, 372
Sea whistle, 373
Sempervirine, 368
Seneca snakeroot, 366
Senegal gum, 138
Senegal senna, 152
Senegenic acid, 366
Senegenin, 366
Senegin, 366
Senna italica, 152
Sennoside A, 147
Sennoside B, 147
Sennoside C, 147
Sennoside D, 147
Señorita, 251
Seredamine, 255
Serepentenine, 255
Serpentina root, 243
Serpentine, 244-245, 252
Serpentine root, 243
Serpentinine, 244-245, 255
Serpent-wood, 234, 242-249
Serpine, 244, 252
Serpinine, 244
Serposterol, 252
Sesame, 381
Sesame oil, 381
Sesamin, 381
Sesamolin, 381
Sesamum indicum, 381
Sesamum orientale, 381
Sesquiterpene, 363, 365
Sesquiterpene tree, 104
Seville orange, 377
Sharp cedar, 27
Shogaol, 360
Short-staple cotton, 378
Siam benzoin, 217-218
Siaresinol, 217
Siaresinolic acid, 217
Sickly-smelling nightshade, 303
Simaroubaceae, 365
Sinigrin, 364
Sisal, 66-71
Sisal agave, 67
Sisalagenin, 68
Sisal hemp, 67
Sitosterol, 21, 94, 369
Slash pine, 12-15, 19-23
Sleeping nightshade, 283

Slippery elm, 361
Smilacaceae, 374
Smilagenin, 374
Smilax aristolochiaefolia, 374
Smilax febrifuga, 374
Smilax ornata, 374
Smilax regelii, 374
Smooth strophanthus, 259-262
Smut of rye, 5
Snakeroot, Seneca, 366
Socotrine aloe, 47
Sodium alginate, 373
Sodium potassium tartarate, 148
Solanaceae, 281, 283, 293, 299, 303, 307, 369
Somali gum, 137
Sophora japonica, 364
Sour orange, 377
Southern pine, 17
Southern yellow pine, 17
Spanish licorice, 155
Spanish psyllium, 331, 332
Sparteine, 364
Spearmint, 381
Spogel, 325
Spur kernels, 5
Spurge family, 193
Spurred rye, 5
Squalene, 376, 380
Squaw tea, 36
Stagshorn clubmoss, 359
Star anise, 374
Stearic acid, 326, 376, 377, 378, 380, 381
Sterculiaceae, 209, 211, 378
Sterculia gum, 211
Sterculia urens, 210-211
Stigmasterol, 332
Stinking nightshade, 303
Storax, 129
Storax family, 217
Storesin, 129, 133
Straw foxglove, 321
Stropeside, 320
Strophanthidin, 265
Strophanthin, 265
Strophanthus gratus, 234, 258-262
Strophanthus hispidus, 266
Strophanthus kombe, 264-267
Strophoside, 265
Struxine, 368
Strychnaceae, 368
Strychnine, 368
Strychnos nux-vomica, 368
Strychnos toxifera, 98
Styracaceae, 215, 217

Styracin, 133, 218
Styrax benzoin, 216-218
Styrax paralleloneurum, 217-218
Styrax sumatranus, 217
Styrax tonkinense, 217, 219
Styrene, 130, 218
Styrocamphene, 130
Suakim gum arabic tree, 141
Succinic acid, 284
Sudan gum arabic, 137
Sugar wrack, 373
Sumaresinol—6-hydroxyoleanolic acid, 218
Sumaresinolic acid, 218
Sumatra benzoin, 216-218
Swamp hellebore, 59
Swamp pine, 13
Sweet birch, 361
Sweet gum, 128-130
Sweet orange, 378
Sweet wood, 155
Swizzle stick, 255
Syrian tragacanth, 143
Syzygium aromaticum, 367

T

Tailed pepper, 360
Talha gum arabic tree, 141
Tall oil, 20
Tangle, 373
Tannin, 166, 208, 278, 326, 343, 360, 361, 362, 363, 364, 366, 368, 371, 380
Teamster's tea, 36
Temisin, 371
Terpenes, 365, 367
Terpinene, 374, 377
1-terpinen-4-ol, 359, 374
Terpineol, 104, 105, 367, 377
Terpin hydrate, 105
Terpinolene, 379
12-0-tetra-decanoyl-phorbol-13-acetate (TPA), 366
Tetrahydroalstonine, 253
Tetramethylputrescine, 294
Tetraphyllicine, 244
Thebaine, 113, 124-125, 244
Theobroma cacao, 378
Theobroma oil, 378
Theobromine, 354, 378
Theophylline, 354
Thiamine, 278, 365, 369
Thiolhistidine, 6
Tholloside, 260
Throatwort, 313
Thujone, 371

Thyme, 276-280
Thyme oil, 278
Thymol, 275, 278, 381
Thymus vulgaris, 275-279
Thymus zygis, 277-278
Thymus zygis var. *floribunda*, 278
Thymus zygis var. *gracilis*, 278
Tickleweed, 59
Tigloidine, 293
Tigogenin, 68
Tigonin, 314
Tinnevelly senna, 147
Tocopherol, 380
Tolu balsam, 161-162
Toluene, 18
Toluifera pereirae, 165
Tolu-resinotannol, 161
Tomentocurine, 98
Tomillo salsero, 277
Totaquina, 343
Toxaphene, 22
Tragacanth, 142-145
Tragacanthin, 143
Tree melon, 223
Tricyclene, 278
Trigonelline, 266, 354
1:3:4′ trihydroxyflavone, 147
3,4,5-trimethoxybenzoyl methyl reserpate, 244
3,4,5-trimethoxycinnamic acid methyl reserpate, 244
Trimethylamine, 6
1,3,7-trimethylxanthine, 354
Tripoli senna, 152
Tropacocaine, 178
Tropine, 178, 294
True veratrum, 59
Tryptophane, 224
Tubocurarine, 98
Turkey red oil, 198
Turkish licorice, 155
Turkish sweet gum, 133
Turlington's balsam, 219
Turpentine, 18-22
Turpentine pine, 17
Tyramine, 6
Tyrosine, 224

U

Ulmaceae, 361
Ulmus fulva, 361
Ulmus rubra, 361
Umbelliferae, 367
Umbelliferone, 156
Umbrella plant, 87

Uncaria gambier, 370
Undecylenic acid, 196
Urginea maritima, 360
Uronic acid, 138, 143, 325-326
Ursolic acid, 278, 368

V

Valepotriate, 370
Valerianaceae, 370
Valeriana officinalis, 370
Valeric acid, 354
Valerine, 178, 370
Valeroidine, 293-294
Valtropine, 294
Vanillin, 105, 130, 133, 162, 166, 218, 367
Vegetable calomel, 87
Vegetable mercury, 87
Velban, 238
Veneno, 251
Veratramarine, 63
Veratramine, 60
Veratridine, 60
Veratrosine, 60, 63
Veratrum album, 62, 63
Veratrum viride, 58-61
Verodoxin, 314
Viborilla, 251
Vinblastine, 238-239
Vincaine, 238
Vincaleukoblastine, 238, 240
Vinca rosea, 237
Vinceine, 238
Vincovin, 238
Vincristine, 238-240
Vinleurosine, 238-239
Virginia pine, 17
Virgin's milk, 219
Vomicine, 368
Vomifoline, 256

W

Wahoo, 201
White hellebore, 59, 63
White mallow, 207
White poppy, 111
White sennar gum, 138

White thyme oil, 278
Whorehouse tea, 36
Wild black cherry, 376
Wild coffee, 201
Wild grape, 97
Wild jalap, 87
Wild lemon, 87
Wild mandrake, 87
Wintergreen, 368
Witch hazel, 364
Witch hazel family, 129
Wolfsbane, 59
Wood marjoram, 277
Woolly foxglove, 319
Wormseed, 371
Wormseed, American, 362

X

Xanthaline, 113
Xylene, 18
Xylose, 325

Y

Yam family, 75
Yamogenin, 68
Yellow bark, 343
Yellow berry, 87, 364
Yellow foxglove, 321
Yellow gentian, 369
Yellow pine, 17
Yellow tang, 373
Yellow thorn, 137
Yerba de San José, 251
Yerba santa, 380
Yohimbine, 238, 244, 252, 255
y-terpinene, 379
y-yohimbine, 244

Z

Zanzibar aloe, 47
Zaragatona, 331
Zingerone, 360
Zingiberaceae, 360, 374
Zingiberene, 360
Zingiber officinale, 360